SpringerWienNewYork

Nasrullah Memon · Jonathan D. Farley
David L. Hicks · Torben Rosenorn
Editors

Mathematical Methods in Counterterrorism

SpringerWienNewYork

Editors

Nasrullah Memon, Ph.D.
The Maersk Mc-Kinney Moller Institute
University of Southern Denmark
Campusvej 55
5230 Odense M
Denmark
memon@mmmi.sdu.dk

David L. Hicks, Ph.D.
Department of Computer Science
and Engineering
Aalborg University Esbjerg
Niels Bohrs Vej 8
6700 Esbjerg
Denmark
hicks@cs.aaue.dk

Jonathan D. Farley, D.Phil. (Oxon.)
Institut für Algebra
Johannes Kepler Universität Linz
4040 Linz
Austria
lattice.theory@gmail.com

Torben Rosenorn
Esbjerg Institute of Technology
Aalborg University Esbjerg
Niels Bohrs Vej 8
6700 Esbjerg
Denmark
tur@aaue.dk

Softcover re-print of the Hardcover 1st edition 2009

SpringerWienNewYork is a part of Springer Science + Business Media
springer.at

Printed by: Books on Demand, Norderstedt, Germany
Printed on acid-free paper
SPIN 12197798

With 124 Figures

ISBN 978-3-7091-1644-7 SpringerWienNewYork

Foreword

Terrorism is one of the serious threats to international peace and security that we face in this decade. No nation can consider itself immune from the dangers it poses, and no society can remain disengaged from the efforts to combat it.

The term *counterterrorism* refers to the techniques, strategies, and tactics used in the fight against terrorism. Counterterrorism efforts involve many segments of society, especially governmental agencies including the police, military, and intelligence agencies (both domestic and international). The goal of counterterrorism efforts is to not only detect and prevent potential future acts but also to assist in the response to events that have already occurred.

A terrorist cell usually forms very quietly and then grows in a pattern – spanning international borders, oceans, and hemispheres. Surprising to many, an effective "weapon", just as quiet – mathematics – can serve as a powerful tool to combat terrorism, providing the ability to connect the dots and reveal the organizational pattern of something so sinister.

The events of 9/11 instantly changed perceptions of the words *terrorist* and *network*, especially in the United States. The international community was confronted with the need to tackle a threat which was not confined to a discreet physical location. This is a particular challenge to the standard instruments for projecting the legal authority of states and their power to uphold public safety. As demonstrated by the events of the 9/11 attack, we know that terrorist attacks can happen anywhere. It is clear that the terrorists operate in networks, with members distributed widely across the globe. To fight such criminals, we need to understand the new "terrain": networks – how they are formed, how they evolve, and how they operate.

The case for the development of mathematical methods and tools to assist intelligence and law enforcement officials is undeniable. The intelligence community is faced with the overwhelming tasks of gleaning critical information from vast arrays of disparate data, choosing the best forms of gathering information, developing new data sources and setting up personnel to effectively function in adverse circumstances. Each of these difficult tasks demands an intensive effort on behalf of the mathematical community to develop techniques and tools to assist in the asymmetric confrontation with secretive terrorist networks.

The important role of mathematics in the scientific advances of the last century, from advanced computing to lasers and optical communications to medical diagnosis, is now widely recognized. The central role of mathematical principles and techniques in assisting intelligence and law enforcement officials is equally striking. From new cryptographic strategies to fast data processing techniques, mathematical ideas have provided the critical link.

We are now faced with an extraordinary situation in the global fight against terrorism, and this book seeks to encourage the mathematical community to bring its full capabilities to bear in responding to this challenge. It presents the most current research from mathematicians and computer scientists from around the world aimed at developing strategies to support counterterrorism and enhance homeland security. These new methods are more important now than ever in order to glean the maximal possible benefit from the tremendous amount of information that has been gathered since 2001 regarding terrorist cells and individuals potentially planning future attacks.

I am confident that this book will help to advance the discourse, and enable its insights to be shared among a broad range of researchers, policy makers, politicians, and the members of intelligence and law enforcement agencies. The articles the book contains can help create momentum in the effort to transform theoretical techniques and strategies into concrete results on the ground.

Finally, I congratulate Nasrullah Memon, co-editors of the book, and the authors for their substantial efforts to address one of the most important crises faced by the international community today.

Brussels, January 2009 *Gilles de Kerchove*

Gilles de Kerchove is the EU Counterterrorism Coordinator and Law Professor at the Catholic University of Louvain and the Free University of Brussels.

Contents

Mathematical Methods in Counterterrorism: Tools and Techniques for a New Challenge

David L. Hicks, Nasrullah Memon, Jonathan D. Farley, and Torben Rosenørn

1 Introduction

Throughout the years mathematics has served as the most basic and fundamental tool employed by scientists and researchers to study and describe a wide variety of fields and phenomena. One of the most important practical application areas of mathematics has been for national defense and security purposes. For example, during the Second World War, the mathematical principles underlying game theory and cryptography played a very important role in military planning. Since that time, it has become clear that mathematics has an important role to play in securing victory in any global conflict, including the struggle faced by national security and law enforcement officials in the fight against those engaged in terrorism and other illicit activities.

Recent events of the past decade have produced an increased interest in and focus upon the area of counterterrorism by a broad range of scholars, including mathematicians. At the same time, government decision makers have often been skeptical about mathematics and statistics, even while faced with the considerable challenges of sifting through enormous amounts of data that might hold critically important clues. Realizing that policy makers were not always receptive, the mathematical

David L. Hicks
Department of Computer Science, Aalborg University, Esbjerg, Denmark, e-mail: hicks@cs.aaue.dk

Nasrullah Memon
The Mærsk Mc-Kinney Møller Institute, University of Southern Denmark, Odense, Denmark, e-mail: memon@mmmi.sdu.dk

Jonathan D. Farley
Institut für Algebra, Johannes Kepler Universität Linz, Linz, Austria, e-mail: lattice.theory@gmail.com

Torben Rosenørn
Esbjerg Institute of Technology, Aalborg University, Esbjerg, Denmark, e-mail: tur@aaue.dk

N. Memon et al. (eds.), *Mathematical Methods in Counterterrorism*,
DOI 10.1007/978-3-211-09442-6_1, © Springer-Verlag/Wien 2009

community has pondered about how best to put what they knew to work in building a more secure world. They felt especially qualified to help decision makers see the important patterns in the haystack of data before them and detect the most important and relevant anomalies.

Though governments have begun to engage the research community through grants and collaborative opportunities, across the sciences, and in particular within the fields of mathematics and statistics, the interesting problems and viable methodologies are still at a very early and speculative stage. The recently increased interest in counterterrorism has driven the research focus towards revisiting and strengthening the foundations necessary to build tools and design techniques capable of meeting the new challenges and producing more accurate results. This book provides a look at some of the latest research results in a variety of specialty topics that are central to this area.

2 Organization

This volume is composed of 21 contributions authored by some of the most prominent researchers currently focused on the application of mathematical methods to counterterrorism. The contributions span a wide variety of technical areas within this research field. In this book they have been organized into the five categories of network analysis, forecasting, communication/interpretation, behavior, and game theory. The remainder of this section provides a brief overview of the contributions in each of those categories.

Section 1: Network Analysis. The first section of the book begins with a contribution by Brantingham, Glässer, Jackson, and Vajihollahi. The authors describe their work on the development of a comprehensive framework and tool to support the mathematical and computational modeling of criminal behavior. They focus on criminal activities in urban environments, but also seek to extend the approach beyond conventional areas and support the application of computational thinking and social simulations to the analysis process in the area of counterterrorism. The next contribution in this section is from Skillicorn and discusses methods to obtain knowledge from graphs that are used to represent and study adversarial settings. It describes that, while graphs are appropriate for use in such analyses, they can also be more difficult to analyse than more traditional representations, and this article presents practical methods to help understand the structures these graphs contain.

The section continues with a contribution from McGough that examines the modeling of terrorist cells. The focus is on discussing and determining the strength of terrorist cell structures, and using the partially order set model and algorithms to do so. The next contribution in the first section is from Zaidi, Ishaque, and Levis. It describes an approach to combine and apply temporal knowledge representation and reasoning techniques to criminal forensics. An emphasis is placed on answering questions concerning time sensitive aspects of criminal or terrorist activities. The section concludes with a contribution by Farley that examines the structure of the "perfect" terrorist cell. In particular it examines two theoretical questions related to the number of cutsets that exist in partially ordered sets that are used to represent that structure.

Section 2: Forecasting. A contribution by Gutfraind begins the second section of the book. It describes a dynamic model that is capable of representing the relevant factors involved as a terrorist organization changes over time. The approach enables those factors and their effects to be considered together, in a quantitative way, rather than individually, and for predictions to be made about the organization based on these analyses. In the next contribution of the section, Rhodes describes methods for constructing social networks of covert groups. The focus is on the use of Bayesian inference techniques to infer missing links, making the approach suitable for cases where only limited and incomplete data is available.

The following contribution, by Pinker, develops a model to represent the uncertainty in the timing and location of terror attacks. A framework is then described for guiding the issuance of terror warnings and the deployment of resources to combat attacks, and balancing the tradeoffs between these two defensive mechanisms. In the final contribution of the section, Ferry, Lo, Ahearn, and Phillips consider detection theory from the traditional military domain and its relationship to network theory. In particular they describe ways in which the detection theory approach might be used to leverage results from network theory as a way to find and track terrorist activities.

Section 3: Communication/Interpretation. The third section begins with a contribution by Hartnell and Gunther that focuses on communication in covert groups. A graph is used as a theoretical model to study the tradeoff between the competing demands of ease of communication and the potential danger for betrayal when members are captured, and a number of related questions. The section continues with a contribution from Woo that examines constraints on terrorist network plots imposed by the intelligence gathering efforts of law enforcement services. It also describes some of the implications inherent in both the intelligence gathering level and the methods that are utilized, and their relation to the important campaign to win the hearts and minds of the larger population.

In the following contribution, Lindelauf, Borm, and Hamers examine a theoretical framework for analysing homogeneous networks, especially in terms of the competing factors of secrecy and operational efficiency. An evaluation and comparison are provided of the different network topologies and their suitability for use for different graph orders. In the final contribution of the section, Ozgul, Erdem, and Bowerman describe two models for representing terrorist groups and analysing terrorist groups. The models are evaluated and compared in terms of their ability to support the semi-supervised detection of covert groups.

Section 4: Behavior. The fourth section starts with a contribution by Sliva, Subrahmanian, Martinez, and Simari in which they examine group behavior. Moving forward from previous research efforts focused on models derived mainly from past group behavior, they describe an architecture and algorithms to predict the conditions under which a group is likely to change their behavior. The next contribution in this section is from Baccara and Bar-Isaac and discusses the impact of methods of interrogation on terror networks. They investigate how the legal limits of interrogation to extract information under which authorities operate relate to and have impact on the degree to which terrorist organizations diffuse or distribute their information in the attempt to increase the efficiency of the organization.

The section continues with a contribution from McCormick and Owen that discusses state-terrorist coalitions. They examine the factors of mutual advantage and mutual trust, how they change and evolve over time, and the impact these changes have upon the behavior of the partners in this type of coalition, and in particular the circumstances that might lead to continued cooperation, or the dissolution of the partnership. A contribution by McGough concludes the section with a look at the modeling of the behavior of terrorist groups. A mathematical model is described to represent terrorist groups and their behavior, and then it is evaluated through experimentation to test its projections, shedding light on the question of theory versus reality.

Section 5: Game Theory. A contribution by Melese begins the fifth section of the book. It describes a game theory approach to simulate terrorist cells. The principal question that is considered in the analysis examines under what conditions a threat of preemptive action by a world leader might successfully deter terrorist organizations or a sovereign state from the acquisition of weapons of mass destruction. The section continues with a contribution from Arce and Sandler in which they examine continuous policy models and describe an extension of them including differentiable payoff functions. They also consider the relationship between terrorist actions and government responses, and the effects they might have on the future support for terrorist organizations, within a game-theoretic context.

In the final contribution of the fifth section, Shapiro and Siegel discuss funding aspects of terrorist organizations. A model of a hierarchical terror organization is used to examine the implications of an arrangement in which leaders delegate financial and logistical tasks to middlemen, and, for security reasons, are not able to closely monitor their actions. A series of policy implications based on the analysis are also discussed.

Section 6: History of the Conference on Mathematical Methods in Counterterrorism. A contribution from Farley in the last section provides a look at the background and history of the Mathematical Methods in Counterterrorism conference series along with some personal observations.

3 Conclusion and Acknowledgements

The science of counterterrorism is still unfolding. As demonstrated by the variety of contributions to this volume, researchers welcome the opportunity to influence and shape the landscape of this important emerging area. The task before them is a challenging one, nothing less than to develop a new kind of mathematics, one that can equip national security and law enforcement officials with the tools they urgently need to face a new challenge – a new kind of war.

The editors would like to gratefully acknowledge the efforts of all those who have helped with the creation of this volume. Firstly, it would never be possible for a book such as this one to provide such a broad and extensive look at the latest research in the field of mathematical methods to combat terrorism without the efforts of all those expert researchers and practitioners who have authored and contributed papers.

Thanks are also due to Gilles de Kerchove, Counter-Terrorism Coordinator for the Council of the European Union, for providing a very motivational Foreword for the book. The support and guidance of a publisher is a critical component to a project such as this one. In this case thanks are due to Stephen Soehnlen and Edwin Schwarz, and their colleagues at Springer. Lastly, but certainly not least, the editors would like to acknowledge the considerable effort and support provided by Claus Atzenbeck with the layout, typesetting, and formatting of the book.

Part I
Network Analysis

Modeling Criminal Activity in Urban Landscapes

Patricia Brantingham, Uwe Glässer, Piper Jackson, and Mona Vajihollahi

Abstract Computational and mathematical methods arguably have an enormous potential for serving practical needs in crime analysis and prevention by offering novel tools for crime investigations and experimental platforms for evidence-based policy making. We present a comprehensive formal framework and tool support for mathematical and computational modeling of criminal behavior to facilitate systematic experimental studies of a wide range of criminal activities in urban environments. The focus is on spatial and temporal aspects of different forms of crime, including opportunistic and serial violent crimes. However, the proposed framework provides a basis to push beyond conventional empirical research and engage the use of computational thinking and social simulations in the analysis of terrorism and counter-terrorism.

1 Introduction

Innovative research in criminology and other social sciences promotes mathematical and computational methods in advanced study of social phenomena. The work

Patricia Brantingham
Institute for Canadian Urban Research Studies (ICURS), School of Criminology, Simon Fraser University, B.C., Canada, e-mail: pbrantin@sfu.ca

Uwe Glässer
Software Technology Lab, School of Computing Science, Simon Fraser University, B.C., Canada, e-mail: glaesser@sfu.ca

Piper Jackson
Software Technology Lab, School of Computing Science, Simon Fraser University, B.C., Canada, e-mail: pjj@sfu.ca

Mona Vajihollahi
Software Technology Lab, School of Computing Science, Simon Fraser University, B.C., Canada, e-mail: monav@sfu.ca

N. Memon et al. (eds.), *Mathematical Methods in Counterterrorism*,
DOI 10.1007/978-3-211-09442-6_2, © Springer-Verlag/Wien 2009

presented here proposes a comprehensive framework and supporting tool environment for mathematical and computational modeling of criminal behavior to facilitate systematic experimental studies of a wide range of criminal activities in urban environments. The focus is on uncovering patterns in the spacial and temporal characteristics of physical crime in urban environments, including forms of crime that are opportunistic in nature, like burglary, robbery, motor vehicle theft, vandalism, and also serial violent offenses such as homicide; they can involve multiple offenders and multiple targets. The principles of environmental criminalogy [1] suggest that criminal events can best be understood in the context of people's movements in the course of their everyday lives: offenders commit offenses near places they spend most of their time, and victims are victimized near places where they spend most of their time. This line of theory and supporting research argues that location of crimes is determined by perception of the environment – separating good criminal opportunities from bad risks [2] – and implies there is a set of patterns and/or rules that govern the working of a social system: one composed of criminals, victims and targets. They interact with each other and their environment, and their movements are influenced by the city's underlying land use patterns and high activity nodes like shopping centers and entertainment districts, street networks and transportation systems.

Computational methods and tools arguably have an enormous potential for serving practical needs in crime analysis and prevention, namely as instruments in crime investigations [3], as an experimental platform for supporting evidence-based policy making [4], and in experimental studies to analyze and validate theories of crime [5, 6]. The approach presented here proposes a formal modeling framework to systematically develop and validate discrete event models of criminal activities; specifically, it focuses on describing dynamic properties of the underlying social system in abstract mathematical terms so as to provide a reliable basis for computational methods in crime analysis and prevention. Besides training and sandbox experiments, our approach aims at intelligent decision support systems and advanced analysis tools for reasoning about likely scenarios and dealing with 'what-if' questions in experimental studies. Building on a cross-disciplinary R&D project in Computational Criminology [7], called *Mastermind* [8], we describe here the essential design aspects of the Mastermind system architecture in abstract functional and operational terms, emphasizing the underlying principles of mathematical modeling in an interactive design and validation context. The description presented here extends and complements our previous work [8, 9, 10] in several ways, opening up new application fields.

Mastermind is jointly managed by the Institute for Canadian Urban Research Studies (ICURS) in Criminology and the Software Technology Lab in Computing Science at SFU and has partly been funded by the RCMP "E" Division over the past three years. Crossing boundaries of research disciplines, the Mastermind project is linked to a wide range of research areas and application fields spanning criminology, computing, mathematics, psychology and systems science. Not surprisingly, any attempt to integrate such diverse views within a unifying computational framework in a coherent and consistent manner faces a number of challenging problems to be addressed. A particular intriguing aspect is finding the right level of accuracy and detail to model real-world phenomena so that the resulting observable behavior is meaningful, at least in a probabilistic sense. This is closely related to the question of how *micro-level behavior* affects *macro-level behavior* and the observable phenomena under study. Another challenging aspect is the question of how to draw the boundaries of any such system, clearly delineating the system from the environment into which it is embedded: that is, what needs be included in the model and what is irrelevant in terms of the resulting behavior of interest?

The nature of modeling something as complex and diverse as crime is an ongoing and potentially open-ended process that demands for an interactive modeling approach – one that embraces frequent change and extensions through robustness and scalability of the underlying mathematical framework. The formal approach taken here builds on modeling and validation concepts using the *Abstract State Machine* (ASM) [11, 12] multiagent modeling paradigm together with CoreASM [13, 14], an open source modeling tool suite[1], as the formal basis for semantic modeling and rapid prototyping of mobile agents and their routine commuting activities through a virtual city they inhabit.

The remainder of this chapter is structured as follows. Section 2 discusses fundamental concepts in Computational Criminology and specific challenges and needs in mathematical and computational modeling of complex social systems. Section 3 introduces the mathematical framework and the tool environment used in the Mastermind project. Section 4 illustrates the main building blocks for modeling criminal activity, namely the representation of the urban environment and the agent architecture, and also summarizes some lessons learned from this project. Section 5 concludes the chapter.

2 Background and Motivation

This section briefly reviews the benefits of applying computational methods to studying crime. We first explain how this new way of thinking and problem solving benefits researchers in criminology. We then discuss related practical requirements of developing software tools in a collaborative research environment.

2.1 Computational Criminology

The use of computational techniques has become well-established and valuable in advancing the boundaries of many research disciplines, such as biology and chemistry. Research in Criminology, like other social sciences, faces the problem of lack of control in running experiments. Computational Criminology aims at pushing

[1] CoreASM v1.0.5 is readily available at http://www.coreasm.org.

these limits through interdisciplinary work with mathematics and computing science. By employing novel technologies and formal methodologies, existing theories of crime can be extended and new applications developed. Computational models allow for running experiments in simplified artificial situations where abstraction is used conveniently and systematically to adjust the influence of different elements under study. This facilitates dealing with the highly complex and dynamic nature of criminal activities. As such, beyond seeing computers as tools, *computational thinking* [15] presents a way of solving problems and designing systems in Criminology.

Conventional research in crime is empirical in nature and tends to use methods that rely upon existing data. In order to analyze the data, mostly statistical methods are used to derive a more abstract view of the data. Nowadays, however, computational methods offer a new way of thinking about the data that leads to new perspectives and new models for analyzing the problems.

Using computer simulations to conduct experiments virtually or to analyze "what-if" scenarios is now commonly practiced in the social sciences. The agent-based modeling paradigm is widely used for describing social phenomena, including criminal events, where individuals are represented by agents interacting with one another in a given environment. Different sub-areas of crime analysis have already benefited from the blending of criminology, computing science and mathematics [4, 16, 17]. For a more detailed review of computational modeling approaches in crime analysis we refer to [8].

Mathematical and computational modeling also introduces a great degree of precision. In the process of modeling (or defining) a theory of crime in abstract computational and mathematical terms, any incompleteness or lack of rigor becomes evident. This demands far more effort on the *theory development* side to complete existing theories, or discard incomplete ones, or even precisely identify the limitations of existing ones. This opens up new opportunities for criminologists to *test* the existing theories beyond conventional means.

There is particular value in expanding the use of computational thinking and social simulations in the analysis of terrorism and counter-terrorism. Terrorism is, as would be expected, a growing area of research in criminology where knowledge gained about human behavior in legal and traditional illegal activities such as robbery, homicide, and burglary is being expanded to include terrorist bombings, kidnapping, execution and general vulnerability to covert actions [18, 19]. Terrorism requires networks, exchange of information, and actions. There is a need for fitting into normal activity patterns until the terrorist act and to maintain safe locations. This is similar, yet different in severity, to many criminal activities where offenders operate in normal environs and search for targets close to established time-space patterns, with a preference for locations where the crime is acceptable or where actions by individuals are not closely watched. For example, shoplifting, motor vehicle theft and robbery occur in high activity shopping areas where diversity is accepted. These high activity and diverse areas are also attractors of terrorist bombing.

In particular, the development and modification of theories that cover terrorism need to push beyond conventional empirical research and engage in methods that allow the logical exploration of alternatives, the modification of contextual background, including cultural and economic differences [20], the interaction between terrorist support networks and anti-terrorism networks [21], covert and adaptive networks, and the ability to explore "what-if" scenarios of alternative policies, and the dynamic nature of simulation models.

2.2 Challenges and Needs

For the social sciences, applying computational techniques helps in overcoming some of the core limitations of studying social phenomena. Social scientists have always been limited by the inextricability of the subject of their research from its environment. Hence, it is difficult to study different factors influencing a phenomena in isolation. Particularly for fields that fall under the umbrella of security, safety and ethical issues can be an obstacle to innovation. For criminologists, it is very difficult to get first-hand evidence of crimes while they are being perpetrated – an observer would most likely be legally required to try to prevent the crime rather than letting it take place. Developing response strategies to unpredictable and dangerous situations is difficult to do in the field, since such situations are unpredictable and by their nature very difficult to control. Computational methods allow us to circumvent these problems by generating scenarios inside a virtual environment. In particular, modeling and simulation allow us to dynamically and interactively explore our ideas and existing knowledge.

Computational thinking about social phenomena, however, means thinking in terms of multiple layers of abstraction [15], which facilitates a systematic study of the phenomena by adjusting the level of detail given to the various factors under study. Computer models of social systems simulate dynamic aspects of individual behavior to study characteristic properties and dominant behavioral patterns of societies as a basis for reasoning about real-world phenomena. This way, one can perform experiments as a mental exercise or by means of computer simulation, analyzing possible outcomes where it is difficult, if not impossible, to observe such outcomes in real life.

2.3 Modeling Paradigm

In the process of interactive modeling of behavioral aspects of complex social systems, one can distinguish three essential phases, namely *conceptual modeling*, *mathematical modeling*, and *computational modeling*, with several critical phase transitions and feedback loops as illustrated in Fig. 1. Starting from a conceptual model that reflects the characteristic properties of the phenomena under study in a direct and intuitive way, as perceived by application domain experts, a discrete mathematical model is derived in several steps, gradually formalizing these properties in abstract mathematical and/or computational terms. This model is then transformed into an initial computational model that is executable in principle; that is, any aspects that have been left abstract provisionally ought to be details to be filled in as the result of subsequent refinement steps. Ideally, any such refinement would be restricted to just adding details as required for running experiments both to help establishing the validity of the formal representation of the conceptual model and for further experimental studies. In reality, however, modeling is a highly iterative and essentially non-linear process with feedback loops within and also across the various phases, potentially affecting the design of the model in its entirety (see Fig. 1).

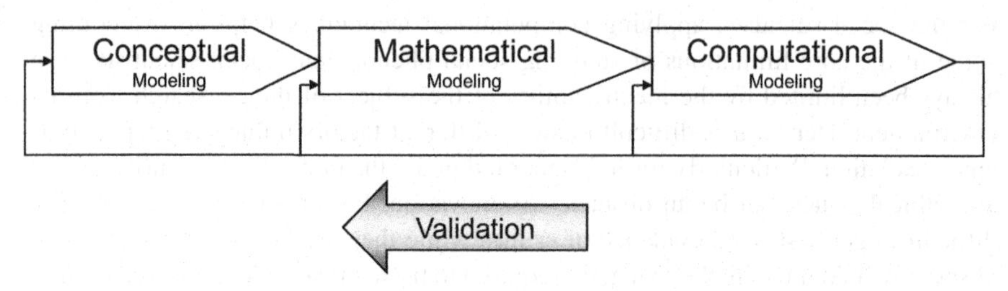

Fig. 1 Phase transitions in social system modeling

Specifically, the role of the mathematical model is to assist in formalizing the conceptual view of the target domain so as to provide an exact description of the characteristic properties as a reliable basis for deriving a computational model. Marking the transition from an informal (or semi-formal) to a formal description, this model serves as "semantic middleware" for bridging the gap between the conceptual and the computational view. As such, it provides a blueprint of the essential understanding, allowing systematic analysis and reasoning about the initial transformation step – typically, the most challenging one. Mathematical precision is essential to accurately characterize the key attributes, ensuring that they are properly established and well understood prior to actually building the computer model.

3 Mastermind Framework

Our approach to modeling complex criminal behavior in the Mastermind project follows the modeling paradigm of Sect. 2.3, emphasizing the need for mathematical rigour and precision. In order to accomodate the highly iterative process of modeling and validation, we build on common abstraction principles of applied computational logic and discrete mathematics using the Abtract State Machine method as the underlying mathematical paradigm. This section provides an overview of the ASM method, the CoreASM tool suite, which allows for rapid protyping, and the Control-State Diagram editor (CSDe), which was designed to facilitate a more interactive modeling approach.

3.1 Mathematical Framework

A central question in computing science is how to precisely define the notion of *algorithm*. Traditionally, Turing machines have been used in the study of the theory of computation as a formal model of algorithms [22]. For semantic purposes, however, this model is utterly inappropriate due to its fixed level of abstraction. The origin of Abstract State Machines (formally called *evolving algebras* [23]) was the idea to devise a generalized machine model so that any algorithm, never mind how abstract, can be modeled at its natural level of abstraction. That is, every computation step of the algorithm essentially has a direct counterpart (usually a single step) performed by the machine model. Theoretical foundations show that both the notion of sequential algorithm and of parallel algorithm are captured respectively by the models of *sequential ASM* [12] and *parallel ASM* [24] in the aforementioned sense. For distributed algorithms (including concurrent and reactive systems), the *distributed ASM* framework provides a generalization of the two other models that is characterized by its asynchronous computation model with multiple computational agents operating concurrently.

This section outlines the basic concepts for modeling behavioral aspects of distributed systems as abstract machine *runs* performed by a distributed ASM. The description builds on common notions and structures from computational logic and discrete mathematics. For further details, we refer to [11, 12].

3.1.1 Concurrency, Reactivity and Time

A distributed ASM, or DASM, defines the concurrent and reactive behavior of a collection of autonomously operating computational agents that cooperatively perform distributed computations. Intuitively, every computation step of the DASM involves one or more agents, each performing a single computation step according to their local view of a globally shared machine state. The underlying semantic model regulates interactions between agents so that potential conflicts are resolved according to the definition of *partially ordered runs* [23].[2]

A DASM M is defined over a given vocabulary V by its program P_M and a non-empty set I_M of initial states. V consists of some finite collection of symbols for denoting the mathematical objects and their relation in the formal representation of M, where we distinguish *domain symbols*, *function symbols* and *predicate symbols*. Symbols that have a fixed interpretation regardless of the state of M are called *static*; those that may have different interpretations in different states of M are called *dynamic*. A state S of M results from a valid interpretation of all the symbols in V and constitutes a variant of a first-order structure, one in which all relations are formally represented as Boolean-valued functions.

Concurrent control threads in an execution of P_M are modeled by a dynamic set AGENT of computational *agents*. This set may change dynamically over runs

[2] For illustrative application examples, see also [25, 26].

of M, as required to model a varying number of computational resources. Agents interact with one another, and typically also with the operational environment of M, by reading and writing shared locations of a global machine state.

P_M consists of a statically defined collection of agent programs $P_{M_1}, \ldots, P_{M_k}$, $k \geq 1$, each of which defines the behavior of a certain *type* of agent in terms of state transition rules. The canonical rule consists of a basic update instruction of the form

$$f(t_1, t_2, \ldots, t_n) := t_0,$$

where f is an n-ary dynamic function symbol and each t_i $(0 \leq i \leq n)$ a term. Intuitively, one can perceive a dynamic function as a finite function table where each row associates a sequence of argument values with a function value. An update instruction specifies a *pointwise* function update: an operation that replaces a function value for specified arguments by a new value to be associated with the same arguments. In general, rules are inductively defined by a number of well defined rule constructors, allowing the composition of complex rules for describing sophisticated behavioral patterns.

A computation of M, starting with a given initial state S_0 from I_M, results in a finite or infinite sequence of consecutive state transitions of the form

$$S_0 \xrightarrow{\Delta_{S_0}} S_1 \xrightarrow{\Delta_{S_1}} S_2 \xrightarrow{\Delta_{S_2}} \cdots,$$

such that S_{i+1} is obtained from S_i, for $i \geq 0$, by firing Δ_{S_i} on S_i, where Δ_{S_i} denotes a finite set of updates computed by evaluating P_M over S_i. Firing an update set means that all the updates in the set are fired simultaneously in one atomic step. The result of firing an update set is defined if and only if the set does not contain any conflicting (inconsistent) updates.

M interacts with a given operational environment – the part of the external world visible to M – through actions and events observable at external interfaces, formally represented by externally controlled functions. Intuitively, such functions are manipulated by the external world rather than agents of M. Of particular interest are *monitored functions*. Such functions change their values dynamically over runs of M, although they cannot be updated internally by agents of M. A typical example is the abstract representation of global system time. In a given state S of M, the global time (as measured by some external clock) is given by a monitored nullary function *now* taking values in a linearly ordered domain $\mathsf{TIME} \subseteq \mathsf{REAL}$. Values of *now* increase monotonically over runs of M. Additionally, $'\infty'$ represents a distinguished value of TIME, such that $t < \infty$ for all $t \in \mathsf{TIME} - \{\infty\}$. Finite time intervals are given as elements of a linearly ordered domain DURATION.

The ASM concept of *physical time* is defined orthogonally to the concept of state transition, flexibly supporting a wide range of time models, also including continuous time [27]. A frequently used model is that of distributed real-time ASM with time values ranging over positive real numbers.

3.1.2 ASM Ground Models

The ASM formalism and abstraction principles are known for their versatility in mathematical modeling of algorithms, architectures, languages, protocols and apply to virtually all kinds of sequential, parallel and distributed systems. Widely recognized ASM applications include semantic foundations of programming languages, like JAVA [28], C# [29] and Prolog [30], industrial system design languages, like BPEL [31], SDL [32], VHDL [33] and SystemC [34], embedded control systems [35], and wireless network architectures [36].[3] Beyond hardware and software systems, this approach has been used more recently in computational criminology [8, 9] and for modeling and validation of aviation security [37, 38]. A driving factor in many of the above applications is the desire to systematically reveal abstract architectural and behavioral concepts inevitably present in every system design, however hidden they may be, so that the underlying *blueprint* of the functional system requirements becomes clearly visible and can be checked and examined by analytical means based on human expertise. This idea is captured by the notion of *ASM ground model* [39] as explained below.

Intuitively, a ground model serves as a precise and unambiguous foundation for establishing the characteristic dynamic properties of a system under study in abstract functional and operational terms with a suitable degree of detail that does not compromise conceivable refinements [40]. A ground model can be inspected by analytical means (verification) and empirical techniques (simulation) using machine assistance as appropriate. Focusing on semantic rather than on syntactic aspects, the very nature of ASM ground models facilitates the task of critically checking the consistency, completeness and validity of the resulting behavioral description. Depending on the choice and representation of the ground model, the transformation from the mathematical to a computational model can be less problematic, whereas the validation of the outcome of the computational phase usually poses another difficult problem.

3.2 *Rapid Prototyping with* CoreASM

CoreASM is an open source project[4] focusing on the design and development of an extensible, executable ASM language, together with a tool environment that supports *high-level design* in application-domain terms, and *rapid prototyping* of executable ASM models [14, 13]. The tool environment consists of a (1) platform-independent extensible engine for executing the language, (2) various plugins that extend the language and the behavior of the engine, and (3) an IDE for interactive visualization and control of simulation runs. The design of CoreASM is novel and the underlying design principles are unprecedented among the existing executable

[3] See also the ASM Research Center at http://www.asmcenter.org.

[4] CoreASM is registered at http://sourceforge.net/projects/coreasm.

ASM languages, including the most advanced ones: Asmeta [41], AsmL [42], the ASM Workbench [43], XASM [44], and AsmGofer [45].[5]

The CoreASM language and tool suite specifically support the early phases of the software and system design process, emphasizing freedom of experimentation and the evolutionary nature of design as a creative activity. By minimizing the need for encoding in mapping the problem space to a formal model, the language allows writing highly abstract and concise specifications, starting with mathematically-oriented, abstract and untyped models, gradually refining them down to more concrete versions with a degree of detail and precision as needed. The principle of minimality, in combination with the robustness of the underlying mathematical framework, makes design for change feasible, effectively supporting the highly iterative nature of modeling complex system behavior.

Executable specifications offer invaluable advantages in model-based systems engineering, serving as a tool for design exploration and experimental validation through simulation and testing [47]. Pertinent to our purpose, they greatly facilitate validating a ground model by executing different scenarios and comparing the resulting behavior with the behavior *expected* by the domain experts. In many cases, observation of system behavior can lead to discovering *new* concepts or elements in the underlying system that may have been previously neglected.

3.3 Interactive Design with Control State ASMs

One of the fundamental principles of our approach is the direct involvement of non-computing experts in the design and development process. Arbitrary design choices made by computing experts not intimately familiar with the social system under study are potentially dangerous and can lead to fatal design flaws due to misconceptions or oversights. However, it is usually difficult for non-computing team members to understand the development process and especially the formal representation of a system. Hence, it is necessary to make development as transparent as possible, for instance, by using visual representation means, such as *ASM control state diagrams (CSD)*[6] as illustrated in Fig. 2. Despite similarity to the more complicated UML activity diagrams, ASM CSDs do not require any special training to understand. Their simplicity allows the interdisciplinary reader to focus on the content of the description rather than the formalism. The accessibility and ease of use of CSDs make them an integral part of our design process. In our experience, the domain experts were able to understand a CSD, and even suggest changes to it, regardless of their technical background. As such, CSDs act as both a means of clarifying communication between development partners and of enabling straight-forward validation.

---

[5] An in-depth introduction to the architecture of the CoreASM engine and its extensibility mechanisms is provided in [13, 46].

[6] Control state ASMs provide "a normal form for UML activity diagrams and allow the designer to define machines which below the main control structure of finite state machines provide synchronous parallelism and the possibility to manipulate data structures" [11].

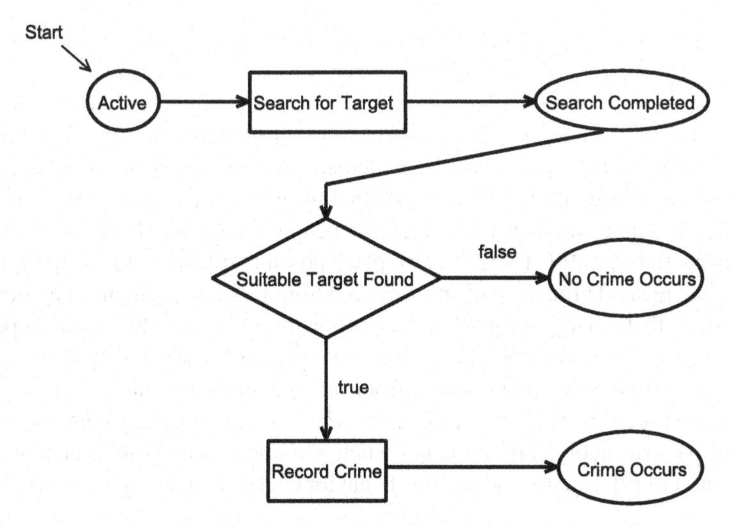

Fig. 2 CSDe: A Control State Diagram editor plugin for the Eclipse development environment, with automatic translation to CoreASM code.

We have further facilitated the involvement of non-computing experts in the development process by the construction of an ASM Control State Diagram editor (CSDe). This editor allows not only the construction and editing of CSDs through a graphical interface, but also automatic translation of the diagrams into the CoreASM language[7].

4 Mastermind: Modeling Criminal Activity

In this section, we describe the scope of the Mastermind project, its development phases, and the core technical aspects of the Mastermind system architecture.

4.1 Overview

Mastermind is a pioneering project in Computational Criminology, employing formal modeling and simulation as tools to investigate offenders' behavior in an urban environment. The goal is to capture the complexity and diversity of criminal behavior in a robust and systematic way. A variety of software development methods were applied and constantly reviewed with respect to their usability, expressiveness

[7] A diagram contains no initial state, so the code may not be immediately executable. However, it will provide a structural foundation for an executable model.

and effectiveness, the result of which has lead to the development of the modeling framework presented in Sect. 2.3.

Crime is understood to be comprised of four main elements: the law, the offender, the target and the location [1]. We construct a multi-dimensional model of crime in order to study the interactions of these elements. Our focus is on the concepts of environmental criminology, which argues that in spite of their complexity, criminal events can be understood in the context of people's movements in the course of everyday routines [1, 48]. Therefore, we place possible offenders in an environment they can navigate. Through their movement within this environment, they develop mental maps that correspond to the ideas of *awareness space* (the places a person knows) and *activity space* (the places a person regularly visits) [1, 49]. In the course of a routine activity, the agents move from one location to another, and may visit potential targets on the way [48]. In its core, Mastermind captures what is suggested by crime pattern theory: crime occurs when a motivated individual encounters a suitable target [49]. Figure 2 shows this behavior captured in terms of a control state ASM.

The main building block of Mastermind is a robust ASM ground model developed through several iterations. To this end, we applied a simple graphical notation for communicating the design (using CSDe) and utilized abstract executable models in early stages of design (using CoreASM). Furthermore, the ground model is refined into more concrete models with specific details systematically added, an example of which is the simulation model of Mastermind implemented in Java. This version provides a responsive user interface and a simulation environment based on real-world Geographical Information System (GIS) data. We also refined the CoreASM executable ground model to derive specific refinements to create more controlled experiments, which allow for a structured analysis of theories in a hypothetical world. Both versions also provide visualization features which are a priority for criminology publications.

The results of our work on the Mastermind project have been well received both by the researchers in academia and law enforcement officials. For additional information on the project and the results, we also refer to [8, 9] and the project website[8]. Next, we describe the main components of the Mastermind architecture, highlighting several key aspects.

4.2 Agent Architecture

The central component of our model is an autonomously acting entity, called a *person agent*, which represents an individual living in an urban environment and commuting between home, work, and recreation locations. Person agents *navigate* within the environment and may assume different roles such as offender, victim, or guardian; depending on the role they exhibit different behaviors.

[8] http://stl.sfu.ca/projects/mastermind/

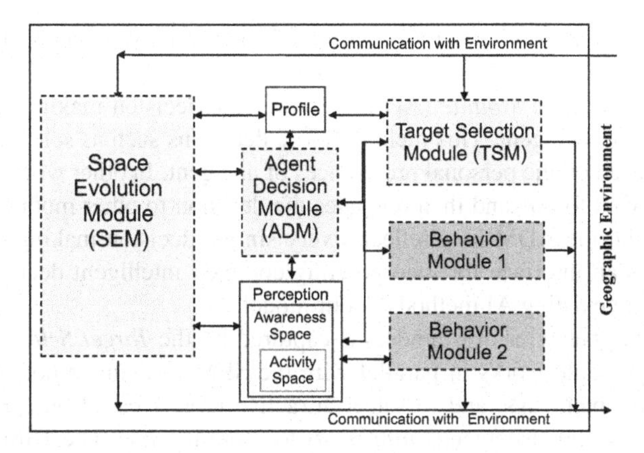

Fig. 3 The architecture of a *person* agent

The agent architecture presented here provides a robust yet flexible design to capture different aspects of an individual's behavior. Intuitively, it is based on a Belief-Desire-Intention (BDI) model [50], providing a structural decomposition of the behavior into different logical components as illustrated in Fig. 3. Each component captures certain aspects of the overall behavior following the classical Divide and Conquer approach.

An agent's personal attributes and preferences are represented by the *profile*. The profile is a repository of all the factors that are specific to an individual agent and have an impact on the behavior under study. These factors include agents' skills, activity nodes, or demographic factors such as age and sex.

To model the urban environment, we follow extant theories of environment in behavioral sciences and divide the environment into two broad categories: the *objective* and the *subjective* environment [51]. The objective environment is external to an agent and encompasses the totality of all things jointly forming the physical reality. The subjective environment, on the other hand, refers to the *perception* of an agent, i. e., a filtered view of the objective environment as an agent perceives it. The perception is modeled as a memory-like repository that is constantly updated as the agent moves in the environment and commits crime. An agent's perception is further divided into two sub-categories [52]. The part of the perception that an agent is aware of through current events, past experiences and interaction with other agents forms the *awareness space* of the agent. The *activity space* is the part of the awareness space that the agent has visited more frequently over a recent period of time. The agent typically has very detailed information about this part of the environment.

A person's navigation behavior is modeled by the *Space Evolution Module* (SEM). It provides a navigation algorithm to move the agent from an origin to a destination considering the particular preferences of the agent. These preferences reflect the importance of different factors in navigation for different types of agent. For instance, teenagers have different priorities in finding their paths compared to working adults. The SEM is also responsible for recording the paths frequently used by agents, which in turn leads to formation of their activity spaces and awareness spaces.

The *Agent Decision Module* (ADM) captures the decision making process that sets the *goals* of the agent. This includes basic decisions such as selecting the next destination based on the personal preferences of an agent. In other words, the ADM decides on "what to do" and then relegates the decision to other modules on "how to do it". While the ADM may reflect a very simple decision making behavior, it is designed as an interface for incorporating complex intelligent decision making behaviors using existing AI methods in our model.

The criminal behavior of offenders is captured by the *Target Selection Module* (TSM). This module works in parallel with the SEM to monitor potential targets on the familiar pathways, and to select attractive targets based on agent-specific selection criteria and also agents' *propensity* to commit crime. The TSM carves out the *crime occurrence space* of an agent at the micro-level, which leads to formation of *crime patterns* at the macro-level.

We like to emphasize the flexibility of this architecture for adding additional behaviors to the model. For instance, a module can be added to model social interactions between agents that lead to formation of *social networks*. Several factors such as common spatio-temporal aspects or common criminal goals may be considered in the evolution of social networks and captured by the module. Similarly, different roles such as victims or police officers can be modeled by adding respective behavior modules.

4.3 Urban Landscape Model

We abstractly model the physical environment as representing some urban landscape with an *attributed directed graph*. This model potentially includes everything from road and rail traffic networks to walkways and hiking trails in city parks. In principle, it may also extend to the layout of public spaces such as shopping malls, underground stations, and even airports and seaports. In the following, we concentrate on street networks, although the same modeling approach applies to virtually any type of urban traffic and transportation system. We gradually define the physical environment model in several steps as follows.

Let $H = (V,E)$ be a directed connected graph representing the detailed street network of some urban area as specified by a city map or, more adequately, by a Geographic Information System (GIS). Let $V = \{v_1,\ldots,v_n\}$ be the set of vertices representing the intersections and other distinguished points of interest located on or next to a road, such as highway exit and entry points, gas stations, recreational facilities and shopping centers. Further, let $E \subseteq V \times V$ be the set of directed edges representing the identifiable road segments; unidirectional road segments are repre-

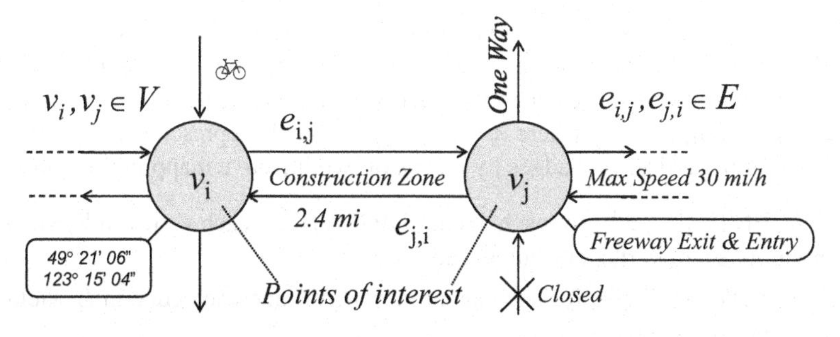

Fig. 4 Geographic Environment

sented by a single edge and bidirectional ones by a pair of oppositely directed edges connecting the same two vertices.[9]

For labeling the edges and vertices of H, let Θ_e and Θ_v denote two disjoint sets of labels, called *edge attributes* and *vertex attributes* respectively. Θ_e splits into two disjoint subsets, Θ_e^{stat} and Θ_e^{dyn}, the edge attributes that are statically defined – like distances, road types and speed limits, for instance – and those that may (and typically do) change dynamically depending on various factors, including weather phenomena affecting road conditions, time of the day affecting traffic conditions, and special events – like road blockages and closures due to construction work, emergency response units, or other factors. In contrast, vertex attributes specify information on locations and characteristic features, such as geographic coordinates, highway exit numbers, as well as other, more specific information related to points of interest.

Next, we define the *geographic environment* as an attributed directed graph $G_{GeoEnv} = (H, \theta)$ by associating a non-empty set of attributes with each of the vertices and edges of H. We therefore introduce a labeling scheme $\theta = (\theta_v, \theta_e)$, with $\theta_e = (\theta_e^{stat}, \theta_e^{dyn})$ consisting of three finite mappings as follows:

1. $\theta_v : V \rightarrow 2^{\Theta_v}$ assigns a finite set of vertex attributes to each vertex in V.
2. $\theta_e^{stat} : E \rightarrow 2^{\Theta_e^{stat}}$ assigns a finite set of static edge attributes to each edge in E.
3. $\theta_e^{dyn} : E \rightarrow 2^{\Theta_e^{dyn}}$ assigns a finite set of dynamic edge attributes to each edge in E.

Figure 4 illustrates the representation of the geographic environment for a simple example consisting of two interconnected points of interest.

G_{GeoEnv} represents the objective urban environment – the physical reality – and serves as the basis for defining an agent's subjective perception of this environment (cf. Sect. 4.2). We model perception by introducing an additional labeling on top of G_{GeoEnv}. The fact that, in general, each agent perceives the geographic environment differently implies that each agent also sees different agent-specific attributes associated with certain edges and vertices of G_{GeoEnv}.

[9] Refining the granularity, one may also represent the individual lanes of a given street network in exactly the same way.

Let Λ_v and Λ_e denote two additional sets of labels for vertices and edges respectively. The urban environment, integrating both the objective environment and the subjective environment for each of the agents, is defined as an attributed directed graph $G_{Env} = (G_{GeoEnv}, \lambda)$ where $\lambda = (\lambda_v, \lambda_e)$ abstractly represents the agent specific labeling of vertices and edges by means of two injective mappings as follows:

- λ_v : AGENT$\times V \rightarrow 2^{\Lambda_v}$, for each agent in *AGENT* and each vertex in V, yields a non-empty set of vertex attributes, and
- λ_e : AGENT$\times E \rightarrow 2^{\Lambda_e}$, for each agent in *AGENT* and each edge in E, yields a non-empty set of edge attributes.

G_{Env} can be seen as a attributed directed graph with *colored attributes*. Each color refers to the specific perception of an individual agent. Λ_v, for instance, specifies the frequency of visits to a location as well as the agent's subjective interest in this location. Λ_e, for instance, specifies the frequency of using a road, reinforcement factors, and intensity of awareness and activity.

Finally, the awareness space and activity space of each agent in any given system state is computed from the abstract representation of the urban environment by means of operations on G_{Env} that extract the subset of edges with an associated intensity above a certain threshold. Likewise, the opportunity space for a certain type of crime is encoded. The crime occurrence space of an agent for a certain type of crime is a subset of the intersection of the opportunity space and the activity space of an agent.

4.4 Space Evolution Module: ASM Model

The main responsibility of the SEM is to model how a person agent navigates the urban environment G_{Env} (cf. Sect. 4.3) during the course of his or her daily routine activities. Intuitively, the SEM moves a person agent in discrete steps from his or her current position on the graph – a vertex or an edge as identified by functions *currentNode* and *currentEdge* – to the destination. It also keeps track of the places visited by the agent, leading to the evolution of the agent's activity space, thus affecting the attribute values of G_{Env}. Such attributes are accessed and manipulated through operations on the graph structure.

The SEM model presented here has gone through several iterations in order to capture different variations of agent navigation behavior in a flexible and robust manner. Abstractly speaking, given an origin and a destination, the SEM finds a "potential" path that reflects the specific preferences of the agent. It then moves the agent on this path, traversing one road segment (edge) at the time. However, at anytime, due to a variety of reasons the agent may divert from this path, e.g. deciding abruptly to take a random road, or being forced to take an alternate road due to the traffic. At that point, it is required to re-route the agent toward the destination by finding a new path. The trip ends when the agent reaches the destination.

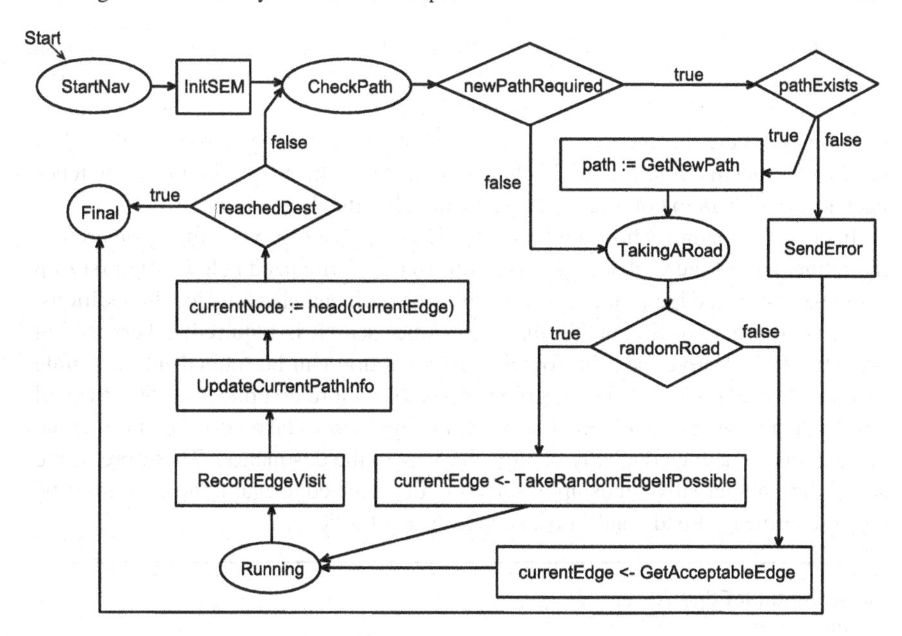

Fig. 5 Control State Diagram of the Space Evolution Module (SEM)

We formalize the operation of the SEM in terms of a control-state ASM, as illustrated in Fig. 5. The *control states* distinguish different *modes* of operation of the SEM. At every step, the SEM is in a certain control state which defines its behavior. Transitions between control states occur when the conditions stated by the respective *guards* hold. *Update rules*, shown in rectangular boxes, define the ASM rules that update the underlying data structure.

The operation of the SEM starts in the *StartNav* mode, where the SEM is initialized and the mode switches to *CheckPath*. In this mode, the SEM first checks if a new path is required (*newPathRequired*), i. e., a new trip is started, or the previously found path is no longer appropriate for any reason. In that case, a new path to the destination is chosen. It is important to note that, at this point, we abstract away from the details of the specific navigation algorithm used for path finding and add the details incrementally in further refinement steps.[10] When a path to the destination is found, the SEM mode changes to *TakingARoad* where a road is chosen to be traversed by the agent. Here, the roads on the existing path are examined on the fly and several possibilities are considered: (1) the agent may decide to take a random road, or (2) the SEM attempts to take the next road on the existing suggested path to the destination. However, this road may not be *acceptable* due to local conditions such as traffic or construction. Thus, (a) the next road on the path is taken if the local conditions are acceptable, or there is no other alternative; (b) otherwise, an

[10] This path, called *suggestedPath*, is calculated using the global information that the agent has about its environment. This includes using a map, or referring to the memory, if the destination is previously known.

alternate road is chosen instead. Here, again we leave the details on how to choose an alternate road abstract: it could be refined to simply choose a random road, or a more complex approach may be used. Once the road is chosen, the mode switches to *Running* where the agent is essentially moving on the selected road. If the agent reaches the destination, the mode changes to *Final*. Otherwise, the mode switches back to *CheckPath* to continue the agent's travel to the destination.

In order to exemplify how the abstract ASM model of Fig. 5 is refined, we present the refinement of the GetAcceptableEdge rule as formalized below. The first step is to take the next edge available on the existing path and check its local conditions. If the conditions are acceptable then no further search is required; otherwise, an acceptable alternative must be found. This operation can be refined into a simple random choice among all the outgoing edges, or a more complex one as presented here. When finding a viable alternative, it is important to take into account that the current unacceptable edge may be the only way to the destination. Therefore, if the search for an alternative ends up in selecting the same edge again, the edge must be taken regardless of bad conditions, such as heavy traffic.

```
GetAcceptableEdge(me) ≡
  return res in
    let e = GetNextEdgeOnPath(me) in
      if acceptable(e,me) then
        res := e
      else
        res := FindAcceptableAlternative(e,me)

FindAcceptableAlternative(e,me) ≡
  return r in
    if attemptedBefore(e,me) then
      r := e
    else
      par
        r := TakeRandomEdgeIfPossible(me)
        attemptedBefore(e,me) := true
      endpar
```

For a complete description of the SEM, including the (sub-)rules left abstract here, the reader is referred to [53].

4.5 Lessons Learned

Mathematical modeling of criminal activity in the form of discrete event models that define the cooperative behavior of multiple autonomous agents in abstract operational and functional terms has proven to be a sensible choice in the cross-disciplinary research context of the Mastermind project. The task of analyzing and reasoning about complex crime activity patterns and their representation in computational form, in an attempt to explain real-world phenomena, requires an amalgamation of expertise from criminology and computer science. Hence, coherence and consistence of a common conceptual view, one that has a virtually identical meaning in both worlds, is crucial for overcoming fatal misconceptions, especially in the transition between modeling phases (cf. Sect. 2.3).

A typical challenge in the early formalization phases is lack of a thorough understanding of the functional requirements, resulting in vague descriptions and fuzzy architectural concepts. Real-life events are not usually thought of in a discrete, mathematical manner that would easily transform into something computable. Striving for more clarity and regularity, any encoding should not only be minimal but also be as direct and intuitive as possible, so as to not add extra weight to the overall problem scope. In light of such practical needs, the relatively simple semantic foundation of the ASM formalism contributes a great deal to the ease of using this approach as a precise analytical means for communicating ideas and for reasoning about design decisions. Viewing states as first-order structures, or simply as abstract *mathematical data structures* being manipulated by autonomously operating computational agents, indeed greatly simplifies the formalization task. A minimal instruction set for describing state transition behavior in CoreASM based on an untyped language, combined with flexible extensibility and refinement techniques, facilitates experimentation by supporting design for change. Finally, the ability to freely choose and mix common modeling styles, e. g., declarative, functional and operational, depending on what appears most suitable in a given application context, is invaluable.

An important part of the modeling exercise is identifying the right level of granularity required for modeling behavior at the micro-level and investigating the impact on the macro-level behavior patterns. To facilitate this proces, we use CoreASM to run experiments in very early stages of design. Through these experiments, we were able to identify key elements which impact the macro-level patterns of behavior, but were left unnoticed at the micro-level. For instance, the specific method used by agents for finding a path to the destination (e.g. completely deterministic vs. random walk) is expected to have a huge impact on macro-level crime patterns. However, the experiments showed that the impact of the *boundaries* of the environment, such as the size of the road network, could be even stronger than individual path finding preferences. Such results re-affirm the benefits of computational models in developing and testing theories of crime.

The Java version of Mastermind is capable of simultaneously processing the daily activities of multiple agents, whose activity spaces and selected targets are displayed on-screen. With its recognizable landscape and dynamic agents, this simulation model turned out to be particularly effective at illustrating concepts in Environmental Criminology to non-criminologists [49]. However, this version of Mastermind was too intimidating for criminologists to be used as a tool in experimentation. Even though the behavior of the agents *appeared* to follow established theory, they could not be confident that the program semantics followed their understanding of the phenomena. Furthermore, if the behavior of the program is not clearly explained, the produced results are not useful in an academic publication: it is impossible to peer-review a black box. Consequently, we focused on developing experiments using the CoreASM version. The CoreASM program code is easier for a non-specialist to read, and it is well-suited for designing *controlled* experiments. The results of our experiments using CoreASM have been more focused and useful for satisfying criminology queries.

5 Concluding Remarks

We propose a novel approach to the modeling of crime patterns and theories in crime analysis and prevention, a key aspect in Computational Criminology, based on the ASM formalism and CoreASM for interactive design and validation of distributed ASM models. Our model is designed in such a way that it is scalable and not only applicable to a broad range of crimes but also to other environments such as airports, ports, shopping centers, and subway stations. Pattern theory and routine activity theory suggest that through a combination of decisions and innate understanding of environmental cues, likely offenders are able to separate good criminal opportunities from bad risks. The nature of this process is highly structured and allows researchers to anticipate likely concentrations of a variety of regular, daily activities, including criminal offences.

Mathematical and computational modeling of crime serves multiple purposes. It has a direct value in law enforcement, in intelligence-led policing, and in proactive crime reduction and prevention. For intelligence-led policing, our model would make it possible to predict likely activity spaces for serial offenders for use in apprehension or developing precautions. For proactive policing, modeling of crime makes it feasible to build scenarios in crime analysis and prevention, and provides a basis for experimental research, allowing experiments that can often not easily be done in a real-world setting. Beyond crime analysis and prevention, one may adopt the Mastermind approach to counter-terrorism, specifically, in critical infrastructure protection, for instance with intruder detection, or in public safety for improving security measures for protecting public spaces.

Although unconventional, the application of the ASM paradigm to social systems turns out to be a promising approach: it nicely combines with the established view of multi-agent modeling of social systems and provides a precise semantic foundation – something multi-agent system modeling is lacking. Our main theoretical result is the abstract behavior model of person agents (based on the *agent architecture*) interacting with their objective and subjective environments which jointly form the *geographic environment*. Our main practical result is the Mastermind system architecture which serves as a platform for the construction and experimental validation of discrete event simulation models.

References

1. Brantingham, P.J., Brantingham, P.L.: Patterns in Crime. New York: Macmillan Publishing Company (1984)
2. Brantingham, P.J., Brantingham, P.L.: Environment, Routine and Situation: Toward a Pattern Theory of Crime. Routine Activity and Rational Choice: Advances in Criminological Theory (1993) 259–294
3. D'Amico, J.: Stopping Crime in Real Time. The Police Chief – The Professional Voice of Law Enforcement (July 2008) URL: http://policechiefmagazine.org/magazine/.
4. Liu, L., Eck, J., eds.: Artificial Crime Analysis Systems: Using Computer Simulations and Geographic Information Systems. Information Science Reference (January 2008)
5. Frank, U., Troitzsch, K.G.: Epistemological Perspectives on Simulation. Journal of Artificial Societies and Social Simulation **8**(4)
6. Groff, E., Birks, D.: Simulating crime prevention strategies: A look at the possibilities. Policing: A journal of Policy and Practice **2**(2) 175–184
7. Brantingham, P.J., Brantingham, P.L., Glässer, U.: Computer Simulation as a Research Tool in Criminology and Criminal Justice. Criminal Justice Matters (58) (February 2005)
8. Brantingham, P.L., Kinney, B., Glässer, U., Jackson, P., Vajihollahi, M.: Mastermind: Computational Modeling and Simulation of Spatiotemporal Aspects of Crime in Urban Environments. In Liu, L., Eck, J., eds.: Artificial Crime Analysis Systems: Using Computer Simulations and Geographic Information Systems. Information Science Reference (2008)
9. Brantingham, P.L., Glässer, U., Kinney, B., Singh, K., Vajihollahi, M.: A Computational Model for Simulating Spatial Aspects of Crime in Urban Environments. In Jamshidi, M., ed.: Proceedings of the 2005 IEEE International Conference on Systems, Man and Cybernetics. (October 2005) 3667–74
10. Brantingham, P.L., Glässer, U., Kinney, B., Singh, K., Vajihollahi, M.: Modeling Urban Crime Patterns: Viewing Multi-Agent Systems as Abstract State Machines. In Beauquier, D., et al., eds.: Proceedings of the 12th International Workshop on Abstract State Machines (ASM'05). (March 2005)
11. Börger, E., Stärk, R.: Abstract State Machines: A Method for High-Level System Design and Analysis. Springer-Verlag (2003)
12. Gurevich, Y.: Sequential Abstract State Machines Capture Sequential Algorithms. ACM Transactions on Computational Logic **1**(1) (July 2000) 77–111
13. Farahbod, R., Gervasi, V., Glässer, U.: CoreASM: An Extensible ASM Execution Engine. Fundamenta Informaticae (2007) 71–103
14. Farahbod, R., Glässer, U., Jackson, P., Vajihollahi, M.: High Level Analysis, Design and Validation of Distributed Mobile Systems with CoreASM. In: Proceedings of 3rd International Symposium On Leveraging Applications of Formal Methods, Verification and Validation (ISoLA 2008). (October 2008)
15. Wing, J.M.: Computational Thinking. Communications of the ACM **4**(3) (2006) 33–35
16. Xue, Y., Brown, D.: A decision model for spatial site selection by criminals: a foundation for law enforcement decision support. Systems, Man, and Cybernetics, Part C: Applications and Reviews, IEEE Transactions on **33** (2003) 78–85
17. Baumgartner, K., Ferrari, S., Salfati, C.: Bayesian network modeling of offender behavior for criminal profiling. In: 44th IEEE Conference on Decision and Control and European Control Conference CDC-ECC'05. (2005) 2702–2709
18. Clarke, R.V.: Technology, Criminology and Crime Science. European Journal on Criminal Policy and Research **10**(1) (2004) 55–63
19. Dean, G.: Criminal profiling in a terrorism context. Criminal Profiling International Theory, Research, and Practice (2007)
20. Elliot, E., Kiel, L.D.: Towards a new vision of complexity. Chaos, Solutions & Fractals **20**(1) (April 2004) 63–68
21. Raczynski, S.: Simulation of the dynamic interactions between terror and anti-terror organizational structures. Journal of Artificial Societies and Social Simulation **7**(2) (2004)

22. Sipser, M.: Introduction to the Theory of Computation. PWS Publishing Company (1997)
23. Gurevich, Y.: Evolving Algebras 1993: Lipari Guide. In Börger, E., ed.: Specification and Validation Methods. Oxford University Press (1995) 9–36
24. Blass, A., Gurevich, Y.: Abstract State Machines Capture Parallel Algorithms. ACM Transactions on Computation Logic **4**(4) (2003) 578–651
25. Glässer, U., Gurevich, Y., Veanes, M.: Abstract Communication Model for Distributed Systems. IEEE Trans. on Soft. Eng. **30**(7) (July 2004) 458–472
26. Farahbod, R., Glässer, U.: Semantic Blueprints of Discrete Dynamic Systems: Challenges and Needs in Computational Modeling of Complex Behavior. In: New Trends in Parallel and Distributed Computing, Proc. 6th Intl. Heinz Nixdorf Symposium, Jan. 2006, Heinz Nixdorf Institute (2006) 81–95
27. Gurevich, Y., Huggins, J.: The Railroad Crossing Problem: An Experiment with Instantaneous Actions and Immediate Reactions. In: Proceedings of CSL'95 (Computer Science Logic). Volume 1092 of LNCS., Springer (1996) 266–290
28. Stärk, R., Schmid, J., Börger, E.: Java and the Java Virtual Machine: Definition, Verification, Validation. Springer-Verlag (2001)
29. Börger, E., Fruja, N.G., Gervasi, V., Stärk, R.F.: A High-level Modular Definition of the Semantics of C#. Theoretical Comp. Sci. **336**(2/3) (May 2005) 235–284
30. Börger, E.: A Logical Operational Semantics for Full Prolog. Part I: Selection Core and Control. In Börger, E., Kleine Büning, H., Richter, M.M., Schönfeld, W., eds.: CSL'89. 3rd Workshop on Computer Science Logic. Volume 440 of LNCS., Springer (1990) 36–64
31. Farahbod, R., Glässer, U., Vajihollahi, M.: An Abstract Machine Architecture for Web Service Based Business Process Management. International Journal of Business Process Integration and Management **1** (2007) 279–291
32. Glässer, U., Gotzhein, R., Prinz, A.: The Formal Semantics of SDL-2000: Status and Perspectives. Computer Networks **42**(3) (2003) 343–358
33. Börger, E., Glässer, U., Müller, W.: Formal Definition of an Abstract VHDL'93 Simulator by EA-Machines. In Delgado Kloos, C., Breuer, P.T., eds.: Formal Semantics for VHDL. Kluwer Academic Publishers (1995) 107–139
34. Müller, W., Ruf, J., Rosenstiel, W.: An ASM Based SystemC Simulation Semantics. In Müller, W., et al., eds.: SystemC - Methodologies and Applications, Kluwer Academic Publishers (June 2003)
35. Beierle, C., Börger, E., Durdanovic, I., Glässer, U., Riccobene, E.: Refining Abstract Machine Specifications of the Steam Boiler Control to Well Documented Executable Code. In Abrial, J.R., Börger, E., Langmaack, H., eds.: Formal Methods for Industrial Applications. Specifying and Programming the Steam-Boiler Control. Number 1165 in LNCS. Springer (1996) 62–78
36. Glässer, U., Gu, Q.P.: Formal Description and Analysis of a Distributed Location Service for Mobile Ad Hoc Networks. Theoretical Comp. Sci. **336** (May 2005) 285–309
37. Glässer, U., Rastkar, S., Vajihollahi, M.: Computational Modeling and Experimental Validation of Aviation Security Procedures. In Mehrotra, S., Zeng, D.D., Chen, H., Thuraisingham, B.M., Wang, F.Y., eds.: Intelligence and Security Informatics, IEEE International Conference on Intelligence and Security Informatics, ISI 2006, San Diego, CA, USA, May 23-24, 2006, Proceedings. Volume 3975 of Lecture Notes in Computer Science., Springer (2006) 420–431
38. Glässer, U., Rastkar, S., Vajihollahi, M.: Modeling and Validation of Aviation Security. In Chen, H., Yang, C., eds.: Intelligence and Security Informatics: Techniques and Applications. Volume 135 of Studies in Computational Intelligence. Springer (2008) 337–355
39. Börger, E.: The ASM ground model method as a foundation of requirements engineering. In N.Dershowitz, ed.: Verification: Theory and Practice. Volume 2772 of LNCS., Springer-Verlag (2003) 145–160
40. Börger, E.: Construction and Analysis of Ground Models and their Refinements as a Foundation for Validating Computer Based Systems. Formal Aspects of Computing **19**(2) (2007) 225–241
41. Gargantini, A., Riccobene, E., Scandurra, P.: A Metamodel-based Simulator for ASMs. In: Proc. of the 14th Intl. Abstract State Machines Workshop. (June 2007)

42. Microsoft FSE Group: The Abstract State Machine Language. (2003) Last visited June 2003, http://research.microsoft.com/fse/asml/.
43. Del Castillo, G.: Towards Comprehensive Tool Support for Abstract State Machines. In Hutter, D., Stephan, W., Traverso, P., Ullmann, M., eds.: Applied Formal Methods – FM-Trends 98. Volume 1641 of LNCS., Springer-Verlag (1999) 311–325
44. Anlauff, M.: XASM – An Extensible, Component-Based Abstract State Machines Language. In Y. Gurevich and P. Kutter and M. Odersky and L. Thiele, ed.: Abstract State Machines: Theory and Applications. Volume 1912 of LNCS., Springer-Verlag (2000) 69–90
45. Schmid, J.: Executing ASM Specitications with AsmGofer. (2005) Last visited Sep. 2005, www.tydo.de/AsmGofer/.
46. Farahbod, R., Gervasi, V., Glässer, U., Ma, G.: CoreASM plug-in architecture. In: Proceedings of the Dagstuhl Seminar on Rigorous Methods for Software Construction and Analysis (LNCS Festschrift), Springer (2008 (to be published))
47. Farahbod, R., Gervasi, V., Glässer, U., Memon, M.: Design exploration and experimental validation of abstract requirements. In: Proceedings of the 12th International Working Conference on Requirements Engineering: Foundation for Software Quality (REFSQ'06), Essener Informatik Beitrage (June 2006)
48. Felson, M.: Routine Activities and Crime Prevention in the Developing Metropolis. Criminology (1987) 911–931
49. Brantingham, P.J., Brantingham, P.L.: The Rules of Crime Pattern Theory. Environmental Criminology and Crime Analysis (2008)
50. Bratman, M.E., Israel, D., Pollack, M.E.: Plans and Resource-Bounded Practical Reasoning. Computational Intelligence **4** (1988) 349–355
51. Koffka, K.: Principles of Gestalt Psychology. Harcourt (1967)
52. Sonnenfeld, J.: Geography, Perception and the Behavioral Environment. In English, P.W., Mayfield, R.C., eds.: Man, Space and the Environment. Oxford University Press, New York (1972) 244–251
53. Brantingham, P.L., Glässer, U., Jackson, P., Vajihollahi, M.: Modeling Criminal Activity in Urban Landscapes. Technical Report SFU-CMPT-TR-2008-13, Simon Fraser University (Aug 2008)

Extracting Knowledge from Graph Data in Adversarial Settings

David Skillicorn

Abstract Graph data captures connections and relationships among individuals, and between individuals and objects, places, and times. Because many of the properties of graphs are emergent, they are resistant to manipulation by adversaries. This robustness comes at the expense of more-complex analysis algorithms. We describe several approaches to analysing graph data, illustrating with examples from the relationships within al Qaeda.

1 Characteristics of Adversarial Settings

In adversarial settings, there are some who are motivated to alter the results of data analysis for their own benefit – often to conceal their existence, properties, or actions, but sometimes also to acquire a direct benefit.

Adversaries have three opportunities to affect the data analysis. These are:

1. Avoid data collection altogether or cause the data to be collected in a way that makes analysis useless (for example, by breaking the connection between data record attributes and some identifying key).
2. Change some data records so that the analysis algorithms will produce incorrect models, models that contain blind spots or hiding places in which the records of adversaries can be concealed.
3. Choose the records and their attributes in such a way that the consequences of the constructed models will be discounted or ignored (as 'computer errors' or 'false positives') – in other words, use social engineering.

All of these attacks use the same basic mechanism – altering some aspect of the collected data records – but the nature of the alteration is different for the different forms of attack. For example, avoiding data collection directly hides records about

David Skillicorn
School of Computing, Queen's University, Kingston, Canada, e-mail: skill@cs.queensu.ca

N. Memon et al. (eds.), *Mathematical Methods in Counterterrorism*,
DOI 10.1007/978-3-211-09442-6_3, © Springer-Verlag/Wien 2009

the adversaries, while attacking the analysis adds records about someone or something else whose indirect effect is to enable the adversaries' records to be hidden.

Mainstream knowledge-discovery technologies work, in a general way, by maximizing the 'fit' between a model and the data from which it is derived (or, equivalently, minimizing some objective or error function). In an adversarial setting, this strategy is an inherently dangerous one for three reasons:

- The regions where the fit between model and data is poor tend to be ignored or characterised as 'noise' – but, in an adversarial setting, these regions may deserve the most attention.
- It hands control to adversaries, who can exploit knowledge of the fitting strategy to achieve their goals by, for example, ensuring that the fit is good in regions where they wish their data to seem innocuous.
- The amount of data that needs to be altered to affect the model is often quite small because of the way 'fit' is defined. For example, minimizing sum of squares error is statistically principled, but makes it possible for a single, unusual data record to completely alter a model.

In adversarial settings, therefore, attention needs to be paid to model-building algorithms that are hardened against attacks by adversaries. Little work in the direction has been done [1]. However, one way to avoid the problem is to use graph or relational data. Such data consists of (usually) pairwise connections between nodes that represent individuals, places, or things; and these connections are labelled with some local affinity. These *local* properties lead to a set of emergent *global* properties of the graph as a whole; and these global properties, in turn, have implications for the nodes and pairwise connections. Because the global properties are emergent, it is extremely difficult to manipulate or alter the structure of the graph by altering local affinities, and even more difficult to do so in a predictable way. Hence, analysis algorithms using graph data are naturally hardened against adversarial manipulation (even, as it turns out, against some forms of collection-avoidance strategies).

However, graph data is undoubtedly more difficult to work with than more-conventional record-based data. The algorithms are conceptually more difficult to understand, and more expensive to execute; and the presentation of the resulting models is also more difficult. Although substantial progress has been made, there remain many open problems.

In this chapter, we will examine some of the known analysis algorithms, and ways to exploit their results to understand typical graph datasets. We will use relational data on al Qaeda as a running example.

2 Sources of Graph Data

Graph data in adversarial settings arises in three main ways:

- *Surveillance*. Watching the activities of any group creates links between its members and particular places they visit, or objects they use. The nodes of the resulting graphs may be of different types, and the edges represent some contact between them. The local affinity between two nodes might, for instance, capture the frequency of contact.

- *Communication*. Any communication from one individual to another creates an arrow between them that is naturally mapped to an edge. The local affinity may capture any one of several properties of the edge such as frequency of use or total volume of use.
- *Correlation*. Given a set of data records, each record can be identified with a graph node, and edges inserted between nodes based on the correlation between the attributes of the records associated with the nodes. There are several variations: correlation or normalized correlation can be used; and correlations may be thresholded so that weak or negative correlations do not generate edges. The main advantage of this approach is that the resulting graph data structure is independent of the number of attributes in the original data records. Thus high-dimensional data is mapped to much lower dimensional data in a way that preserves much of its structure. This is especially true when the apparent high dimensionality is an artifact of the data capture: for example, a coloured image may contain a million pixel attribute values, but the objects in the image are still 3-dimensional and can be precisely described with many fewer attributes.

All three data-capture mechanisms produce graph data in which there are, say, n nodes corresponding to objects of interest, and a set of edges between them, each weighted with a positive value representing the local affinity. In practice, the number of edges is much, much smaller than the possible $n(n-1)/2$.

Graph data is usually represented by an *adjacency matrix*, a matrix with n rows and n columns, where the ijth entry is either the value of the local affinity of the edge between node i and node j, or 0 is nodes i and j are not connected. We will assume that edges are undirected, so the ijth and jith entries will always be the same. The iith entries are 0.

3 Eigenvectors and the Global Structure of a Graph

It may seem at first surprising that eigenvectors are helpful ways to understand the structure and properties of a graph, because the usual presentation of eigenvectors is in terms of a matrix acting as an affine transformation of some geometric space. The following intuition, however, illustrates the connection to graphs. Suppose that each node of a graph begins with a fixed quantity of some substance, which we will think of as importance. Plausibly, one becomes important by having affinity with other important people. So suppose each node distributes its importance to its neighbours by sending, to each of its k neighbours, $1/k$th of its total. Each node will distribute all of its importance to its neighbours, but it will receive some back as its neighbours, in turn, distribute their importances to it. The amount of outgoing and incoming importance at a node will not, in general, be the same because it depends on how richly connected its neighbours are. A node connected only to a well-connected neighbour will send it all of its importance but receive back only a small fraction.

After one round of importance redistribution, well-connected nodes will have increased their total importance while poorly-connected ones will have decreased theirs. The change reveals something about the *local neighbourhood* of each node. If the process is repeated, the importance at each node reflects properties of an increasingly large region around the node. Somewhat surprisingly, given some mild conditions on the connectivity of the graph and the local affinities, the redistribution process converges to a steady-state distribution of the importance values. The magnitudes of these values capture the global importance of each node, derived from the initial identical importances via the global connection structure of the graph. The importance values are, in fact, the principal eigenvector of the adjacency matrix of the graph (and the distribution process is the well-known power method of computing the principal eigenvector). The associated eigenvalue which, in a geometric view represents a kind of stretching or scaling, has no direct interpretation.

An eigenvector decomposition of the graph's adjacency matrix produces other eigenvectors as well (up to $n-1$ of them, in general, not n because, if the connections of the first $n-1$ nodes are known, then all of the remaining connections are also known). Each of these eigenvectors has a similar interpretation in terms of quantities passing along the edges of the graph. The orthogonality of the eigenvectors implies that all of these flows can be thought of as independent – the actual graph edges act as conduits for up to $n-1$ non-interacting flows of different properties. This is quite a realistic way to model what happens in a relational graph – the edges usually do not exist for only a single reason and are not used for only a single purpose. The eigendecomposition of the adjacency matrix separates these different modes of use of the graph (and orders them, since the magnitudes of the eigenvalues indicate the relative importance of the separate flows).

As we shall see, there are many different ways in which eigenvectors can be useful in determining the properties of graphs and their nodes.

4 Visualization

One obvious way to understand the structure and properties of a graph is to create a rendering (a drawing) and use our human visual strengths to examine it. We are extremely good at seeing patterns in our visual field, so this approach is a good symbiosis of human strengths and algorithm strengths [2].

There are, however, several problems with visualization as a strategy. These are:

- *Scalability*. It is easy to render a graph with a few hundred nodes, but when the size of the graph increases, some nodes necessarily occlude other nodes, no matter how clever the rendering. This can be mitigated, to some extent, by using powerful rendering technology. For example, using screens on more than one surface makes it possible to view the rendering from 'inside' rather than looking at it from outside; and something of the same effect can be achieved using glasses that provide a 3-dimensional view and the ability to move within the rendering. Both of these techniques increase the size of graph that can be understood by an order of magnitude, but do not solve the deeper problem.

- *Rendering the irregularities.* Most graph-rendering algorithms concentrate on rendering the main structure(s) of the graph well, pushing the small, idiosyncratic structures to the margins of the rendering. In most situations, this is the obviously right thing to do. However, in an adversarial setting, it is the small, idiosyncratic structures that are likely to be most interesting. It is not straightforward to alter rendering algorithms to make the irregularities more obvious because such irregularities are not patterns in the usual sense, so the algorithm cannot anticipate what kind of structures it should look for.
- *Training the observer.* Although humans are good at finding structure in images, it seems difficult to provide guarantees about how well an observer will do given a particular graph; and also to train observers to become better at detecting irregularities that might be interesting.

5 Computation of Node Properties

There is a long tradition of analysing graph data by using the edges, and so the global structure of the graph, to calculate properties of the individual nodes. Such properties may not be obvious from the attributes of the nodes, perhaps not even from their immediate neighbourhoods.

5.1 Social Network Analysis (SNA)

One major family of techniques that uses this approach is *Social Network Analysis* (SNA). Its origins are in the analysis of small family and social groups, and so the nodes are almost always individuals and the edges are relationships between them [3, 4, 5]. A set of node measures have been developed, describing two fundamental kinds of properties:

- How much each node is in the 'middle' of the graph, assuming, in some weak sense, that nodes close the to middle are more powerful and important in the group.
- How significant each node is as a conduit, perhaps for information or influence.

Within these two categories, there are a large number of different measures. As a concrete example, in any organisation, those with significant strategic power tend to be those with a large span of control, direct or indirect, and this tends to place them far from the boundary of the graph of relationships in the organisation. The graph describing communication in the organisation may look completely different from the official organisational chart, with strong local affinities between personal assistants, rather than their bosses, and between informal groups, such as smokers, who meet each other outside several times a day to smoke.

For organisations that operate using normal human behaviour, social network analysis can reveal properties of the individuals within the organisation. However, in an adversarial setting, it has significant weaknesses.

A group that is aware that it may be subjected to social network analysis will take steps to organise itself in such a way that its principals will not be easily discovered using such techniques. This is not entirely possible – the use, in the analysis, of the global structure of the graph puts the spotlight on nodes that are globally significant, even though they may not appear so locally. However, one obvious way for a principal to hide is to use someone else as a front, and communicate only with them (relying on them to communicate to and from the rest of the organisation). It is very difficult to distinguish such an *éminence grise* from an unimportant hanger-on – the supreme leader and the brother-in-law may look very similar.

5.2 Principal eigenvector of the adjacency matrix

We discussed above how the flow of an attribute such as importance around a graph can be calculated as an eigenvector of an appropriate adjacency matrix. This idea is the basis for the well-known PageRank algorithm used by Google to rank web pages [6]. The importance value calculated depends on the number of pages that point, directly or indirectly, to each page, a measure that has sometimes been called *authority*.

Of course, links in the web give rise to *directed* edges, which makes the process slightly more complicated. A node, or a region of the graph, with no outgoing edges eventually accumulates all of the importance. The standard way to avoid this problem is to allow some importance to leak from every node to other nodes in each round of the calculation – a process that, for example, allows Google to impose some extra ranking on web pages by constraining which those other nodes will be.

The same basic algorithm can be used to compute authority, a kind of centrality measure, in human networks. Even in a mixed network, where the nodes represent people, places, and things, the calculation of the authority of each node may hint at opportunities for disruption or increased surveillance.

The precise way in which local affinities, the edge weights, are extended to global affinities can be altered by changing the matrix appropriately. For example, normalizing the rows of the matrix allows it to be considered as a stochastic transition matrix and allows the machinery of random walks to be applied. In this view, the importance score of a node becomes the fraction of time a random walker spends at that node.

6 Embedding Graphs in Geometric Space

The strength of graph data, that the properties of every part of the graph depend on all of the other parts, also make them difficult to work with.

For data consisting of records with attributes, there is a natural way to represent the data geometrically. Suppose that there are n records, each with m attributes. Each record can be placed as a point in an m-dimensional space, spanned by the attributes, according to the values of its m attributes. The (global) similarity of two records can then be estimated by the reciprocal of their distance apart in this geometric space.

In a geometric space, the distance between a pair of objects remains the same regardless of the rest of the data; in a graph data space, the addition or removal of a single node can alter the distances between *all* of the remaining nodes. This makes every computation global and tends to make the running times of analysis algorithms worse than quadratic.

Fortunately, it is possible to embed graph data into a geometric space in a way that respects the local affinities [7, Chapter 4]. This provides the best of both worlds – the sophistication of graph data, with its resistance to manipulation, and the ease of computation associated with data embedded geometrically.

There are two complications to this embedding. The first is that a graph space and the corresponding geometric space are 'inside out' with respect to one another. Consider a 'central' node in the graph that is connected to many other nodes. Its row in the adjacency matrix contains many non-zero entries. This row, considered in a geometric sense, is far from the origin. Similarly, a 'peripheral' node with very few connections will have few non-zero entries in its row and so is close to the origin. An embedding, therefore, must take the rows of the adjacency matrix and 'invert' their magnitudes as rows to reflect their proper relationships – placing well-connected nodes close to the origin and poorly-connected nodes far away from it.

The second complication is deciding how to extend local affinity to global affinity. This decision cannot be made simply as a mathematical abstraction – it depends on the meaning associated with local affinities. For example, if local affinity represents ease of communication, such as the time it takes, then it is natural to extend it globally using the sum of local affinities along paths. However, if local affinity represents influence, then it is natural to extend it globally by considering not just the length of the paths connecting each pair of nodes, but also *how many* alternate paths exist between each pair. This latter extension, which corresponds to an electrical resistance model, is often the most useful, and the one we will assume in what follows.

6.1 The Walk Laplacian of a graph

The embedding of a graph space in a geometric space is accomplished by transforming its adjacency matrix appropriately, handling the two complications described above [8].

The best known way to do this is to replace the adjacency matrix by its *walk Laplacian*. In the first step, the diagonal entries are set to the sum of their corresponding rows, and the off-diagonal entries are negated. Then the entries in each row are divided by the diagonal entry, so that the diagonal of the matrix is all 1s, and the off-diagonal entries are negative numbers between 0 and -1.

The walk Laplacian corresponds to embedding the graph data in $n-1$ dimensions in such a way that the distances (for example, Euclidean distances) between pairs of connected points agree with their local affinities; and nodes that are central to the graph are placed close to the origin.

This is useful progress – for the geometric distance between unconnected points reflects their global affinity (using the multiple-pathway extension of local affinity); and (the reciprocal of) distance from the origin can be used as a surrogate for centrality measures.

However, n is typically extremely large, so an embedding into an $(n-1)$-dimensional space is not as useful as it might be – distances are not well-behaved for sparse points in a high-dimensional space. Hence we would like to reduce the dimensionality, both because it makes analysis more tractable, and because, for most real data, the manifold occupied by the data is of much lower dimension than the number of records or nodes. Hence much of the apparent dimensionality must be illusory.

6.2 Dimensionality reduction

We have already explained how an eigendecomposition reveals some of the global structure of a graph. It is also well-known that eigendecompositions (such as *Principal Component Analysis* (PCA) and *Singular Value Decomposition* (SVD) [9]) can be used for dimensionality reduction. They do this by transforming the space in which the data points lie so that axes are placed along directions of maximal variation. This helps both to see the important ways in which the data varies, and the unimportant ways. These latter can then be removed, preserving the main structure in the dataset while removing 'noise'.

An eigendecomposition of a matrix that embeds graph data into a geometric space produces a double win. It provides an allocation of up to $n-1$ orthogonal importance measures to the nodes of the graph, together with scores (the eigenvalues) indicating how significant each measure is; and it allows the dimensionality of the space in which the data is embedded to be reduced so that analysis techniques that work well in geometric spaces (for example, clustering or nearest-neighbour computations) can be applied effectively.

For simplicity, we will use the singular value decomposition, since its usual implementation orders the axes of the transformed space in decreasing order of significance (variation). Given a walk Laplacian matrix, A, of size $n \times m$ its singular value decomposition is

$$A = USV'$$

where U and V are orthogonal matrices, S is a diagonal matrix with non-increasing non-negative entries, and the superscript dash indicates transposition. If A is $n \times m$, then U is $n \times m$, S is $m \times m$, and V is $m \times m$. The rows of U can be understood as the coordinates of each object (row of A) with respect to new axes defined by the rows of V', stretched by the corresponding entries of S. The entries of S are called the *singular values*, the columns of U the left singular vectors, and the columns of V the right singular vectors. (Of course, in this setting A will be $n \times n$.)

Because the singular values are arranged in decreasing order, the right-hand side of the decomposition can be truncated at some k, with the $k+1$st singular value providing an estimate of how much structure is being ignored by such a truncation. The resulting decomposition is:

$$A \approx U_k S_k V_k'$$

where U_k is $n \times k$, S_k is $k \times k$ and V is $n \times k$.

The rank of the matrix A is at most $n-1$, so the nth singular value is necessarily zero. Other singular values towards the end of the diagonal of S may also be zero – the number of zero entries indicates the number of connected components in the graph.

We consider the left singular vectors, that is the columns of the U matrix, which correspond to eigenvectors. Because of the 'twist' inserted by the embedding, the *rightmost* columns of U (with non-zero singular values) correspond to the principal eigenvectors of the adjacency matrix, and so to the most significant structures in the graph. On the other hand, the singular value decomposition does not know that the matrix A arose from a graph, and so the leftmost columns of U correspond to the most significant variation in the rows of A considered as a geometry. Recalling that nodes with few edges are placed far from the origin, these columns capture information about unusual individual nodes and small regions connected to them. In fact, the singular value decomposition can be understood as revealing information about structure and variability of the graph at all scales, from the global to the local.

One useful interpretation, motivated by the connection between the walk Laplacian and the continuous Laplace-Beltrami operator, is that each column of U represents a mode of vibration of the graph when struck by a virtual mallet parallel to one of the axes of the transformed space. The magnitude of each entry in the column indicates how significantly that node is involved in the corresponding vibrational mode.

6.3 The rightmost eigenvectors

The rightmost columns of U describe modes in which almost all nodes vibrate, but only by small amounts. For the rightmost column with a non-zero singular value, dividing the points whose entries are positive from those that are negative is guaranteed to cut the graph into two connected components and, in fact, this gives a good clustering of the graph.

As a running example, we use a graph dataset that captures publicly known relationships among the members of al Qaeda up to 2004. These relationships include family relatedness, friendship and known collaboration on a mission. The data was collected by Marc Sageman[1]. We will use two forms of the data. The first is a simple adjacency matrix describing the local affinities between the members. The second also includes some demographic and descriptive information (and so is not a square matrix). Figure 1 shows a plot of the adjacency matrix, where blue dots indicate non-zero entries. There are 366 members, and 2174 edges connecting them.

The members have been arranged in groups, based on their participation in attacks. This makes it easier to understand the analysis, since individuals who appear close together in the rows and columns of the matrix probably are related in interesting ways. Note, though, that the analysis algorithms make no use of the ordering or rows and columns and would find the same connections no matter how the matrix was arranged.

It is clear from the figure that al Qaeda consists of more or less independent groups, the most densely connected being the South East Asian group in rows from 150–200. Each group is quite sparsely connected to the leadership, who appear in the early rows.

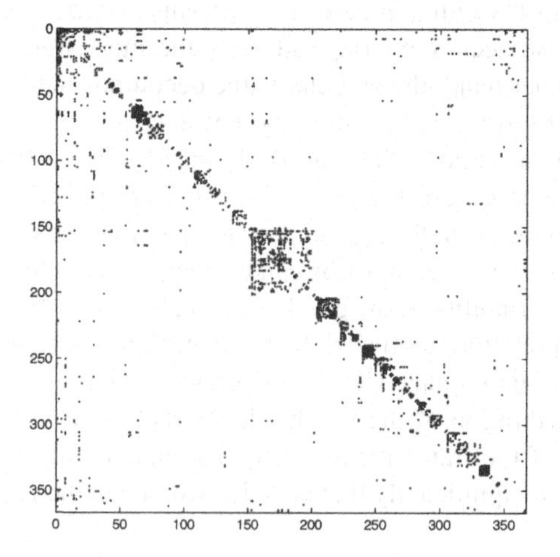

Fig. 1 The adjacency matrix of the al Qaeda relationships.

Plotting the nodes of the graph using the values of the rightmost and next-to-rightmost columns of U typically provides a good rendering for the graph. This is shown for the example dataset in Fig. 2. The rendering shows the surprising fact that al Qaeda is not naturally one unified group, but consists of two quite distinct

[1] To whom I am grateful for access to the data.

subgroups. The vertical arm are those members from the Moroccan and Algerian units; while the horizontal arm is the rest of al Qaeda.

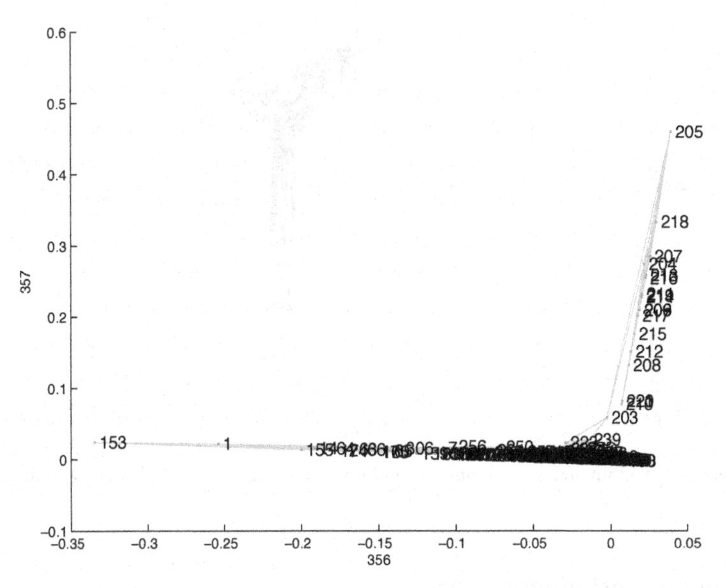

Fig. 2 Plot of the 357th and 356th columns of U provides a good rendering of the graph.

From our discussion of the association of importance with the principal eigenvector, we can see that this is not just a good rendering of the global structure of the graph. Those points at the extremes of each dimension are those with (respectively) the maximum and minimum importance values, and this can help to focus interest on them, independently of how they are rendered. The point labelled 205 in the figure is Abdelilah Ziyad while the point labelled 153 is Hambali, who is actually more significant than bin Laden, being both better-connected and important because he connects the Arab part of al Qaeda to the South East Asian part. The point labelled 1 is Osama bin Laden.

The plot based on the next two columns of U is shown in Fig. 3. The vertical arm towards the bottom of the figure is the Jemaah Islamiya group in Indonesia, the group towards the top left is the French group (the point labelled 240 is Christopher Caze), and the group towards the top right is the group that carried out the Casablanca bombings. The point labelled 306 is Imad Eddin Barakat Yarkas, the suspected leader of al Qaeda in Spain, and clearly a connection between the North African, European, and central parts of al Qaeda.

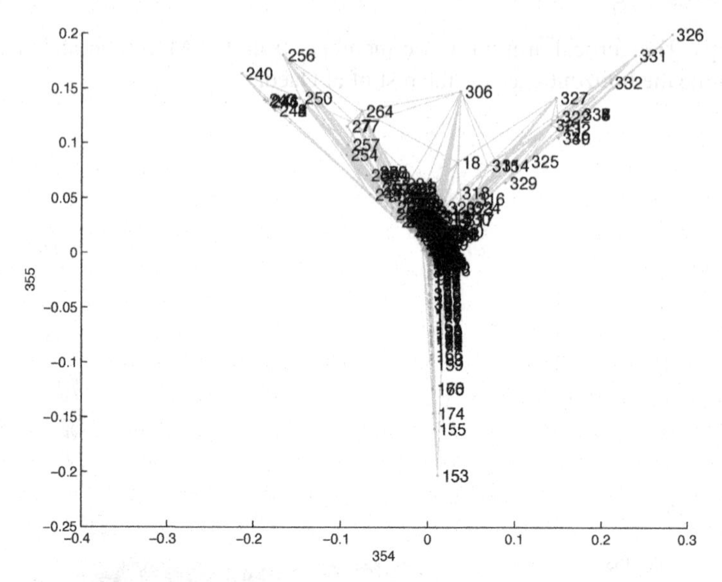

Fig. 3 Plot of the 355th and 354th columns of U reveals the next level of clustering in the group.

6.4 The leftmost eigenvectors

The leftmost columns of U describe modes in which vibrations tend to be centred around single nodes, whose neighbourhoods, for example, number of neighbours, is unusual. This is because these columns represent nodes that vary the most, when their rows in the walk Laplacian matrix are regarded as sets of attributes. In other words, nodes vibrate strongly when their connection pattern to the rest of the graph is different to that of most nodes. Such nodes are interesting in adversarial settings, because much connection structure is mediated by normal human interactions, so outliers may indicate unusual actions and relationships. Some of these may, of course, be because of unusual individuals who are eccentric or idiosyncratic in some way, but nevertheless these columns can suggest places for further investigation.

The plot in Fig. 4 picks out four unusual members of the group, based on their neighbourhoods (really the structure of the entire graph from their perspectives). These are: the point labelled 1 (Osama bin Laden); the point labelled 153 (Hambali); the point labelled 200, Jack Roche, surely an unusual member of the group; and the point labelled 36, Saad bin Laden, Osama bin Laden's son.

Plots of single columns of U can also be revealing, as they show which nodes are involved in each particular vibrational mode. For the example dataset, the two modes associated with columns 1 and 2 are shown in Fig. 6.4. This figure reveals more detail about the members involved in each vibrational mode but, at the same time, make it harder to pick out the most important members.

Other left-hand columns of U can also indicate important nodes, and so important individuals. Figure 6 picks out another four unusual members. The point labelled 2

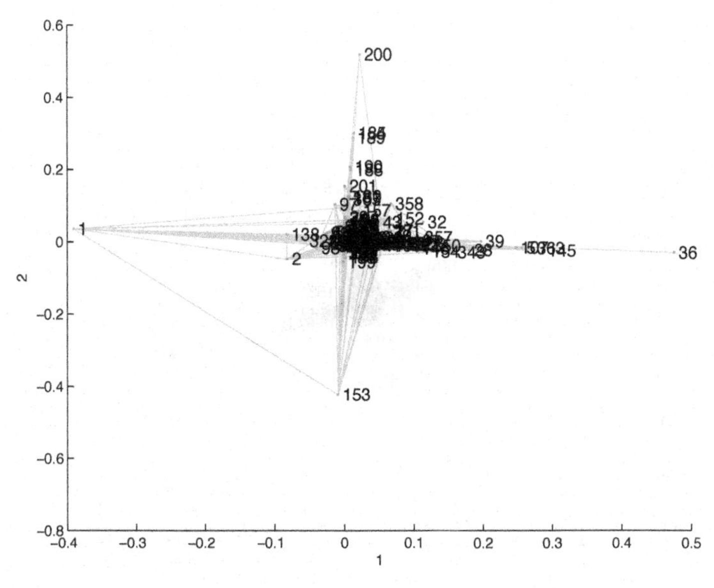

Fig. 4 A plot based on the first two columns of U, revealing four members whose connections to the rest of the group are unusual.

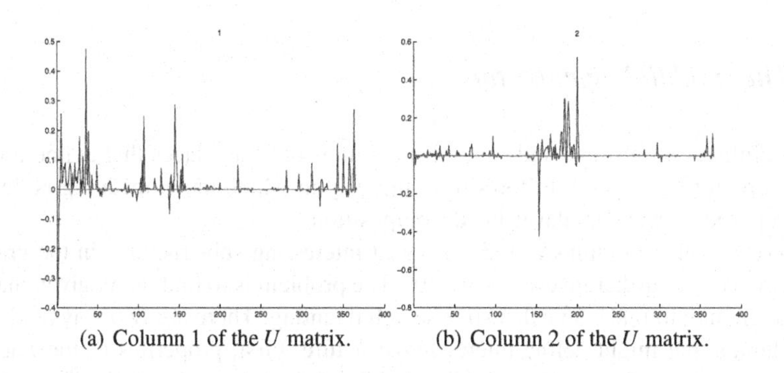

(a) Column 1 of the U matrix. (b) Column 2 of the U matrix.

Fig. 5 Magnitudes of entries in two columns

is Ayman al-Zawahiri, and the point labelled 32 is Abd Aziz al-Jamal, his aide. The point labelled 14 is Khalid Sheikh Mohammed and the point labelled 53 is Ahmed Omar Sheikh. Without any knowledge of al Qaeda, these plots focus attention on a set of significant individuals, some of whom are leaders and some of whom are simply unusual within the group context.

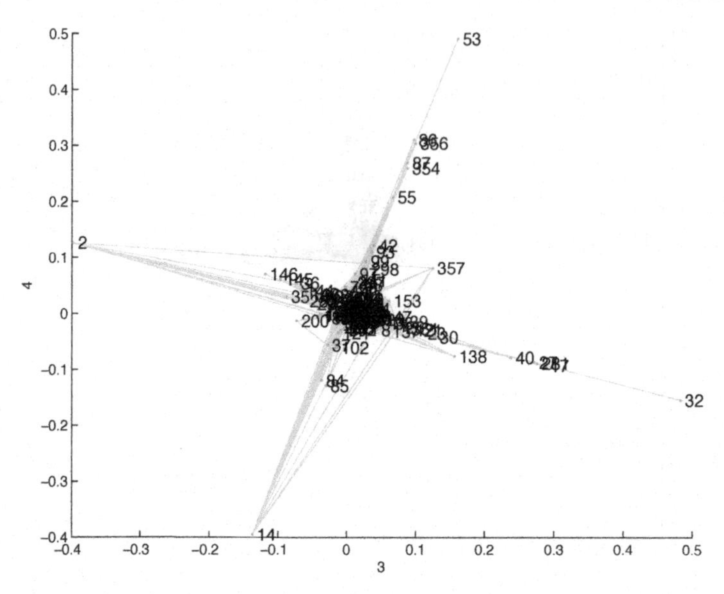

Fig. 6 A plot based on the next two columns of U, revealing four more members whose connections to the rest of the group are unusual.

6.5 The 'middle' eigenvectors

The usefulness of leftmost and rightmost columns of U as places that define interesting structure in a graph follows from the way in which the embedding is done, and properties of the singular value decomposition.

However, other columns can also suggest interesting substructures in the graph that may require further investigation [10]. The problem is to find them, given that n is large enough to make an exhaustive search daunting. There are two ways to look for columns that might define interesting structure. First, properties of the Laplacian suggest that it is columns near the middle of the matrix that typically contain useful structures (but this depends, to some extent, on how strongly the graph contains clusters). Second, interesting structures involve vibrations of only a few nodes, while the remainder stay at rest; whereas the more random vibration patterns tend to have almost all nodes vibrating. Hence calculating the mean of the absolute values in a column and looking at columns where this value is small is a useful way to find interesting, small structures.

This is shown for the example dataset in Fig. 7. Those places where the value is low are vibrational modes of low energy, which tend to be those in which only a few nodes are involved. Some of the low-energy columns are those numbered 46, 48, 50, 101, 121, 142, 155, and 188.

The plot of a single column of U can be thought of as illuminating the members of a kind of cluster – not in the usual sense since its members need no even be connected to one another in the graph. Rather the members of a cluster share

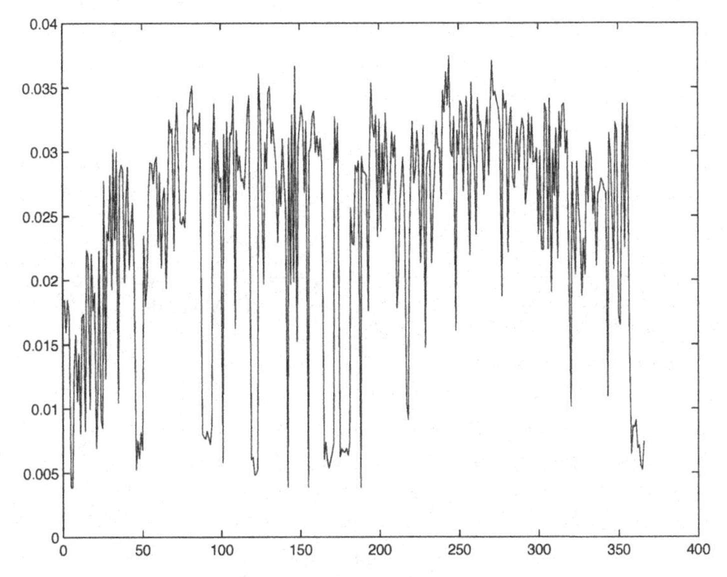

Fig. 7 Mean absolute value of the columns of U.

neighbourhoods in the graph with common properties when those neighbourhoods are integrated over the entire graph.

The single-column plots for the low-energy vibration modes are shown in Fig. 8. Most of these modes pick out small groups, in some cases brothers. Not surprisingly, brothers are connected in the relationship graph – but so are many other pairs of individuals. Notice also how often those in the same mode are close together in the matrix – a sign that they have a practical connection since that is how the matrix rows were sorted. However, these same individuals would have been selected no matter how the matrix rows were ordered, since the algorithms make no use of the ordering.

Returning to the importance interpretation of the magnitudes of the entries of U, we can also think of a column of U as labelling the nodes by their importances with respect to a particular flow associated with this mode. Each such flow is orthogonal to the other flows. It is not easy, and perhaps not helpful, to try an assign a meaning to each flow. Nevertheless, it can be evocative to see which nodes are associated by each flow, particularly when most nodes are neutral with respect to it (that is, they do not vibrate in this mode).

It can be particularly helpful to plot the magnitudes of each node using a right-hand column along one axis and another interesting column along the other. Now the position of points vertically is an indication of their absolute importance, while the horizontal position indicates their vibrational sense according to the interesting column. These two properties, rendered simultaneously, can reveal properties of the graph that are otherwise hard to see.

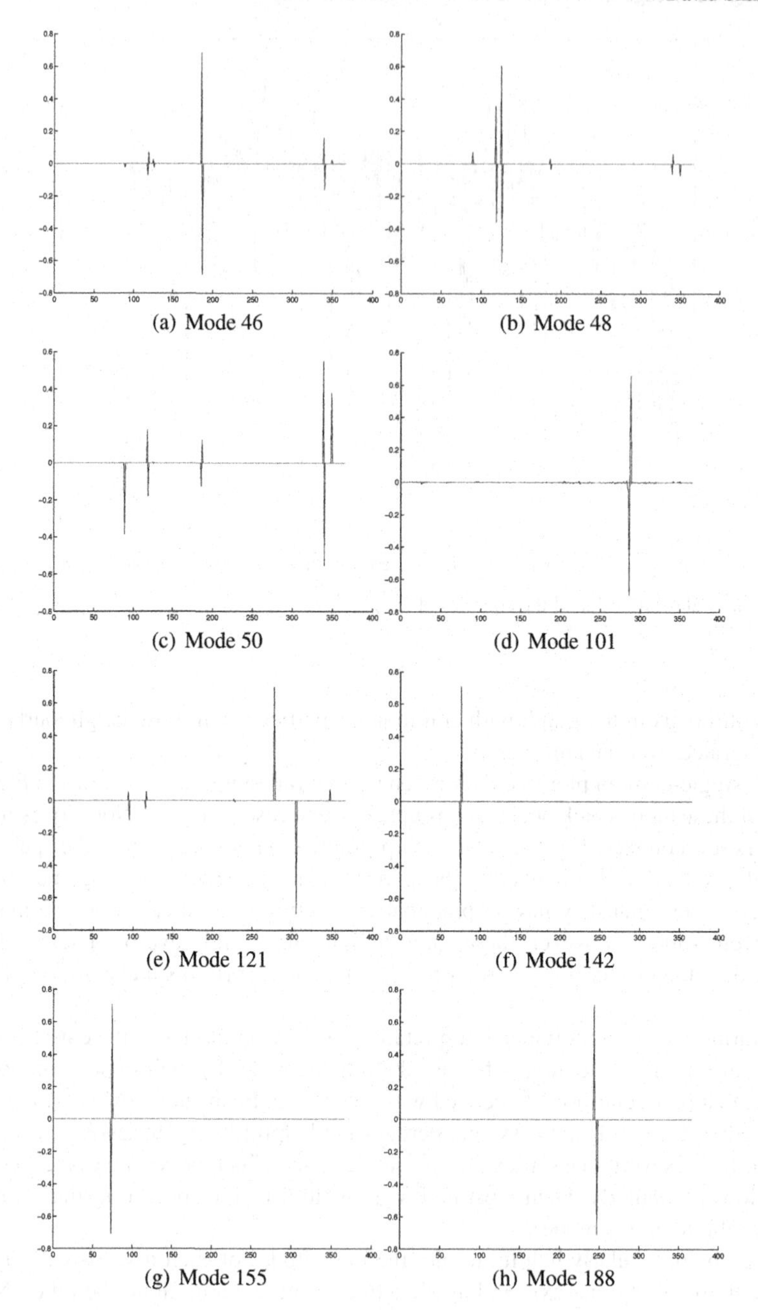

Fig. 8 Low-energy modes in 'middle' columns of U.

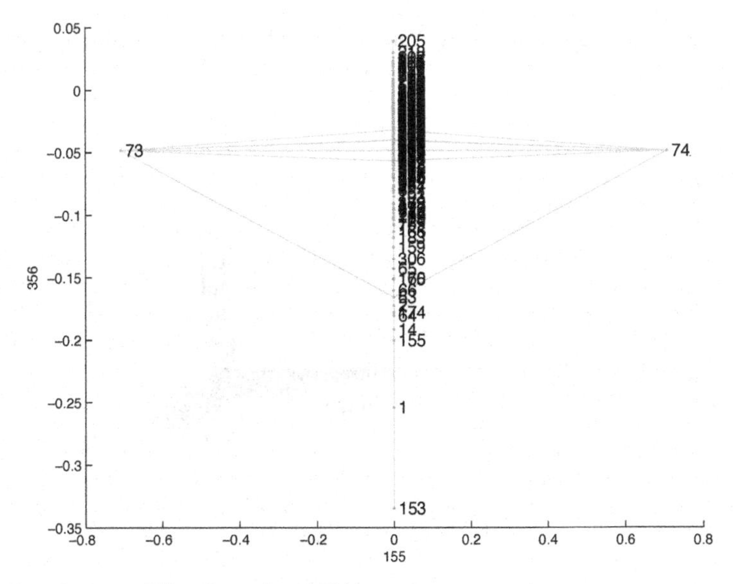

Fig. 9 Plot of column 155 against column 356.

Figure 9 shows such a plot for column 356 (which represents the conventional hierarchy of al Qaeda) against column 155, one of the interesting vibrational modes involving the points labelled 73 and 74, Wail and Waleed al-Shehri. The plot shows that this vibrational mode is only of moderate importance within the hierarchy of al Qaeda; but also that they both have a connection to the point labelled 63, Mohammed Atta.

6.6 Working in a lower-dimensional space

Some subset of the rightmost columns of U can be interpreted as the positions in a geometric space of points corresponding to each node. Conventional knowledge-discovery algorithms can be applied in this geometric space, including clustering and visualization.

Such a plot using columns 355–357 is shown in Fig. 10. From this figure, it is clear that al Qaeda can be thought of as composed of four distinct parts. The arms extending to the left is the South East Asian branch, with Hambali as the dominant individual. The branch extending to the lower left is the Arabic branch, with bin Laden as the dominant individual. The branch extending downwards is less coherent and represents North African branches. The branch extending upwards represents the Moroccan branch, with Abdelilah Ziyad as the dominant figure. Note the point labelled 7 which represents Abu Zubaydah, who is a crucial linking figure between several disparate groups (also visible in some of the previous plots); and the point labelled 306 which represents Imad Yarkas, alleged to be a Spanish member of al Qaeda.

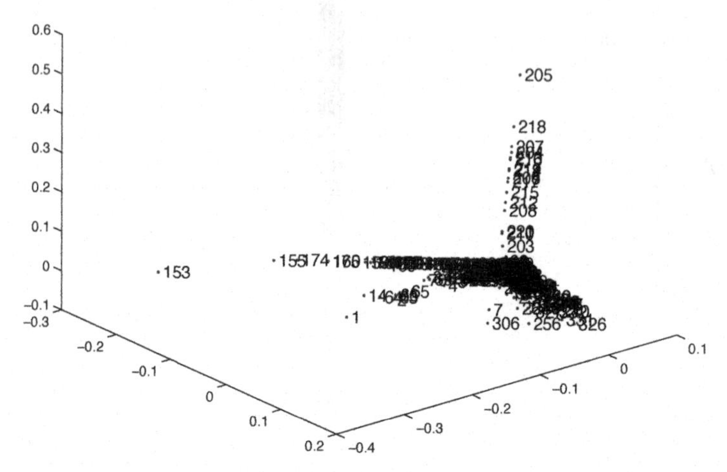

Fig. 10 3-dimensional plot using columns 355–357 of U.

6.7 Overlays of eigenvectors and edges

The plot produced by the rightmost columns of U is often used as a good rendering of a graph. Since these columns capture the large-scale structure of the graph, nodes that are directly connected in the graph should be rendered close together (of course, not for every possible pair of nodes and not all of the time because a rendering is a 2-dimensional embedding of the original graph). When directly connected nodes plot far apart, this signals that their global affinity – which determines their position in the plot – does not agree with the their local affinity – which dictates that they should be close. Discrepancies such as these suggest that some nodes, and perhaps their neighbourhoods, should receive extra scrutiny.

For example, one possible way for an adversarial group to conceal themselves in graph data is to make themselves look, as a group, the way that other groups look. However, it is much harder to make their group fit into the larger-scale structure of groups of groups and so on, because they cannot control the other groups. Making a structure look normal at one scale but failing to do at other scales is one important way in which discrepancies between plotted position and graph position can arise.

It is also possible for two points that plot close together *not* to be connected by an edge in the data. This is also suggestive since it indicates that two nodes have strong global affinity, and so might be expected to have strong local affinity too. This suggests that there may indeed be an edge between them which has failed to be captured by the data collection; or that they are trying explicitly to conceal the existence of a direct link by, for example, communicating using cutouts.

Figure 11 shows a plot using columns 351 and 352 of U. This figure illustrates both of these properties. Consider the point labelled 348 (Mourad Merabet). This point is far from the only points to which it is directly connected (277 and 279), suggesting that this individual is globally dissimilar to his only directly-connected neighbours. On the other hand, this point is quite close to 276 (Merouane ben Ahmed) with whom he has no direct connection. Looking at the demographic data, which was not used in the computations so far, shows that this is plausible. The two men share national origin and background, although their participation in al Qaeda operations has been quite different.

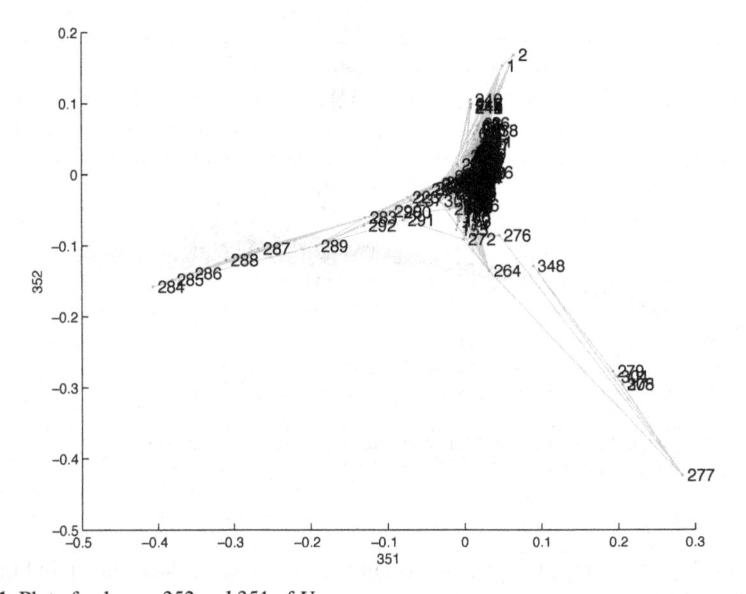

Fig. 11 Plot of columns 352 and 351 of U.

6.8 Using correlation rather than connection

The analysis so far has used only local affinity data between pairs of individuals. However, it is possible instead to construct a graph in which the correlation between individuals, suitably thresholded, is the local affinity.

For the example dataset, this can be done in two ways: first, treat the rows of the adjacency matrix as sets of attributes and calculate the correlation between the *connection patterns* of each pair of individuals; and second, use the version of the dataset with demographic and personal details as well as connections, and compute the correlations between the entire descriptions of individuals.

When the correlation approach is applied to the al Qaeda connection data, the resulting plot is similar to that from the direct analysis, although it has become more obvious how weakly the vertical arm is connected to the rest of the group. This is shown in Fig. 12.

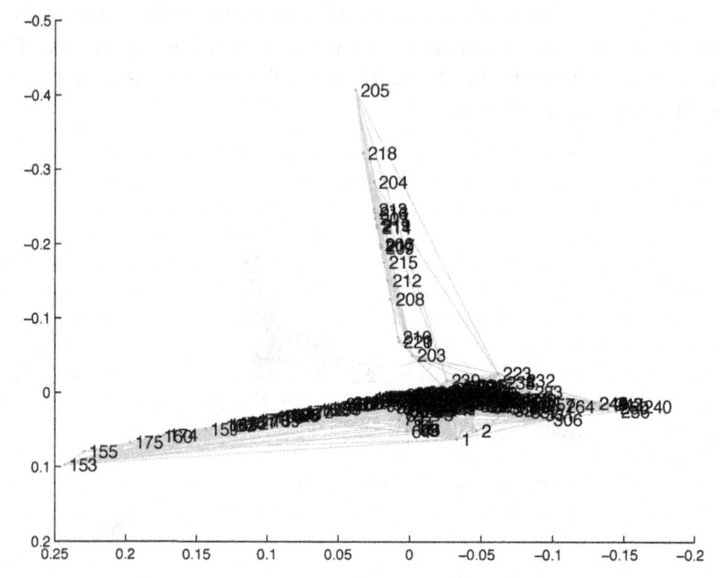

Fig. 12 3-dimensional plot of the rightmost columns of U derived from the correlations between individuals based on connections.

The analogous plot using the demographic and connection data, shown in Fig. 13, shows that the European group, to the right of the figure, is well connected to the rest of the group, but is far more unusual than any of the other subgroups. The other subgroups remain clearly visible, although the two groups whose arms are at the bottom of the plot (the Arab and North African groups) are much more homogeneous (and so less spread out along a line) than the other subgroups.

7 Summary

Graph or relational data is a natural way to understand many kinds of systems, but is particularly attractive in adversarial settings because of its resistance to manipulation. This resistance arises because the global properties of a graph emerge in complex ways from the local connection structures and affinities. One side effect is that graph data is harder to analyse than attributed data.

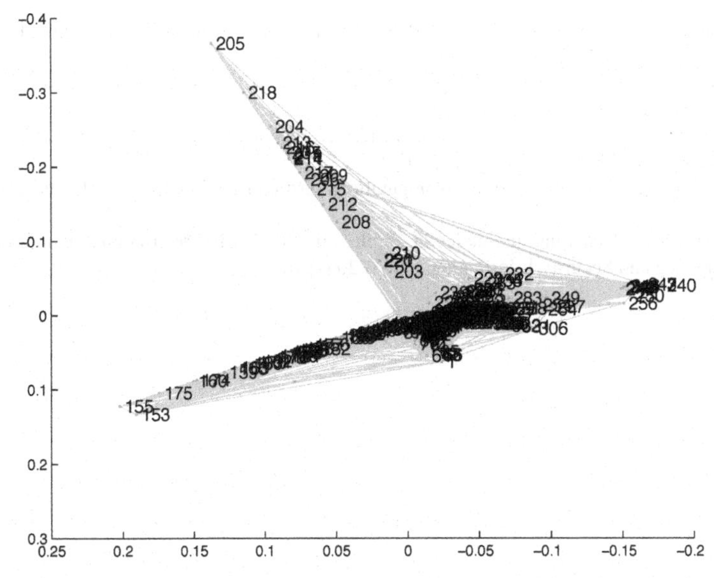

Fig. 13 3-dimensional plot of the rightmost columns of U derived from the correlation between individuals based on connections and demographic data.

Techniques based on eigenvectors of various matrices that represent the connections and properties of graphs have been studied extensively by mathematicians, and have been used as sophisticated ways to cluster data. We have instead concentrated on the practical use of eigenvector-based techniques to understand structures – particularly the unusual structures that are of most interest in adversarial settings. Eigenvectors from all parts of the spectrum have something to say about substructures within graphs, from the large pieces that form good clusters, to small anomalous regions, to individual nodes with unusual neighbourhoods.

References

1. Dutrisac, J., Skillicorn, D.: Subverting prediction in adversarial settings. In: 2008 IEEE Intelligence and Security Informatics. (2008) 19–24
2. Thomas, J., Cook, K., eds.: Illuminating the Path: The Research and Development Agenda for Visual Analytics. IEEE Press (2005)
3. Watts, D., Strogatz, S.: Collective dynamics of 'small-world' networks. Nature **393** (1998) 440–442
4. Jensen, D., Neville, J.: Data mining in social networks. Invited presentation to the National Academy of Sciences Workshop on Dynamic Social Network Modeling and Analysis (November 2003)
5. van Meter, K.: Terrorists/liberators: Researching and dealing with adversary social networks. Connections **24**(3) (2002) 66–78

6. Bryan, K., Leise, T.: The $25,000,000,000 eigenvector: The linear algebra behind Google. SIAM Review **48**(3) (2006) 569–581
7. Skillicorn, D.: Understanding Complex Datasets: Data Mining with Matrix Decompositions. CRC Press (2007)
8. von Luxburg, U.: A tutorial on spectral clustering. Technical Report 149, Max Plank Institute for Biological Cybernetics (August 2006)
9. Golub, G., van Loan, C.: Matrix Computations. 3rd edn. Johns Hopkins University Press (1996)
10. Skillicorn, D.: Detecting anomalies in graphs. In: 2007 IEEE International Conference on Intelligence and Security Informatics. (2007) 209–216

Mathematically Modeling Terrorist Cells: Examining the Strength of Structures of Small Sizes

Lauren McGough

Abstract This paper aims to discuss the strengths of different terrorist cell structures, using the partially ordered set ("poset") model of terrorist cells to define the strength of terrorist cell structures. We discuss algorithms implemented in a program examining the structures of posets of seven elements, and the patterns observed in this analysis. We then discuss implications of these findings, and their applicability to government strategic operations - namely, the possibilities for future expansion and use of the algorithms to produce structures fitting certain parameters, and the caution which must be exercised in following previous suggestions that terrorist cell structures can be assumed to be trees, since structures containing "V" structures and more than one leader can be more secure than trees, according to our findings. We end with several questions that require future investigation in order to increase applicability to strategic operations, and suggestions for how such questions may be approached.

1 "Back to Basics": Recap of the Poset Model of Terrorist Cells

In the paper, "Breaking Al Qaeda Cells: A Mathematical Analysis of Counterterrorism Operations (A Guide to Risk Assessment and Decision Making)", Farley presents a model of terrorist cells using order theory [2]. This model is based on the idea that terrorist cells are best modeled not as graphs, but as partially ordered sets, or "posets," which capture not only the connections between people within terrorist cells, but also the intrinsic hierarchical structure that is likely present in terrorist networks.

Poset visual representations are usefully intuitive: the people within the cell are represented as nodes on a graph, and lines connecting nodes represent that two peo-

Lauren McGough
Phoenix Mathematics, Inc., 35 Northfield Gate, Pittsford, NY 14354, e-mail: lauren@phoenixmath.com

N. Memon et al. (eds.), *Mathematical Methods in Counterterrorism*,
DOI 10.1007/978-3-211-09442-6_4, © Springer-Verlag/Wien 2009

ple are connected in real life (for example, through communication lines that allow the person of higher status to pass information and commands to the person of lower status in the hierarchy). To capture the hierarchy of the network, the people of higher status are depicted closer to the top of the diagram, with the foot soldiers of the cell represented by the nodes on the bottom of the diagram and the leaders of the cell represented by nodes at the top level of the diagram.

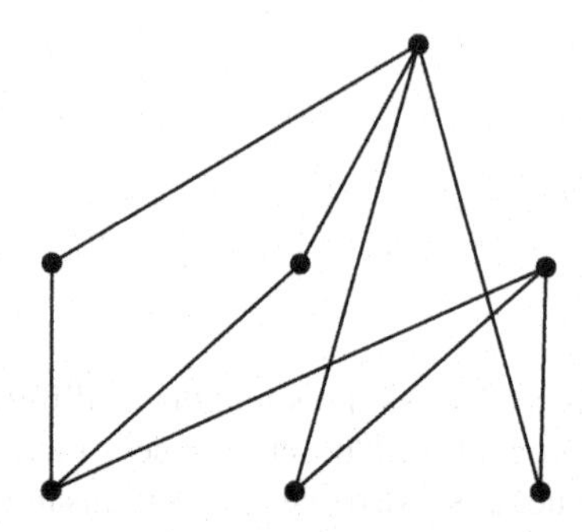

Fig. 1 Example of a Poset – If this poset represented a terrorist cell, this cell would have two leaders, two middle-men, and three foot soldiers, with three levels of hierarchy.

After having established posets as a useful tool in modeling the structures of terrorist cells, we are concerned with several concepts that allow us to use this model as a tool not only for representation but also for interpretation of counterterrorist operations. First, we consider "maximal" and "minimal" nodes. Maximal and minimal nodes are merely the nodes at the top and bottom of the poset structure, respectively. Thus leaders of the cell are represented by maximal nodes, and foot soldiers, of lowest hierarchical status, are represented by minimal nodes.

Next we consider "maximal chains". A saturated chain can be thought of as a path from one node to a node directly "below" it, moving only along the lines between nodes. A maximal chain is a saturated chain from a maximal element to a minimal element. In terrorist cell representations, each maximal chain represents a path by which information may pass from a leader to a foot soldier.

One can then make the assumption that a terrorist attack can occur when a leader of a terrorist cell is able to pass commands to a foot soldier within that cell. Using this assumption, we can then develop a method for the quantitative analysis of counterterrorist operations. First, one may notice, under this assumption, it is logical that one might want to render a terrorist cell disabled by removing people from the cell (killing or capturing these people) in such a way as to cut all of the maximal chains of the structure, thereby preventing all commands from being sent from leaders of the cell to foot soldiers. We define such a collection of nodes (a collection that intersects every maximal chain) as a cutset of size k, where k is the number of nodes contained within the cutset (the number of people removed from the cell).

Using the idea of a cutset, if one knows the structure of a given terrorist cell, and the number of people removed from the cell in a given action against the cell, but not the positions of the removed people within the intact cell structure, one may calculate the likelihood that the action rendered the cell ineffective. If n is the total number of people within a terrorist cell P with a certain known structure, k is the number of people removed from the cell in a given action, and $Cut(P,k)$ is the number of cutsets of size k in poset P, then the probability, Pr, that the cell has been dismantled is expressed by the formula:

$$Pr = \frac{Cut(P,k)}{{}_nC_k}, \tag{1}$$

where
${}_nC_k = \frac{n!}{k!(n-k)!}$ and $r! = r(r-1)(r-2)\ldots(2)(1)$ for all natural numbers r.

This is the number of cutsets of size k divided by the number of possible ways to remove k nodes from a set of size n. [2]

Using these definitions and the resulting formula, it is possible for government operatives to determine the effectiveness of a given action. It is also possible to define the concept of the "strength" of terrorist structures, and to use this definition to analyze different structures in terms of strength in order to determine which structures are strongest for a given size, in order to allow military strategists to make better-informed decisions about a cell especially when its exact structure is unknown. We do just this in the sections that follow.

2 Examining the Strength of Terrorist Cell Structures – Questions Involved and Relevance to Counterterrorist Operations

After one has defined a model, such as the poset model, for representing terrorist cells, certain questions arise regarding that model. Namely, using this representation and the resulting formula for calculating the probability that a cell has been dismantled, one may ask, what structures are the most stable? This question is relevant in assessing the difficulty of dismantling a terrorist cell, especially a cell of unknown or partially unknown structure. For example, often government planning agencies may only know certain parameters that a cell structure fits – its approximate size, or the structure of a part of the cell. This makes the poset model and the formula above difficult or impossible to use, as the cutset formula depends on complete knowledge of the structure of the cell.

In this paper we look at posets of seven elements, and consider, given a specific size (such as seven elements), what structures are the most stable for that size. We also discuss the algorithms used to analyze such a problem. The observations made with small posets may be applicable to posets of larger sizes, and the algorithms used may be improved for use by government planning agencies in order to determine possible or worst-case scenario cell structures, given certain parameters for the structure of the cell.

3 Definition of "Strength" in Terms of the Poset Model

Throughout our analysis, we look at two different ways of defining strength. Both may be useful in the fight against terrorism, depending on the situation.

By saying that one poset is stronger than another, we mean that is more difficult to break by removing people from the structure. In order to determine strength, we use the idea of a "cutset vector" as described in Farley's "Toward a Mathematical Theory of Terrorism: Building the Perfect Terrorist Cell, I" [3]. A cutset vector is a row vector of n entries (where n is the size of the poset in question) whose first entry represents the number of cutsets of size 1 for the specified structure, the second entry represents the number of cutsets of size 2 for that structure, and, in general, whose ith entry represents the number of cutsets of size i for the structure being analyzed. We first define strength in the following way:

Definition 1 (Strength (1)). We say that a poset P of size n and cutset vector $(c_1,\ldots,c_n)$ is *strictly stronger in the first sense* than a poset K of size n and cutset vector $(d_1,\ldots,d_n)$ if and only if

$$\sum_{i=1}^{n} c_i < \sum_{i=1}^{n} d_i \tag{2}$$

We realize, however, that sometimes this definition is not ideal. Another definition for strength may be concerned with the idea that one poset is stronger than another poset if it is more difficult to dismantle that poset by removing fewer than half of its members than it is to dismantle another poset in this manner. Therefore we developed the second definition of strength, which we also use for our analysis:

Definition 2 (Strength (2)). We say that a poset P of size n and cutset vector $(c_1,\ldots,c_n)$ is *strictly stronger in the second sense* than a poset K of size n and cutset vector $(d_1,\ldots,d_n)$ if and only if

$$\sum_{i=1}^{\lfloor \frac{n}{2} \rfloor} c_i < \sum_{i=1}^{\lfloor \frac{n}{2} \rfloor} d_i \tag{3}$$

where $\lfloor r \rfloor$ denotes the greatest integer less than or equal to r for all real r.

The $\frac{n}{2}$ may seem arbitrary, but one could argue (although this should be confirmed) that it could be replaced by any "significant" fraction of n. This second definition tries to capture the idea that we do not need to necessarily worry about capturing all of the terrorists in a cell, just a large fraction of them.

If in either case the sums mentioned above are equal, then we say the posets are of equal strength in the first or second sense, respectively. Note that these definitions are different than those of Farley; they have the advantage that all posets then lie in a linear (quasi)order. One might also consider the so-called dominance order of combinatorics.

These definitions provide us with a way to compare the strength of any two posets of the same size. We examine posets of size seven using both definitions, computing which posets are strongest under each definition. We find that for the most part, the two definitions do not provide very different results for the properties of the strongest posets (suggesting a kind of "Church-Turing thesis"), and it is easy enough to compare the strength of any two posets of the same size using either method in a computer program.

4 Posets Addressed

We chose to address the structures of posets of seven members, as there are 2,045 posets of size seven, which is a reasonable amount with which to work. (We then make observations that we can generalize to posets of larger sizes, as posets representing terrorist networks are usually of larger sizes than 7.)

First, we used the definition of strength mentioned in Definition 1 to find the five-hundred most structurally secure posets of size seven (in the first sense). We then looked at the posets of size seven using the "strength of small cutsets" definition to find the five-hundred most structurally secure posets, and we looked at how the structures of these compared to the structures of the strongest posets of the first kind. We did not limit the set of posets we considered to trees or structures without isolated vertices, as we wanted to see how such structures fared in terms of security, in part to see if it is safe or reasonable to make the assumption, as some have previously suggested it is fitting to do, that terrorist networks are most likely trees. We leave the problem of working more specifically with posets fitting stricter parameters for the future.

5 Algorithms Used

We used John Stembridge's Maple package for posets to generate our data, and we developed an algorithm for finding the cutset vector of a poset. Then, using each definition of strength, we determined in each set of posets which posets were the five-hundred strongest posets, and we looked at these posets to find patterns and common themes in their structures. Though some aspects of the algorithms we used were not ideal in terms of time and memory usage, the algorithms could be modified and improved. They could be very useful to people interested in government planning and military strategic operations, as they could determine, given certain parameters, which are the most structurally secure posets that fit those parameters for either definition of strength.

The algorithm we used was as follows:

First, we created an array of dimensions 1 by 2045, that held all of the posets of size 7. After doing this, we created another array which held the maximal chains of each poset.

For easier manipulation of the maximal chains, we then created a “hash table” for each maximal chain of each poset. This “hash table” was an array of seven entries, and each entry was a 0 or 1, corresponding to whether that element was contained within the maximal chain (for example, if a maximal chain contained element 4 but not element 5, the fourth entry in the array was a 1, while the fifth entry in the array was a 0). We created the hash tables by testing the membership of each element within each maximal chain of each poset. We then placed the hash table arrays in a table of lists, where each poset was represented by a list of arrays in the table, and each array represented a maximal chain of the poset. The creation of the hash tables was the most time-consuming step, as it required looping over each element for each maximal chain for each of the 2045 posets to test that element's membership. However, the arrays simplified the next step of calculating the cutset vectors.

In preparation for calculating the cutset vectors by testing each subset to see if it was a cutset, we created a table, of seven arrays. The ith array in this category simply held all of the combinations of i elements taken out of 7 (for example, the second array held all possible combinations of two elements taken from seven elements). Each element of each array was accessed using three subscripts, where the first subscript indicated which array one was in, the second indicated which combination, and the third indicated which element of that combination, where the combinations were in increasing order (as the elements were specified by the numbers 1 to 7). We also created an empty array, CutsetVector, of dimension 2045 by 7, where each row stood for the cutset vector of a different poset, and the columns indicated which entry of the cutset vector one was accessing.

We then used six procedures to calculate the cutset vector of each poset, where each procedure calculated a different entry of the cutset vector (the seventh entry was assigned to equal 1, as the nth entry of a cutset vector is always 1 for an n-element poset). Each procedure went through each member of each subset for each maximal chain of a poset. If one of the subset's members intercepted the maximal chain, then the procedure increased the cutset vector entry that it calculated for the poset it was using as input, and continued to the next maximal chain. However, if all of the subset's members did not intercept the maximal chain, than the procedure exited the loop that went through all of the maximal chains with that subset, did not increase the cutset vector entry of that poset, and went onto the next subset. We then applied each procedure to each poset using the map command. For example, see Fig. 2, the code for the procedure that calculated the cutset vector's fifth entry for each poset.

After calculating the cutset vector for each poset, we calculated either the sum of all of the entries of the cutset vector for each poset, or the sum of the first three entries of the cutset vector of each poset, depending on which definition of strength we were using. We then converted these sums to a list, and sorted the list. We determined the five-hundredth entry of the list, and then selected all of the posets with cutset vector sums less than or equal to that number. We did this using the select command and a procedure designed to compare that poset's cutset vector sum to the five-hundredth entry, returning true if the poset's cutset vector sum was less than or equal to the five-hundredth entry, and returning false if it was not. This was all

```
CutsetVector5 := proc (P :: list)
local cutsetvectorvalue , p, j, val1, val2, val3, val4, val5;
cutsetvectorvalue := 0;
for p from 1 to 21 do
        val1 := subsets [5][p][1];
        val2 := subsets [5][p][2];
        val3 := subsets [5][p][3];
        val4 := subsets [5][p][4];
        val5 := subsets [5][p][5];
    for j from 1 to nops (P) do
        if P[j][val1] = 0 then
            if P[j][val2] = 0 then
                if P[j][val3] = 0 then
                    if P[j][val4] = 0 then
                        if P[j][val5] = 0 then
break;
                        end if;
                    end if;
                end if;
            end if;
        end if;
        end do;
    if j > nops (P) then
  cutsetvectorvalue := cutsetvectorvalue + 1;
    end if;
end do;
return cutsetvectorvalue :
end proc;
```

Fig. 2 The code used for calculating the fifth entry of the cutset vector for each poset.

possible because for each array, the first dimension stood for the poset one was accessing, and the arrays and lists were compatible, such that the 189th entry of one array, which corresponded to the 189th poset, also corresponded to the same poset in another array. After having selected the "top five-hundred" (or slightly more, given that there were repeats), we printed the Hasse diagram of each of these posets along with its identifying number, cutset vector and cutset vector sum so that we could examine the structure of approximately the top 25 percent of the total number of posets.

6 Structures of Posets of Size 7: Observations and Patterns

One of the most fundamental observations that result from this analysis is that posets with "V" structures, several "leader" maximal nodes, disconnected pieces and isolated vertices are the strongest poset structures of size 7, regardless of whether one uses the first or second definition of strength. Though posets with isolated vertices and disconnected parts do not translate into the language of terrorist networks, as a terrorist who does not communicate with anybody within a network can not be thought of as part of that network, and disconnected parts would not be thought of as one network but as several networks, the "V" structures and multiple leaders of strong cell structures lead to an important observation; namely, that one must be careful when making guesses, for example, that terrorist networks are organized as trees (where a tree is a poset with only one maximal node and no "V" structures), for terrorist networks organized as trees are significantly less secure than networks that break the requirements of a tree, at least when no restrictions are placed on the structures of the networks besides size.

The most secure structure was, trivially, the antichain (seven disconnected vertices with no relations between them), but this structure does not translate into the war on terrorism, as no terrorist network organized as an antichain could be practically thought of as a "network."

The most structurally secure network under the first definition of strength that did not have any isolated vertices is found in Fig. 3. One can see that this structure is not a tree. This structure, or any similar structure, may be unrealistic for a real terrorist structure, as each of the foot-soldiers knows information about each of the leaders, making the risk of betrayal high in case of capture. It is, however, important to note that this structure is very "strongly connected," as each node on one level is connected to all of the other nodes of the other level. Therefore, according to the data, cells where members are connected to many other members are most mathematically secure in that they are least likely to be disabled through removal of their members. This does satisfy logical intuition, as discussed later.

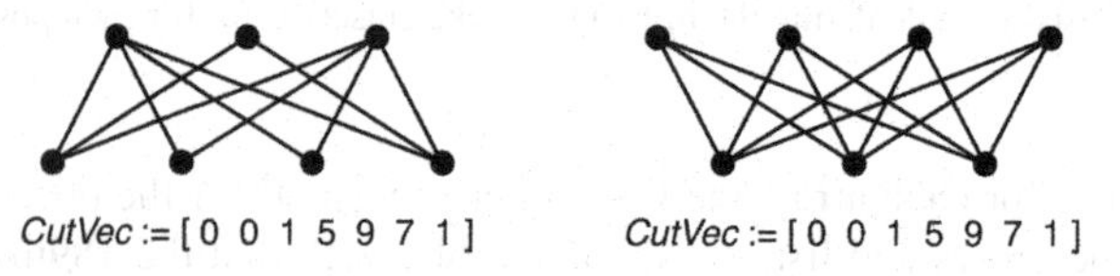

Fig. 3 The most structurally secure posets with no isolated vertices of size 7 according to the first definition of strength. They are duals of one another.

No trees lacking isolated vertices and/or disconnected pieces were present in the "Top 500" strongest posets using the first definition. Many of the strongest networks, however, did follow certain patterns of structure, differing only by the numbers and placements of connections between posets. One pattern present was that of networks with two levels of hierarchy: a level with three nodes, and a level with four nodes. The most secure of these networks was the network where each of the nodes on one level was connected to all of the nodes of the other level, as already discussed. Less strong posets of the same pattern of nodes (a level of 3 and a level of 4) followed as the level of "connectedness" between nodes decreased. Other prominent patterns included a pattern of three levels, one with one node, one with two nodes, and one with four nodes (think of the binary tree), a pattern of three levels that was set up in a "1-3-3" fashion, a "2-2-3" pattern, a "1-4-2" pattern, and a two-leveled "2-5" pattern; some of these are shown in Fig. 4. Posets of these patterns were the most numerous posets lacking disconnected parts and isolated vertices (shown in Fig. 5), and appeared many times in forms differing only in the connections between the nodes.

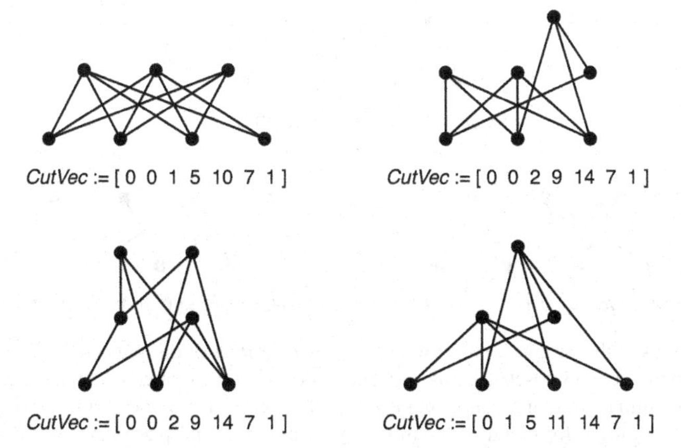

Fig. 4 Examples, in decreasing order of strength, of strong posets in the first sense with no isolated vertices. It is evident that many of the posets follow specific patterns of similarities in structure, offering differing only by several "connections" between people.

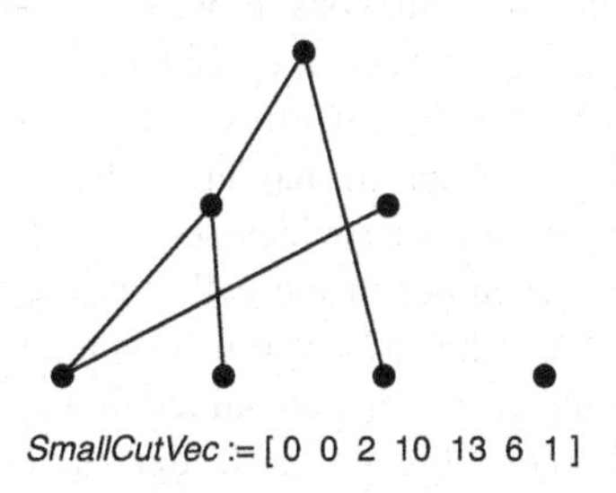

Fig. 5 Example of a poset that is strong in the first sense and contains an isolated vertex. This poset structure and similar ones do not translate as representations of terrorist cells.

Results obtained using the second definition of strength were similar to those obtained using the first definition of strength, but more posets were equivalent to one another using this definition, as this definition took into account only the first three entries of the cutset vectors. The structural patterns found in the "Top 500" posets using this definition were similar to the ones found using the other definition, and the posets in both lists often coincide. Many posets with isolated vertices and disconnected pieces were present in this list, as in the previous case, and of the posets lacking disconnected pieces and isolated vertices, the same structural patterns prevailed. The "3-4" structure was highly prominent, as were the 1-3-3 structures, 2-2-3 structures, 1-2-4 structures, 1-4-2 structures, and 2-5 structures. (One might think we are merely listing all partitions of 7, but in fact we are not.) As in the previous case, the specific strength of each of these structures depended on the connections between the elements within the structures.

These results are not necessarily surprising, though they do bring up important questions that need to be raised about how to analyze counterterrorist actions if a terrorist cell's structure is not fully known. The most important idea that is raised is

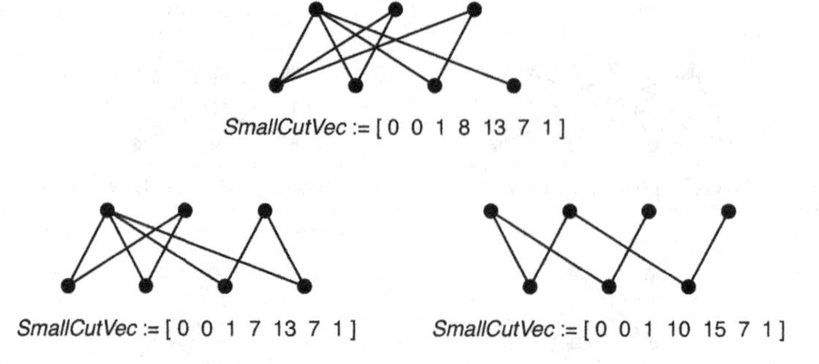

Fig. 6 Examples of the strongest posets with no isolated vertices or disconnected pieces in the second sense. There were 33 posets lacking isolated vertices that were of equal strength to these, and all 33 were of similar structure, containing two "levels", one of three nodes and one of four nodes (with either one on the top), and containing varying amounts and placements of "connections" between those nodes.

that nontrees are, according to this analysis, more secure structures than trees when no restrictions are placed on the posets one considers.

This idea of the strength of "V" structures and connecting as many leaders to each foot soldier makes sense. After all, having such a connected cell where each person reports to numerous leaders provides security, especially in the case that a middleman or upper-level member of the cell is removed. Whereas removing a leader or middleman from a tree has a large effect on the network, as all of the people who were part of maximal chains that contained that middleman are cut off from the cell by the removal of that middleman or leader, the removal of such a middleman from a cell with many connections and "V" structures is not as devastating, as people who were "below" that middleman and connected to him either directly or indirectly may still be connected to other middlemen who have not been captured. (Perhaps one could formulate a metric for the "vulnerability" of a poset based on the number of people who are cut off when a subset is removed. This would be related to but different from the notion of k-connectivity.) The "V" structures that others have speculated would be unreasonable or unlikely structures within a cell are actually structures that could, if contained within terrorist networks, provide strength to the cell that would make it very difficult to break. To illustrate this point, the top 500 posets of size seven under the second definition had at most 7 cutsets of size three or smaller, and, as mentioned before, none of them were trees. The "V" structures of these cells provided built-in support and protection against attacks, making dismantling the cells through removal of few members a difficult task indeed. As the "V" structure strength is not dependent on the size of the cell, we suspect that this result holds for cells of all sizes. We therefore present the following problem:

Problem 6.1

Prove that poset structures containing at least one "V" substructure are generally stronger (by either the first or second definition) than posets of similar structure not containing such V's.

Proof of such a result would provide theoretical insight to the results observed and discussed here, and would prove that the results discussed here regarding posets of size seven apply to posets of any size, and therefore are applicable to the war on terror.

7 Implications and Applicability

This analysis has applications to the struggle against terrorism both in the algorithms used and the results obtained.

The algorithms used could be improved, expanded, and implemented into software that could be useful to government planning agencies and military strategists. Most helpful would be a program that answers the question: given certain parameters, what are the strongest possible poset structures that fit those parameters? This would be helpful when government officials do not know the complete structure of a terrorist cell but only aspects of that structure, and they want to assess, for example, the difficulty of breaking that cell, or the likelihood that they have already dismantled that cell. It is best in such situations that government planning officials use the most secure possible structure that fits their parameters, as such information allows them to calculate a lower bound on the likelihood that they have dismantled a cell, and therefore the analysis is the most conservative. The algorithms used here could easily be expanded and enhanced so as to be useful for a variety of situations and parameters that strategists may encounter.

The results obtained raise two possible interpretations as far as their applicability to the struggle against terrorism. First, the results bring up the question of whether we can really hypothesize that terrorist cells are, as some have suggested, trees. According to this analysis, this hypothesis may be flawed, and to assume that such networks are trees may, in fact, be an impractical and unsafe assumption, as cell structures containing, for example, "V" structures, are safeguarded against dismantling through the removal of members of the cell in ways that trees are not.

If, however, one stands firmly on the idea that cells are most likely trees, as the "V" structures do provide instability in the risk of betrayal if members are captured, then these results are good news for counterterrorist operations. The results show that trees are not the most stable structures of a given size when no restrictions are placed on the structures, and therefore, removing people from the cells by, in one manner or another, cutting off their communication with other members of a cell could have a strong effect on the cell, even if fewer than half of the members are removed.

In truth, though, these results most likely imply that the strongest and most practical terrorist structures are some middle ground between trees and highly connected structures where underlings are connected to many of the people of higher status than themselves. Analysts of counterterrorist operations need to be aware of the idea that "V" structures are mathematically more secure than tree structures, but they are also less practically secure, and take this into account when finding an assumed model for a terrorist cell of unknown structure. One question that is raised is where exactly is this middle ground between theoretical and practical security? This is a question that may be answered through computer or real-life simulations such as the simulation described in "Simulating Terrorist Cells: Experiments and Mathematical Theory" [5].

8 Ideas for Future Research

One could use the algorithm described in Sect. 5 as a basis for development of a more widely applicable program that could take in certain parameters, such as size, on a terrorist cell of partially known structure, and return the strongest or weakest possibilities that meet those parameters.

Another area that needs further work is an expansion of the Maple poset package used to generate the data used in this analysis. The package contains a library of all of the posets of up to 10 vertices, but the posets dealt with in counterterrorism studies are often larger than this. In creating a program for use in counterterrorist operations, one would most likely need to have a larger library of posets. One could even create a Maple package useful for counterterrorism analyses, and include functions such as one that calculates the cutsets of size k in a poset, and one that calculates the probability that a cell is disrupted after an action against that cell. Such programs would undoubtedly be useful in research on the mathematics and theory of counterterrorist operations.

Further questions include theoretical analysis and proof of the results described here, such as the question presented in Problem 6.1. Also, one could do similar analyses with posets of larger sizes, or with a restricted set of posets, such as trees, posets where each element has a restricted number of lower covers, or posets lacking disconnected parts and isolated vertices, and see how the results compare.

Another idea that could inspire further investigation is the seeming dichotomy between what is practically safe and what is mathematically secure – for example, the "V" structures in some posets are practically dangerous, given betrayal risks, but mathematically secure. One could organize realistic or computer simulations of terrorist cells in order to assess the risk presented by such structures, and make conclusions about which terrorist cell structures are strongest in the real world, so that people may better assess the task at hand in fighting such cells.

9 Conclusion

According to our analysis of posets of size seven, certain patterns in node set-up and connections between those nodes create the most stable posets, where stability refers to the difficulty in breaking the cell by removing people from its lines of connection. Trees were not found to be the most stable structures when no restrictions were placed on the structures of the posets we were dealing with.

In light of this analysis, one must always keep in mind that mathematical models do not fully represent real situations, and though these results are useful in finding the representation of the most stable structure, further studies must be conducted to determine what restrictions one must place on such models, and how well results from one set of posets fitting certain parameters carry over to other posets with other parameters. We also must remember that more research is needed in terms of which structures contain the most practical value. It is necessary to compare mathematical value in strength to practical considerations in order to develop a complete picture of the networks that we are trying to disable in the war on terrorism.

In the end, such analyses are useful for improving strategies against terrorist cells, so that the war against terrorism may be fought more efficiently, with fewer casualties and less resources used.

Acknowledgments

The author would like to acknowledge Jonathan Farley for suggesting this topic of research in his paper “Toward a Mathematical Theory of Counterterrorism: How to Build the Perfect Terrorist Cell, I” and for proofreading this paper.

References

1. Davey, B. A., Priestley, H. A.: Introduction to Lattices and Order, Second Edition. Cambridge University Press, Cambridge (2002)
2. Farley, J. D.: Breaking Al Qaeda Cells: A Mathematical Analysis of Counterterrorism Operations (A Guide for Risk Assessment and Decision Making). Studies in Conflict and Terrorism. 26, 399–411 (2003)
3. Farley, J. D.: Toward a Mathematical Theory of Counterterrorism: How to Build the Perfect Terrorist Cell, I. (2006)
4. Hartnell, B. L.: The Optimum Defense against Random Subversions in a Network. Tenth Southeastern Conference on Combinatorics, Graph Theory and Computing. 493–499. Florida Atlantic Univ., Boca Raton, Fla. (1979)
5. McGough, L. R.: Simulating Terrorist Cells: Experiments and Mathematical Theory. (2005)

Combining Qualitative and Quantitative Temporal Reasoning for Criminal Forensics*

Abbas K. Zaidi, Mashhood Ishaque, and Alexander H. Levis

Abstract The paper presents an application of temporal knowledge representation and reasoning techniques to forensic analysis, especially in answering certain investigative questions relating to time-sensitive information about a criminal or terrorist activity. A brief introduction to a temporal formalism called Point-Interval Logic is presented. A set of qualitative and quantitative temporal facts is taken from the London bombing incident that took place on July 7, 2005, to illustrate the use of temporal reasoning for criminal forensics. The information used in the illustration is gathered through the online news sites. A hypothetical investigation on the information is carried out to identify certain time intervals of potential interest to counter-terrorist investigators. A software tool called Temper that implements Point-Interval Logic is used to run the analysis and reasoning presented in the paper.

1 Introduction

While a sequence of events may unfold linearly in time, information about it comes in segments from different locations at different times, often overlapping and often with small contradictions. This phenomenon occurs at a centralized information

Abbas K. Zaidi
System Architectures Lab, George Mason University, Fairfax, VA 22030, e-mail: szaidi2@gmu.edu

Mashhood Ishaque
System Architectures Lab, George Mason University, Fairfax, VA 22030, e-mail: mishaq01@eecs.tufts.edu

Alexander H. Levis
System Architectures Lab, George Mason University, Fairfax, VA 22030, e-mail: alevis@gmu.edu

* An earlier version of the paper has appeared in the proceedings of Descartes Conference on Mathematical Models in Counterterrorism, Center for Advanced Defense Studies, Washington DC, 2006.

N. Memon et al. (eds.), *Mathematical Methods in Counterterrorism*,
DOI 10.1007/978-3-211-09442-6_5, © Springer-Verlag/Wien 2009

gathering node where information can be analyzed and fused both because of the different times sensors make their information available and because of the paths that the data or information takes to reach the centralized node. As the US Department of Defense and the Intelligence Community move towards an information sharing paradigm in which data and information are published by all entities and subscribers can access them, the need to develop algorithms that acknowledge this paradigm and even exploit it gains in importance. In colloquial terms, the question is how quickly can we connect the dots when different dots arrive in random order and at random times. We need to know when we have enough dots (given the information that they carry) to declare that we can make a useful inference – that they have been connected.

This problem has particular significance when one tries to reconstruct the sequence of events that led to an observable effect and, especially, to identify the time interval during which some critical activity has taken place. This can be thought of as forensic analysis of a set of given data. The problem becomes more challenging, if the process is undertaken while pieces of information are arriving and there is a time sensitive aspect to it, i. e., useful inferences need to be made as quickly as possible.

Consider, for example, information regarding events surrounding some criminal activity or an act of terrorism to be unfolding in no specific order. The information gathered at some instant in time, in turn, may be incomplete, partially specified, and possibly inaccurate, or inconsistent, making it difficult for investigators and counter-terrorism experts to piece together the events that can help resolve some of the investigative questions. In situations where the gathered information contains qualitative temporal references (e. g., Event A occurs during Event B) to some events of interest, the investigators (and tools supporting them) are not even equipped with standard representation for a temporal language that can capture these qualitative temporal relations. The time-sensitive information, the information about the timing of events surrounding a criminal/terrorist act, even if provided with qualitative temporal references, may contain hidden patterns or temporal relations that can help identify missing links in an investigation. This calls for a formal, computer-aided approach to such an analysis.

The growing need for a formal logic of time for modeling and analyzing temporal information has led to the emergence of various types of representations and reasoning schemes, extensively reported in the research literature. This paper demonstrates the use of one such formalism, called Point-Interval Logic (PIL), in addressing the problems listed above; how it can be used to create temporal models of situations arising in forensics and help investigators operating in real time answer interesting questions in a timely manner. An earlier student paper by Ishaque et al.[1] first proposed and demonstrated the application of PIL to criminal forensics.

This paper is organized as follows: Section 2 gives a very brief overview of some of the influential work on temporal reasoning and knowledge representation. Section 3 presents an informal description of Point-Interval Logic (PIL). This section gives a description of the logic's syntactic and semantic structure, presents a graphical representation for the temporal statements, and illustrates inference mechanism and algorithm for deciding consistency. The software implementation of the approach, called Temper, is also discussed in this section. A hypothetical investigation on the information taken from London Bombing of July 07, 2005, is carried out with the help of Temper to identify certain time intervals of potential interest to counter-terrorism experts. It also shows that the approach not only operates on the arriving sequence of data but, as a result of the diagnostics produced by the inference engine, it also identifies in a visual manner the type of information that is needed to disambiguate the inferences.

2 Temporal Knowledge Representation and Reasoning

The earliest attempts at formalizing a time calculus date back to 1941 by [2], and 1955 by [3]. Since then, there have been a number of attempts on issues related to this subject matter, like topology of time, first-order and modal approaches to time, treatments of time for simulating action and language, etc. The development of some of these formalisms has matured enough to attract comparative analysis for the computational aspects of these calculi and their subclasses. A number of researchers have attempted to use temporal reasoning formalisms for planning, plan merging, conditional planning, and planning with uncertainty. Other applications of temporal logics include specification and verification of real-time reactive planners, and specification of temporally-extended goals and search control rules.

As noted by Kautz [4], the work in temporal reasoning can be classified in three general categories: algebraic systems, temporal logics, and logics of action. The work on the temporal reasoning within the framework of constraint satisfaction problems was initiated with the influential Interval Algebra (IA) of [5] (formalized as an algebra by [6]) and was followed by Point Algebra (PA) of [7]. They also proved that the problem of determining consistency in IA is NP-Complete. The nature of temporal variables, i. e., point and/or interval, and different classes of constraints, namely qualitative and/or quantitative constraints, have led to a characterization of different types of temporal formalisms: qualitative point, quantitative point, qualitative interval, quantitative interval, and their combinations. A temporal relation (or constraint) between two variables X_i and X_j is represented by the expression $X_iC_{ij}X_j$, where $C_{ij} = \{r_1, r_2, \ldots, r_n\}$; r_i's are basic relations and the expression translates to $(X_ir_1X_j) \vee (X_ir_2X_j) \vee \ldots \vee (X_ir_nX_j)$. A relation $r_i \in C_{ij}$ is *feasible* for the time variables if and only if there exists a *solution* holding the relation between the two variables. A *solution* to such a temporal problem consists of finding *consistent* relations among all the variables involved. The notion of consistency is defined with respect to specific semantics for each type of temporal problem and with the help of *transitivity/composition tables*. A computationally less expensive notion of *local-consistency* is employed in most implementations to incrementally achieve global consistency. In most cases, enforcing local consistency can be done in polynomial time. Almost all the temporal formalisms use some form of graph representation, e. g., constraint networks, distance graphs, timegraphs, point graphs, etc., for representing/formulating the temporal problem under consideration. Some of these graphs are merely used for representing the variables and temporal relations between them, whereas others exploit the graph-theoretic properties to enforce consistency and process queries.

Point-Interval Logic (PIL) is a specialization of Pointisable Algebra [8], which is the first and the simplest tractable subclass identified, containing all the basic temporal relations, [9]. This class is characterized by the fact that the temporal relations in it can be represented by specifying relations between the *start/end* points of intervals. A time interval X is defined with the help of a pair of its starting point sX and end point eX. All allowable temporal relations can be represented as constraints between these start and end points associated with time intervals. Polynomial time algorithms for processing Pointisable algebra have been developed [10], [11]. The work on PIL originated from an earlier attempt on temporal knowledge representation and reasoning by Zaidi [12]. A graph model, called Point Graph (PG), is shown to represent the temporal statements in this approach. An inference engine based on this Point Graph representation infers new temporal relations among system intervals, identifies temporal ambiguities and errors (if present) in the system's specifications, and finally identifies the intervals of interest defined by the user. Zaidi and Levis [13] further extended the point-interval approach by adding provisions for *dates/clock* times and time distances for points and intervals. This extension allowed the assignment of actual lengths to intervals, time *distances* between points, and time stamps to points representing actual time of occurrences, whenever such information is available. A temporal model may change during and/or after the system specification phase. Support for an on-the-fly revision (add, delete, modify) was added to Point Graph formalism in [14]. Zaidi and Wagenhals [15] consolidated the results of the previous work on the logic and its application to the modeling and planning time-sensitive aspects of a mission and extended the approach further. The extension allows for a larger class of temporal systems to be handled by incorporating an enhanced input lexicon, allowing increased flexibility in temporal specifications, providing an improved verification and inference mechanism, and adding a suite of analysis tools.

3 Point-Interval Logic

This section presents a brief introduction to the PIL formalism with the help of illustrative examples. A more technical and detailed description can be found in [15], [16], and/or [13].

3.1 Language and Point Graph Representation

The lexicon of the Point-Interval Logic (PIL) consists of the following primitive symbols:

Points: A point X is represented as $[pX, pX]$ or simply $[pX]$.

Intervals: An interval X is represented as $[sX, eX]$, where sX and eX are the two end points of the interval, denoting the 'start' and 'end' of the interval, such that $sX < eX$.

Point Relations: These are the relations that can exist between two points. The set of relations R_P is given as:
$R_P = \{<, =, \leq\}$ or $R_P = \{before, equals, precedes\}$

Interval Relations: These are the atomic relations that can exist between two intervals. The set of relations R_I is given as:
$R_I = \{<, m, o, s, d, f, =\}$ or
$R_I = \{before, meets, overlaps, starts, during, finishes, equals\}$

Point-Interval Relations: These are the atomic relations that can exist between a point and an interval. The set of relations R_{PI} is given as:
$R_{PI} = \{<, s, d, f\}$ or $R_{PI} = \{before, starts, during, finishes\}$
The symbol ? is used to represent an unknown relationship.

Functions: The following two functions are used to represent quantitative information associated with intervals:
The *Interval length function* assigns a non-zero positive real number to a system interval.
Length $X = d$, where $X = [sX, eX], d \in \Re^+$
This function is also used to assign lower and upper bounds to an interval length. The two bounds can also be seen as representing *at least* and *at most* temporal relations.
Length $X \geq d$, where $X = [sX, eX], d \in \Re^+$ (d is a lower bound on length)
Length $X \leq d$, where $X = [sX, eX], d \in \Re^+$ (d is an upper bound on length)
The *stamp function* similarly assigns a non-negative real number to a point, or lower and upper bounds to it. The two bounds can also be seen as representing *no later than*, and *no earlier than* temporal relations.
Stamp $p = t$, where $t \in \Re$ $(= \Re^+ \cup \{0\})$
Stamp $p \leq t, t \in \Re$
Stamp $p \geq t, t \in \Re$

Figure 1 shows the syntactic and semantic structure of PIL expressions. Note that each relationship between intervals or an interval and a point can be constructed with the help of inequalities between their start and end points, and by assigning values to expressions involving these points.

Qualitative Relations

CASE I—X and Y both points: X = [px] and Y = [py]

1. $X < Y$ $px < py$
2. $X = Y$ $px = py$
3. $X \leq Y$ $px \leq py$

CASE II— X and Y both intervals with non-zero lengths:
X = [sx, ex], Y = [sy, ey] with sx < ex and sy < ey

1. $X < Y$ $ex < sy$
2. $X\ m\ Y$ $ex = sy$
3. $X\ o\ Y$ $sx < sy$, $sy < ex$, $ex < ey$
4. $X\ s\ Y$ $sx = sy$, $ex < ey$
5. $X\ d\ Y$ $sx > sy$, $ex < ey$
6. $X\ f\ Y$ $sx > sy$, $ey = ex$
7. $X = Y$ $sx = sy$, $ex = ey$

CASE III— X is a point and Y is an interval: X = [px] and Y = [sy, ey]

1. $X < Y$ $px < sy$
2. $X\ s\ Y$ $px = sy$
3. $X\ d\ Y$ $sy < px < ey$
4. $X\ f\ Y$ $px = ey$
5. $Y < X$ $ey < px$

Quantitative Relations

X is a point and Y is an interval: X = [px] and Y = [sy, ey]

1. Length $Y = d$ $ey - sy = d$
2. Length $Y \geq d$ $ey - sy \geq d$
3. Length $Y \leq d$ $ey - sy \leq d$
4. Stamp $X = d$ $px = d$
5. Stamp $X \geq d$ $px \geq d$
6. Stamp $X \leq d$ $px \leq d$

Fig. 1 PIL Expressions and Their Semantics

A graph construct called Point Graphs (PG) is used as an underlying structure to represent statements in PIL. In a PG, a node represents a point (or a *composite* point) and an edge between two points represents one of the two temporal relations, *before* and *precedes*, between the two. Two or more points $p_i, p_j, \ldots, p_n$ are represented as a composite point $[p_i; p_j; \ldots; p_n]$, or a single node in a PG, if all are mapped to a single point on the timeline. The statements in PIL can be converted to an equivalent PG representation with the help of the corresponding analytic inequalities shown in Fig. 1. In addition, the quantitative temporal information, modeled using the length and the stamp functions, is represented as node and arc inscriptions on the PG. All the verification, revision, and inference algorithms work by manipulating this Point Graph representation of the set of PIL statements. Figure 2 shows a set of PIL statements and the corresponding Point Graph representation.

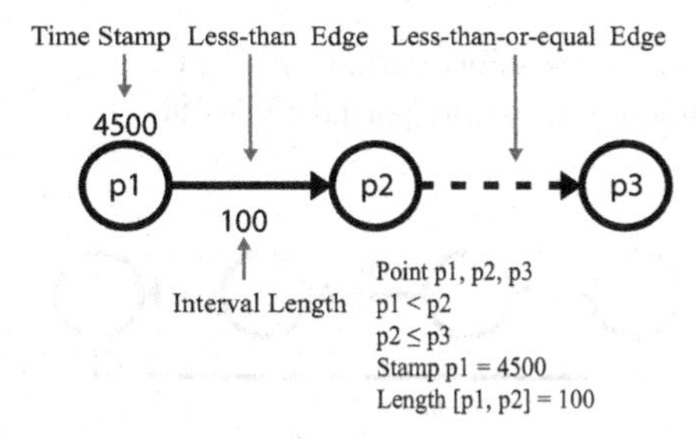

Fig. 2 Point Graph Representation of a Set of PIL Statements

The Point Graph (PG) in Fig. 2 illustrates how the inequalities shown in Fig. 1 for each qualitative temporal relation can be translated to a corresponding equivalent graph structure. The figure also shows graphical objects representing some of the quantitative temporal relations. The graphical representation of the remaining temporal relations requires introduction of virtual time point(s) or *virtual nodes* in a PG. A virtual node is like any other node in a PG except for the fact that there is no temporal variable (point, start of interval, or end of interval) associated with it. It, therefore, does not have a unique identifier or *name* associated with it. Figures 3 and 4 illustrate the PG representations of the quantitative temporal cases not covered by the structure in Fig. 2.

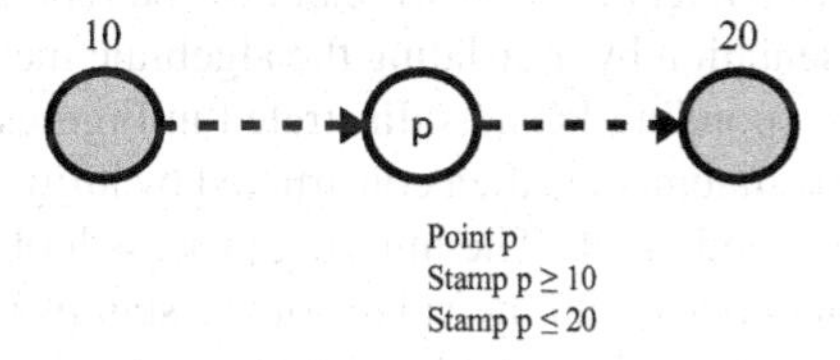

Fig. 3 PG Representation of Lower and Upper Bounds on Stamps

A formal definition of Point Graphs is given as follows:

Definition 1 (Point Graphs). A Point Graph, PG (V,E_A,D,T) is a directed graph with:

- V: Set of vertices with each node or vertex $v \in V$ representing a point on the real number line. Points $p_i, p_j, \ldots, p_n$ are represented as a composite point $[p_i; p_j; \ldots; p_n]$ if all are mapped to a single point on the line.
- E_A: Union of two sets of edges: $E_A = E \cup E_{\leq}$, where
  - E: Set of edges with each edge $e_{12} \in E$, between two vertices $v1$ and $v2$, also denoted as $(v1,v2)$, representing a relation '<' (*before*) between the two vertices-$(v1 < v2)$. The edges in this set are called LT edges;
  - $E_{\leq}$: Set of edges with each edge $e_{12} \in E_{\leq}$, between two vertices $v1$ and $v2$, also denoted as $(v1,v2)$, representing a relation '$\leq$' (*precedes*) between the two vertices – $(v1 \leq v2)$. The edges in this set are called LE edges.
- D: Edge-length function (possibly partial): $E \rightarrow \Re^+$
- T: Vertex-stamp function (possibly partial): $V \rightarrow \Re$

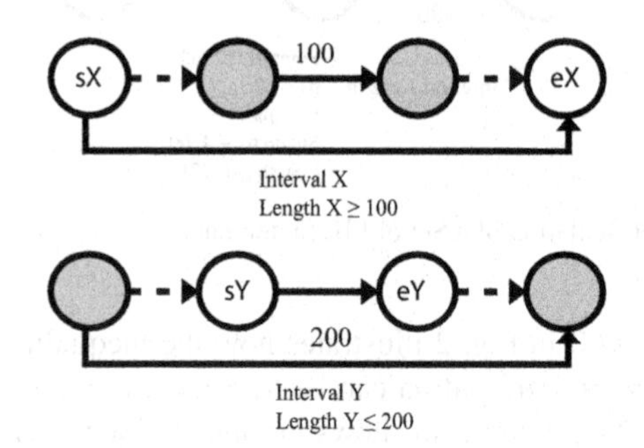

Fig. 4 PG Representation of Lower and Upper Bounds on Interval Lengths

3.2 *Operations on Point Graphs*

A relation R_i between two intervals X and Y can now be represented by an equivalent Point Graph representation by translating the algebraic inequalities/expressions shown in Fig. 1 to corresponding PGs, as illustrated in Figures 2, 3 and 4. The PG representing a set of PIL statement is then constructed by *unifying* individual PGs to a (possibly) single connected graph. The unifying process looks at the labels of the nodes (except for virtual nodes) and the values of the stamps associated with them to identify equalities. The nodes identified as being equal to one another are merged into a single node with a composite label. The *unified* PG is then *folded* with the help of lengths on edges. This folding process establishes new relations among system intervals, inferred through the quantitative analysis of the known relations specified by interval lengths and stamps. Figure 5 illustrates the two operations, unification and folding, on an example set of PIL statements. A more technical and detailed description of the two processes can be found in [15, 16] or [13].

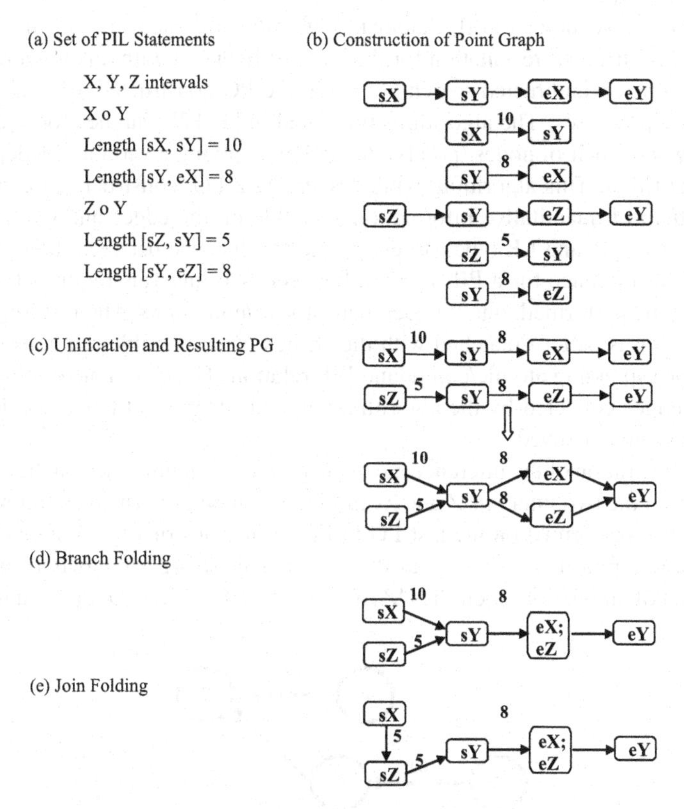

Fig. 5 Steps in PG Construction

3.3 *Inference*

Two points, $p1$ and $p2$, on a real number line are related to each other by one of the following three algebraic relations: '$<$' (*before*), '$=$' (*equals*), and '$\leq$' (*precedes*). A relation R_i between two intervals X and Y, denoted as XR_iY can, therefore, be represented as a 4-symbol string '*abcd*' made of elements from the alphabet $\{<,=,>,\leq,\geq,?\}$, where the first (left-most) symbol '*a*' represents the algebraic relation between sX and sY, second symbol '*b*' between sX and eY, third symbol '*c*' represents relation between eX and sY, and fourth '*d*' between eX and eY. The '?' is added to incorporate incomplete information. The inequalities in Fig. 1, the definition of an interval, i. e., $X = [sX, eX]$ implies '$sX < eX$', and a set of basic PIL axioms [15] are used to construct this 4-symbol string representation for all possible temporal relations, i. e., both basic and compound relations, in Pointisible logic. A complete list of temporal relations and the corresponding string representations is provided in [15].

The PG representation of PIL statements helps the inference mechanism of PIL to construct the string representation for the pairs of intervals with unknown relations by performing an undirected search [1, 17] in the PG constructed after unification and folding processes. The algorithms presented in [1, 17] searches for undirected paths between a pair of nodes (points) in a PG by applying a variant of a depth-first search algorithm. This algorithm, while traversing a PG, constructs an expression with the help of quantitative information available on the edges and the nature of the edges, i. e., LT and LE types, to determine the distance between the two nodes (points). An inference for a PIL relation between two intervals requires four such searches to be performed, one for each pair of start/end points. The resulting string representation is pattern-matched with the strings of all possible relations to identify the corresponding atomic/compound PIL relation. The quantitative information from the edges collected by the algorithm helps identify quantitative relations between the points involved.

As an illustration of the inference mechanism of PIL, an inference on the relationship between the two intervals Z and X, in Fig. 5, can be performed as follows: The four searches performed on the last PG in Fig. 5, to construct the 4-symbol string representation, return '$>,<,>,=$' as the output. This string when pattern-matched with the list of all relations identifies 'ZfX' (or, Z *finishes* X) as the inferred relation.

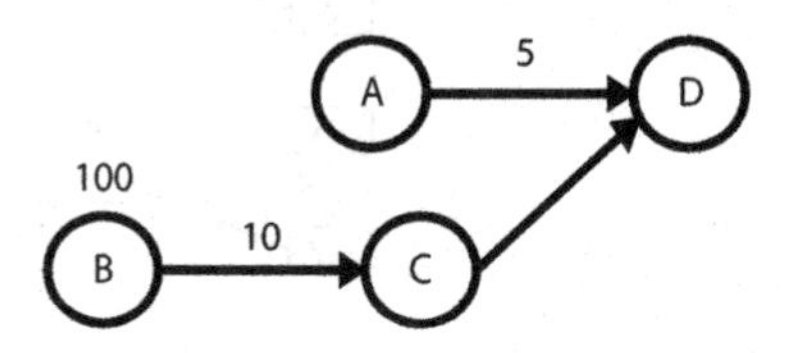

Fig. 6 Inference in a PG

As mentioned earlier, the search algorithm in PIL performs an undirected traversal of a PG to make inferences. The need for an undirected search arises from the presence of quantitative information in PGs, i. e., edge lengths. Figure 6 illustrates a situation where an algorithm that looks for directed paths between a pair of nodes will not be able to infer the temporal relationship between points 'A' and 'B', since there is no directed path either from 'A' to 'B' or from 'B' to 'A'. A search from node A' to B', using the algorithm in [16] returns '$5-\delta-10$' as the output expression, where δ represents the unknown, non-zero length on LT edge between nodes 'C' and 'D', and the negative signs represent the backward orientation of the edges traversed from 'A' to 'B'. The expression represents the distance from point 'A' to point 'B' on a timeline. The expression when evaluated results in a negative value, i. e., '$5-\delta-10<0$', thus establishing the temporal relationship '$B<A$' (or, B *before* A).

In addition to the inference algorithm used to infer temporal relations between a pair of points/intervals, a couple of similar algorithms are presented in [17] to calculate time stamps on points and lengths for intervals. The two algorithms try to identify an exact value for the stamp or the length. In case where exact value cannot be ascertained, the algorithms try to calculate an upper and/or lower bounds for the value. For example in Fig. 6, a query for Stamp [D] is returned by the following: 'Stamp [D]> 110'.

The time complexity of the inference algorithms in PIL have a time complexity of $O(mn+n^2)$, where m is the number of edges and n is the number of nodes in a PG. The worst-case complexity, therefore, is $O(n^3)$ (in case of a complete graph when m is to the order of n^2) with a much better average performance observed during empirical studies.

3.4 Deciding Consistency

The inference mechanism described above may result in erroneous and inconsistent results provided the system of PIL statements, represented by the PG, contains *inconsistent* information. The inference, on the other hand, is guaranteed to yield valid assertions given a consistent PIL system and corresponding PG representation. The following theorem characterizes inconsistency in PIL.

Theorem 1 (Inconsistency in PIL [15]). *A system's description in PIL contains inconsistent information iff*

1. *for some intervals X and Y, and atomic PIL relations R_i and R_j, both XR_iY and XR_jY, $i \neq j$, or XR_iY and YR_jX (with the exception of '=' relation) hold true;*
   or
2. *for some intervals and/or points, the system can determine two string representations such that at least one pair of the algebraic inequalities representing relationships between the corresponding points represents an inconsistency. Let the two string representations be 'abcd' and 'uvwx', where a, b, c, d, u, v, w, and $x \in \{<,=,>,\leq,\geq,?\}$. One of the (un-ordered) pairs of corresponding inequalities, i. e.,*
   (a,u), (b,v), (c,w), or $(d,x) \in (<,=),(<,>),(<,\geq),(=,>),(>,\leq)$;
   or
3. *for a point $p1$, the system calculates two different stamps;*
   or
4. *for some points $p1$ and $p2$, '$p1 < p2$', the system can determine two different lengths for the interval $[p1,p2]$.*

The verification mechanism of PG representation identifies these inconsistent cases by using a path-searching algorithm, [18]. The path-searching algorithm employs techniques by [19] and Warshall's algorithm [20] to identify the erroneous cases. The inconsistencies identified in the theorem, above, manifest themselves in one of the two forms in the PG representation: (a) cycles, and (b) multiple paths between a pair of nodes in a PG with infeasible path lengths. An example of this second case is illustrated in Fig. 7. The details of actual algorithms used to decide consistency in a PIL system can be found in [16].

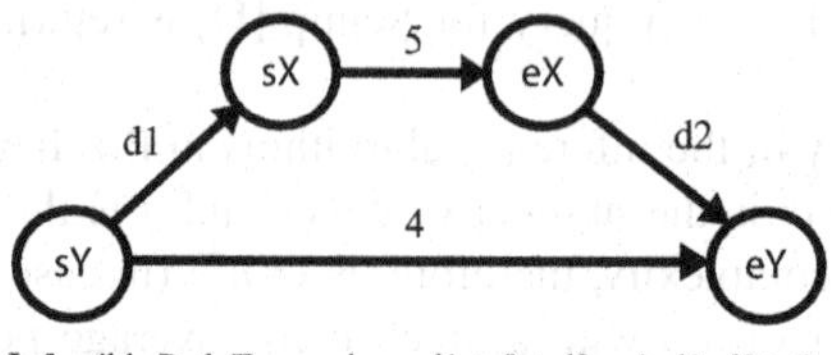

Infeasible Path Expressions: d1 + 5 + d2 = 4; d1, d2 > 0

Fig. 7 An Inconsistency in a PG

Verification is the most costly step of the Point Graph construction process. It uses Warshall's algorithm [20] for searching paths and has time complexity of $O(n^3)$, where n is the number of nodes in a Point Graph. It must be noted that the $O(n^3)$ time complexity is only for detecting cycles and inconsistent paths in the Point Graph. Reporting all cycles or inconsistent paths can have an exponential time complexity, since there can be an exponential number of paths or cycles in a Point Graph, which may not be the case for most real-world PIL systems under study.

The total time complexity of Point Graph construction is $O(n^3)$, where $O(n^3)$ is due to the verification mechanism. But if it is known a priori that the set of PIL statements is consistent, the verification step can be avoided and the Point Graph can be constructed fairly quickly. This and several other techniques have been employed in the software implementation of the approach, called Temper, to make representation and reasoning algorithms more efficient. A brief introduction of Temper follows in the next subsection.

3.5 Temper

A software tool called Temper (**Tem**poral Program**mer**) implements the inference mechanism of Point-Interval Logic along with its verification and revision mechanisms. A screen shot of Temper's user interface is shown in Fig. 8. Temper provides a language editor, shown in Fig. 9, to input PIL statements and a query editor, shown in Fig. 10, to run various queries on the constructed Point Graphs. It has a graphical interface to display the Point Graphs and also a text I/O interface to display information and results of the analysis (Fig. 8). In the PG shown in Fig. 8, each point is represented as a node, and each interval is represented by two nodes connected by a LT or LE edge. Each LT edge is represented by a solid arc and the length, if available, appears adjacent to the arc. Each less-than-or-equal ($\leq$) or LE edge is represented by a dotted arc. The stamp on each point appears inside the node representing the point. A special type of node, called virtual node, is used to represent *at*

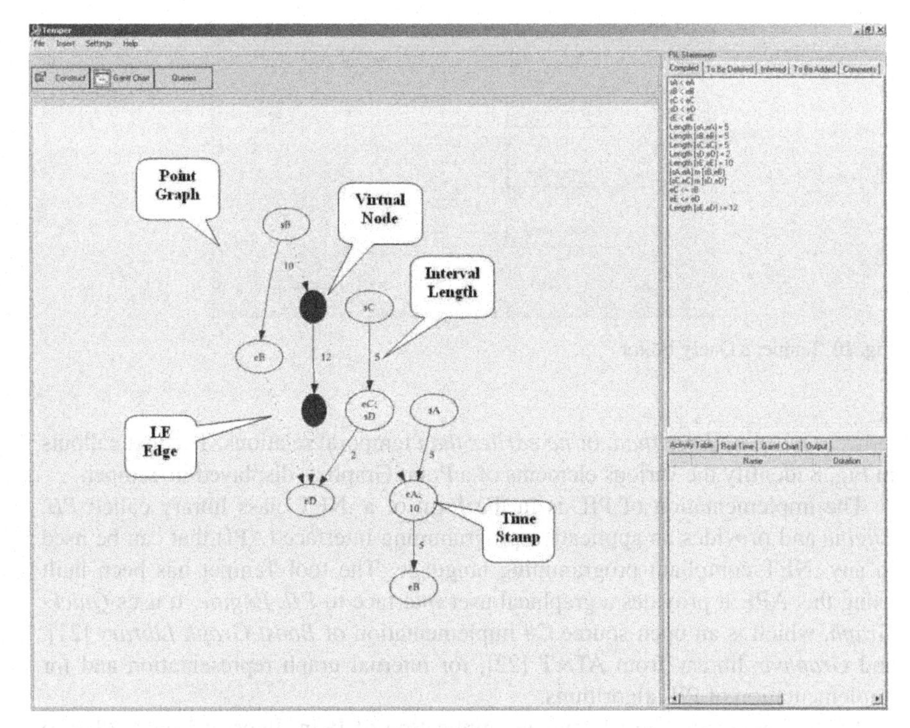

Fig. 8 Temper User Interface with a PG Visualization

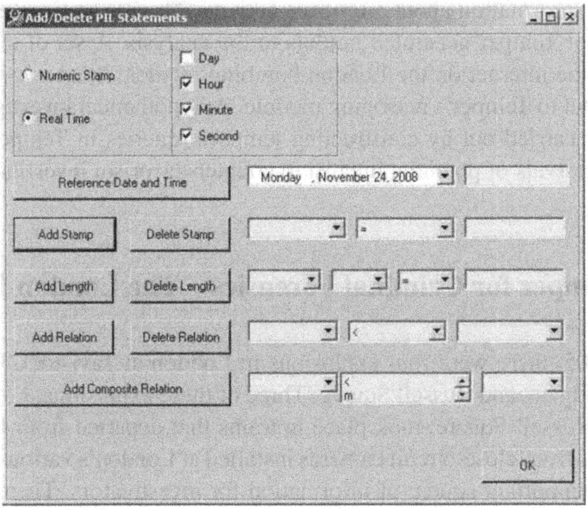

Add/Delete PIL Statements

Numeric Stamp
Real Time
Day
Hour
Minute
Second
Reference Date and Time
Monday , November 24, 2008
Add Stamp
Delete Stamp
Add Length
Delete Length
Add Relation
Delete Relation
Add Composite Relation
OK

Fig. 9 Temper's Language Editor

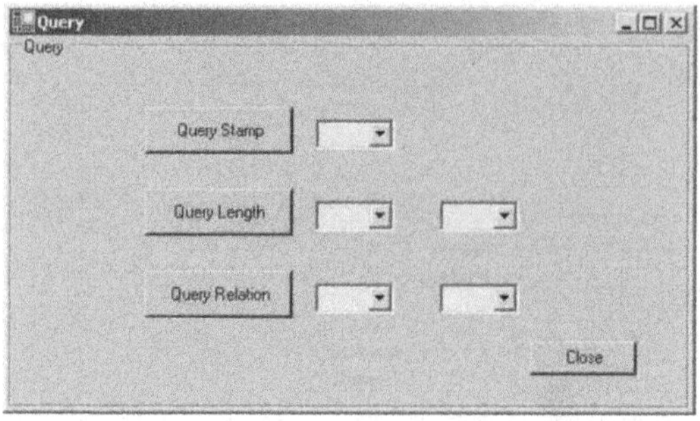

Fig. 10 Temper's Query Editor

least, at most, no later than, or *no earlier than* temporal relations. The text callouts in Fig. 8 identify the various elements of a Point Graph as displayed in Temper.

The implementation of PIL is in the form of a .NET class library called *PIL Engine* and provides an application programming interface (API) that can be used in any .NET compliant programming language. The tool Temper has been built using this API. It provides a graphical user interface to *PIL Engine*. It uses *QuickGraph*, which is an open-source C# implementation of *Boost Graph Library* [21], and *Graphviz* library from AT&T [22], for internal graph representation and for implementation of PIL algorithms.

The following section presents an application of Point-Interval Logic (PIL) to criminal forensics, especially in answering certain investigative questions relating to time-sensitive information about a terrorist activity. The illustrations are presented with the help of Temper generated outputs to the analysis. A set of temporal facts is taken from the internet on the London bombing incident that took place on July 7, 2005, as input to Temper's reasoning module. A hypothetical investigation on the information is carried out by constructing temporal queries in Temper to identify certain time intervals of potential interest to counter-terrorism investigators.

4 Using Temper for Criminal Forensics – The London Bombing

On July 7, 2005, there were four explosions in London at Tavistock Square, Edgware Road, Aldgate, and Russell Square. Three of these explosions, Edgware Road, Aldgate, and Russell Square, took place in trains that departed from King's Cross station. Images from close-circuit cameras installed at London's various railway stations were an important source of information for investigators. There were hours of images available from these cameras and the task of investigators was to analyze these images to identify possible suspects. The large number of such images, although desirable, can make an investigation that requires searching through them in a timely manner very time consuming. Time pressure was also created because there was need to identify the perpetrators quickly enough to apprehend any ones that survived.

In this section, we demonstrate how temporal reasoning, in general, and especially Temper can be used to restrict the size of a potential interval for which to analyze images (by making sense of the available temporal information) and thus speeding up the investigation. Since Temper has the ability to handle both qualitative and quantitative constraints, both types of information regarding the incident and/or the surrounding events can be input to it. Temper also offers the additional advantage of the verification mechanism that can be invoked to check the consistency of the available temporal information. This can be very useful when temporal information may originate from multiple (and possibly unreliable) sources. In this example, we demonstrate the capabilities of Temper by modeling a set of temporal information items related to the incident and by trying to identify the exact time or the shortest possible interval during which one of the ill-fated trains left from King's Cross station for Edgware. The graphical visualization of the temporal relations shows clearly where additional data are needed to establish temporal relations that will facilitate developing responses to specific queries.

The journey of the three trains from King's Cross station can be represented as PIL intervals. The journey of these trains ended in explosions. We also know the lower bounds on the travel times of these trains after their departures from King's Cross station, based on the distances of the sites of the explosions from King's Cross station. The train from King's Cross to Edgware must have traveled for at least 5 minutes. Similarly trains to Aldgate and Russell Square must have traveled for 4 and 5 minutes, respectively. Table 1 shows how this information can be represented as PIL statements. These PIL statements are typed or read into Temper using its language editor. Figure 11 shows the corresponding Point Graph in Temper.

Once the temporal information has been inserted, Temper can be used to draw inferences about the event(s) of interest, i. e., the instant when one of the trains left King's Cross station for Edgware. We run a query, using the query editor of Temper, for the time stamp of the point "sTrain_King_Cross_to_Edgware" which represents the departure of the train from King's Cross station to Edgware. Figure 12 shows the query in Temper. The inference algorithm in Temper returns a '?' or 'unknown' as the result of the query, since the temporal information available to the inference mechanism is not enough to draw any meaningful relationship for the temporal event under investigation.

For the illustrated case, Temper cannot infer anything about the stamp of the event based on the information provided so far. This creates the need for information pull. Suppose, further investigation reveals that the explosion near Edgware took place between times 8:40 and 8:52 (the explosion is considered to be an instantaneous event so the range 8:40 to 8:52 does not represent duration but the uncertainty in determining the actual occurrence time). Similarly the explosions near Aldgate and Russell Square occurred between 8:45 and 8:50, and between 8:40 and 8:50 respectively. Table 2 shows how this information can be represented as PIL statements. These PIL statements are added to the initial temporal model to get the Point Graph of Fig. 13. Once again, the query for the time stamp of the point "sTrain_King_Cross_to_Edgware" is executed (Fig. 12). This time, Temper is able to determine an upper bound for the stamp of the event: its inference algorithm returns 'Stamp [sTrain_King_Cross_to_Edgware] $\leq$8:47', i. e., the train from King's Cross to Edgware must have left no later than 8:47.

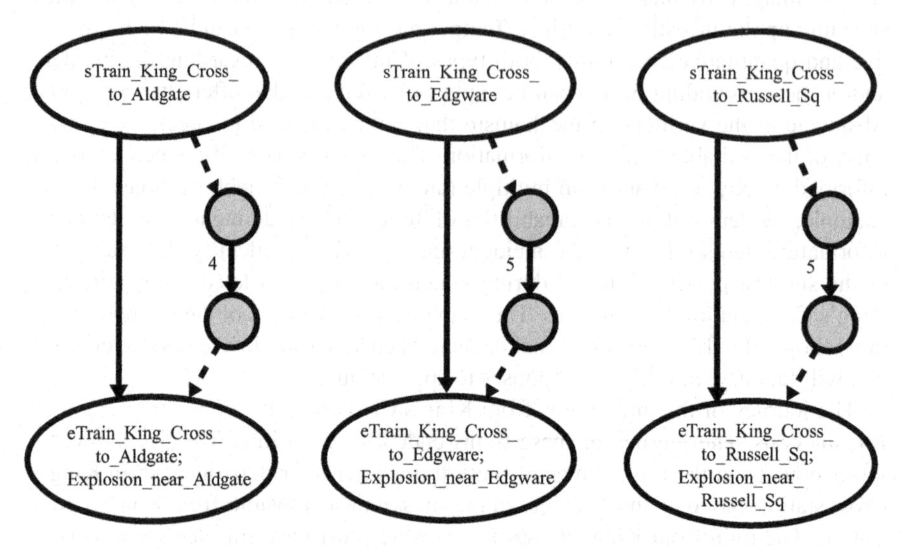

Fig. 11 Point Graph for London Bombing Scenario in Table 1

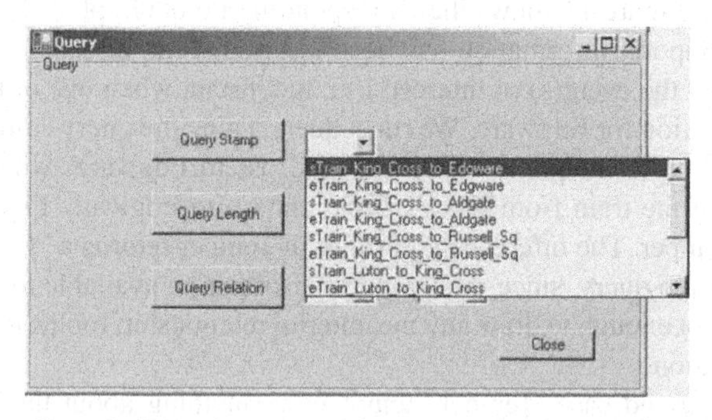

Fig. 12 Running a Query in Temper

As indicated earlier, the verification mechanism of Temper can detect the inconsistencies in the available temporal information. The ability to detect inconsistency can be very useful, especially when the information from different sources is combined into a single model of a situation under investigation. Suppose, we input to Temper the information that the train from King's Cross to Edgware left at time 8:48 (represented by the PIL statement: Stamp[sTrain_King_Cross_to_Edgware] = 8:48). Clearly, this statement is in conflict with the previously added PIL statements.

Table 1 PIL Statements for London Bombing Scenario

Temporal Information	**PIL Statements**
Train traveling from King's Cross to Edgware	*Interval* Train_KingX_Edgware
Train traveling from King's Cross to Aldgate	*Interval* Train_KingX_Aldgate
Train traveling from King's Cross to Russell Square	*Interval* Train_KingX_Russell_Sq
Explosion at Edgware	*Point* Explosion_Edgware
Explosion at Aldgate	*Point* Explosion_Aldgate
Explosion at Russell Square	*Point* Explosion_Russell_Sq
Explosion at Edgware ended the journey of train from King's Cross to Edgware	Explosion_Edgware *f* Train_KingX_Edgware
Explosion at Aldgate ended the journey of train from King's Cross to Aldgate	Explosion_Aldgate *f* Train_KingX_Aldgate
Explosion at Russell ended the journey of train from King's Cross to Russell Square	Explosion_Russell_Sq *f* Train_KingX_Russell_Sq
Train from King's Cross to Edgware traveled at least for 5 time units	*Length* [Train_KingX_Edgware] ≥ 5
Train from King's Cross to Aldgate traveled at least for 4 time units	*Length* [Train_KingX_Aldgate] ≥ 4
Train from King's Cross to Russell Square traveled at least for 5 time units	*Length* [Train_KingX_Russell_Sq] ≥ 5

Table 2 Additional PIL Statements for London Bombing Scenario

Temporal Information	**PIL Statements**
Explosion at Edgware happened no earlier than 8:40	*Stamp* [Explosion_Edgware] $\geq$8:40
Explosion at Edgware happened no later than 8:52	*Stamp* [Explosion_Edgware] $\leq 8:52$
Explosion at Aldgate happened no earlier than 8 : 45	*Stamp* [Explosion_Aldgate] $\geq 8:45$
Explosion at Aldgate happened no later than 8 : 50	*Stamp* [Explosion_Aldgate] $\leq 8:50$
Explosion at Russel Sq. happened no earlier than 8 : 40	*Stamp* [Explosion_Russel_Sq] $\geq 8:40$
Explosion at Russel Sq. happened no later than 8 : 50	*Stamp* [Explosion_Russel_Sq] $\leq 8:50$

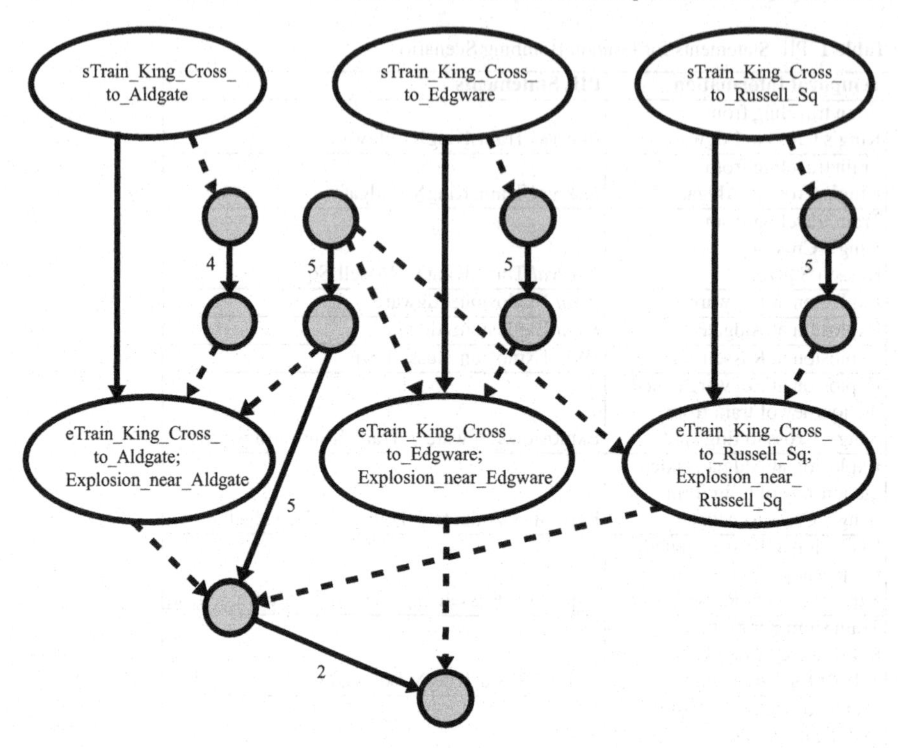

Fig. 13 Revised Point Graph for London Bombing Scenario

Temper detects this inconsistency, and identifies the portion of the Point Graph that contains the contradiction (inconsistent paths shown with an extra boundary around each node in the paths in Fig. 14). Note the two inconsistent paths (from node with stamp 8:40 to node with stamp 8:52): one path with length exactly equal to 12 minutes and the other path having a length of at least 13 minutes. We fix this inconsistency by deleting the last statement added, "Stamp [sTrain_King_Cross_to_Edgware] = 8:48". Temper's revision algorithm employs an efficient use of internal data structures to identify the portion of the Point Graph that is affected by the delete operation and only redraws that affected part to reconstruct the new PG representation. This saves a lot of computational effort required to reconstruct the Point Graph for an entire set of PIL statements from scratch every time there is a need to make modifications in the temporal information. A detailed description of the revision algorithm can be found in [14].

Suppose that the investigators have also identified four suspects who were spotted entering the Luton railway station at time instant 7:20. The investigators believe that these suspects took a train from Luton to King's Cross station, and at King's Cross station they boarded the trains in which the explosions took place. The next train from Luton to King's Cross departed at 7:48 and reached King's Cross at time instant 8:42. Obviously, if these suspects were in fact the bombers, the train from

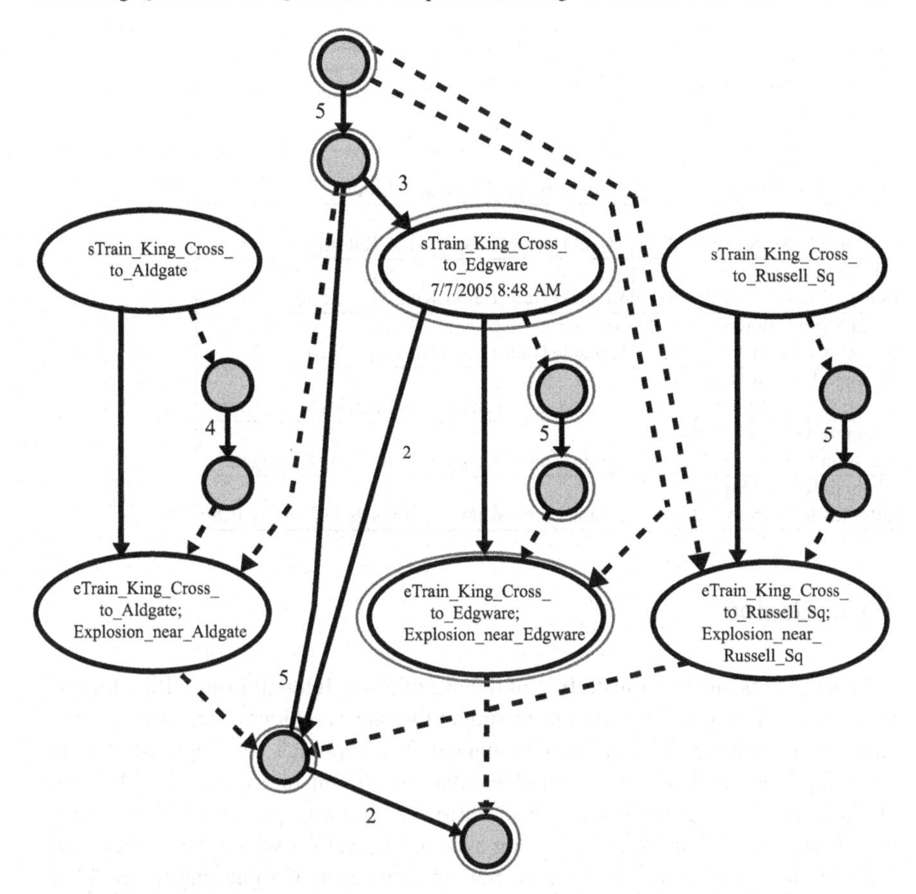

Fig. 14 Inconsistency in the Point Graph

Luton should have reached King's Cross before any train in which there was an explosion left King's Cross station. This information can be represented by PIL statements as given in Tab. 3 and the resulting PG is shown in Fig. 15. Note that Tab. 3 contains both qualitative and quantitative PIL statements.

The query for the time stamp of the point "sTrain_King_Cross_to_Edgware" is executed once more (Fig. 12). This time, Temper is able to determine both an upper bound and a lower bound for the stamp of the event, its inference algorithm returns: '8:42 $<$ Stamp[Train_King_Cross_to_Edgware] $\leq$ 8:47', i. e., the train must have left King's Cross station after time instant 8:42 and no later than 8:47. Note that the lower bound is strict. Thus by applying the inference mechanism of Point-Interval Logic to the analysis of available temporal information, the approach has identified the bounds of the interval that we were interested in. The images need to be analyzed for this interval only; this improves the timeliness of the labor intensive image analysis process.

Table 3 Additional PIL Statements for London Bombing Scenario

Temporal Information	**PIL Statements**
Train traveling from Luton to King's Cross station	*Interval* Train_Luton_KingX
Suspected spotted entering the Luton station	*Point* Suspects_Spotted_at_Luton
Suspected spotted at Luton at time instant 7 : 20	*Stamp* [Suspects_Spotted_at_Luton] = 7 : 20
Train from Luton to King's Cross left at 7 : 48	*Stamp* [sTrain_Luton_KingX] = 7 : 48
Train from Luton to King's Cross arrived at 8 : 42	*Stamp* [eTrain_Luton_KingX] = 8 : 42
Train to Edgware left after the train from Luton	eTrain_Luton_KingX < Train_KingX_Edgware
Train to Aldgate left after the train from Luton	eTrain_Luton_KingX < Train_KingX_Aldgate
Train to Russell Sq. left after the train from Luton	eTrain_Luton_KingX < Train_KingX_Russell_Sq

5 Conclusion

This paper presented an illustration of the use of Point-Interval Logic (PIL) for creating temporal models of situations arising in forensics and for helping investigators answer relevant questions in a timely manner. The approach was demonstrated using Temper, which is a software implementation of Point-Interval Logic (PIL), and the London bombing incident as the scenario. The example presented demonstrates how Temper can be used in identifying a small interval for which the images from close-circuit cameras should be analyzed. The reader may argue that the problem could have been solved manually as well; that is true in the case of a small example like the one presented in this paper; however, in general situations the set of temporal statement may be too large for a human to handle. (The example set of PIL statements used in this paper is intentionally kept small for the sake of presenting the ideas and types of analysis that can be performed, instead of presenting the actual solution to the problem posed.) This presence of large set(s) of incomplete, inaccurate, and often inconsistent data calls for a computer-aided approach to such an analysis. Temper can combine temporal information from multiple sources, detect inconsistencies and identify the specific source(s) with inconsistent information. Temper can, therefore, be used to compare witness accounts of several individuals on the same incident for overlaps and inconsistencies–another useful application for forensics.

Acknowledgements

The work was carried out with support provided by the Air Force Office of Scientific Research under contract numbers FA9550-05-1-0106.

References

1. Ishaque, M.; Zaidi, A. K; and Levis, A. H.: On Applying Point-Interval Logic to Criminal Forensics, Proc. Of 2006 Command and Control Research and Technology Symposium (CCRTS). San Diego, CA. (2006)
2. Findlay, J. N.: Time: A treatment of some puzzles, Aust. J. Phil., vol. 19.216–235 (1941)
3. Prior, A. N.: Diodoran modalities, Phil. Quart., vol. 5. 205–213 (1955)
4. Kautz, H.: Temporal Reasoning, In The MIT Encyclopedia of Cognitive Science, MIT Press, Cambridge. (1999)
5. Allen, J. F.: Maintaining Knowledge About Temporal Intervals, Communications of ACM, 26. 832–843 (1983)
6. Ladkin, P. B. and Maddux, R. D.: On binary constraint problems. Journal of the Association for Computing Machinery. 41(3):435–469 (1994)
7. Vilain, M. and Kautz, H.: Constraint-propagation algorithms for temporal reasoning. In Proceedings of the Fifth National Conference on Artificial Intelligence. 377–382 (1986)
8. Ladkin, P. B., and Maddux, R. D.: Representation and reasoning with convex time intervals, Tech. Report KES.U.88.2, Kestrel Institute. (1988)
9. Vilain, M., Kautz, H., and Van Beek, P.: Constraint propagation algorithms for temporal reasoning: a revised report. In Readings in Qualitative Reasoning about Physical Systems, San Mateo, CA, Morgan Kaufman. 373–381 (1990)
10. Gerevini, A. and Schubert, L.: Efficient Temporal Reasoning through Timegraphs. In Proceedings of IJCAI-93.(1993)
11. Drakengren, T. and Jonsson, P.: Eight Maximal Tractable Subclasses of Allen's Algebra with Metric Time, Journal of Artificial Intelligence Research, 7.25–45 (1997)
12. Zaidi, A. K.: On Temporal Logic Programming Using Petri Nets. IEEE Transactions on Systems, Man and Cybernetics, Part A. 29(3):245–254 (1999)
13. Zaidi, A. K., and Levis, A. H.: TEMPER: A Temporal Programmer for Time-sensitive Control of Discrete-event Systems. IEEE Transaction on Systems, Man, and Cybernetic. (2001) 31(6):485–496
14. Rauf, I. and Zaidi, A. K.: A Temporal Programmer for Revising Temporal Models of Discrete-Event Systems. Proc. of 2002 IEEE International Conference on Systems, Man, and Cybernetics, Hemmamat, Tunisia.(2002)
15. Zaidi, A. K., and Wagenhals, L. W.: Planning Temporal Events Using Point-Interval Logic. Special Issue of Mathematical and Computer Modeling (43)1229–1. (2006)
16. Ishaque, S. M. M.: On Temporal Planning and Reasoning with Point Interval Logic, MS Thesis, CS, George Mason University, VA. (2006)
17. Ishaque M.; Mansoor F.; and Zaidi A. K. An Inference Mechanism for Point-Interval Logic, *The 21st International FLAIRS Conference*, Association for the Advancement of Artificial Intelligence, Coconut Grove, FL. (2008)
18. Ma, C.: On Planning Time Sensitive Operations, MS Thesis, SE, George Mason University, VA. (1999)
19. Busacker, R. G., and Saaty, T. L.: Finite Graphs and Networks: an introduction with application. New York.: McGraw-Hill. (1965)
20. Warshal,l S.: A theorem on boolean matrices, Journal of the ACM, 9, 1,11–12. (1962)
21. Boost Graph Library. Information Available at: http://www.boost.org/
22. Grahpviz. Information Available at: http://www.graphviz.org/

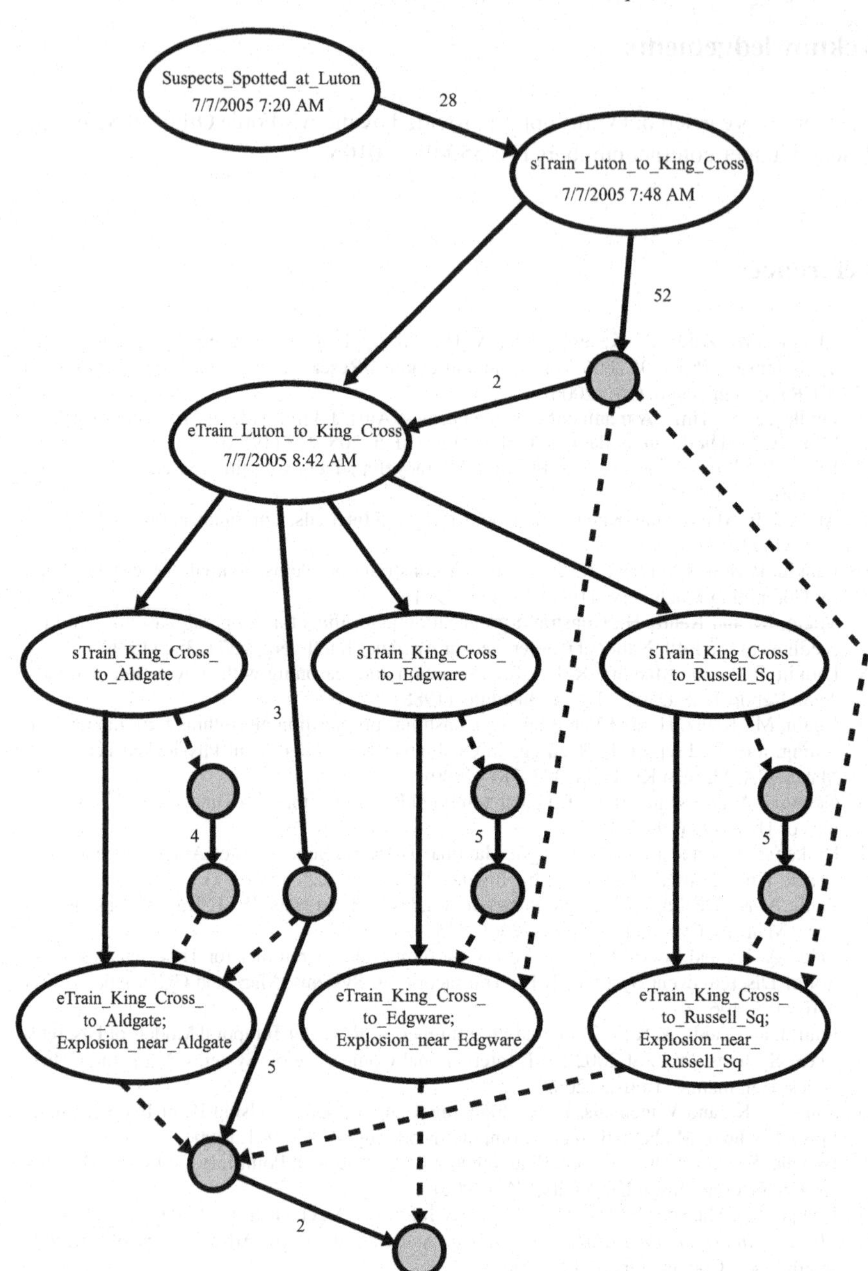

Fig. 15 Revised Point Graph for London Bombing Scenario

Two Theoretical Research Questions Concerning the Structure of the Perfect Terrorist Cell

Jonathan David Farley

Abstract Two questions of theoretical interest regarding the number of cutsets in a poset are presented.

We model terrorist cells as *partially ordered sets*, or *posets*. For background, we refer the reader to Davey and Priestley 2002, Farley 2003 and Farley 2007, where some material from this article previously appeared.

The set of numbers $\{1,2,3,4\}$ is a poset; equivalently, we could be looking at four military officers, a captain, a major, a colonel, and a general. (We call such simply ordered posets *chains*.) In fact, as with numbers, the notation "$a \leq b$" is used to indicate that a is either the same person as b ($a = b$), or else a is a subordinate of b ($a < b$), albeit not necessarily an immediate subordinate.

Let z be a non-negative integer. A poset is z-ary if no member has more than z immediate subordinates. A *tree* is a connected poset whose diagram does not contain the "V" shape. Equivalently, a tree is a poset with a single leader (called the *root*) such that no member has more than one immediate superior. Thus the posets in Figure 1 are trees, but the poset of Figure 2 is not. The technical definition is that a connected poset is a *tree* if, for any member p, the set of superiors of p forms a chain.

Terrorist plans are formulated by the nodes at the top of the organization chart or poset (the leaders or *maximal* nodes); they are transmitted down via the edges (see Figures 1, 2, and 3) to the nodes at the bottom (the foot soldiers or *minimal* nodes), who presumably carry out those plans. The message, we assume, only needs to reach one foot soldier for damage to result. For example, suppose the poset represents a

Jonathan David Farley
Institut für Algebra, Johannes Kepler Universität Linz, 4040 Linz, Austria, e-mail: lattice.theory@gmail.com

N. Memon et al. (eds.), *Mathematical Methods in Counterterrorism*,
DOI 10.1007/978-3-211-09442-6_6, © Springer-Verlag/Wien 2009

courier network. Only one messenger needs to succeed in parlaying the message; but the message must get through. We endeavor to block all routes from the maximal nodes (any one of those nodes) to the minimal nodes (any one of them) by capturing or killing some subset of the agents. Note that the agents we remove need not be maximal *or* minimal. Such a subset is called a *cutset.* A complete chain of command is called a *maximal chain* (not to be confused with a "maximal node").

If k terrorists are killed or captured at random, the probability we have found a cutset (and hence disrupted the cell, according to our model) is the number of cutsets of size k divided by the total number of subsets of size k.

We want to determine the structure of the perfect terrorist cell. Given all our assumptions, this translates into the following mathematical question: What z-ary connected partially ordered set with M maximal nodes and a total of n elements (members) has the fewest cutsets of size k?

It helps to use the following notation: Let Cutsets $(a_1, a_2, a_3, a_4, a_5, \ldots)$ denote the number of cutsets of size $1, 2, 3, 4, 5, \ldots$ respectively. The appendix lists all 63 five-member posets along with how many cutsets and minimal cutsets they have of each size. The labeling scheme indicates the number of superior-subordinate pairs ("non-trivial comparability relations") in a poset. For instance, the 5-member poset 6g in Figure 3 has 6 superior-subordinate pairs: ab, ac, bc, dc, ae, and de. (Note that this is different from the number of edges in the graph.) Starting from the n-member antichain, you can obtain all n-member posets by adding one superior-subordinate pair at a time (up to nC2, for the n member chain). For instance, Figure 4 shows that one can add a superior-subordinate pair to the 4-member poset 3c in three different ways. The extra pair is shown in bold in each of the three augmented posets. (The bold line between diagrams indicates that the number of cutsets is *decreasing* when you would expect it to increase. See below.)

In general, we merely list the posets you get by removing a superior-subordinate pair or by adding one. For instance, in the case of the 5-member poset 6g, by removing a pair you can get 5c, 5d, 5f, 5g, or 5h; by adding a pair you can get 7b, 7d, 7f, or 7h.

As an aside, note that in almost every case, the number of cutsets increases when a superior-subordinate pair is added. The cases where the numbers *decrease* are underlined for 4- and 5-member posets (i. e., for 5-member poset 6g we write "5c,5d,5f,5g,5h / 7b,7d,7f,7h").

We can write down all of the cases where the number of cutsets decreases when a superior-subordinate pair is added. For 4-member posets:

2b → 3c
3c → 4b

For 5-member posets:

2b → 3d
3b → 4b
3b → 4d
3d → 4f

3e → 4g
3e → 4i
4b → 5c
4d → 5c
4e → 5d
4e → 5g .
4g → 5h
4i → 5h
5c → 6b
5d → 6g
5f → 6g
5g → 6g
5h → 6k
6c → 7b
6e → 7f
6h → 7d
6j → 7h
7b → 8b
7h → 8e

Problem 1. Let P and Q be n-member posets such that Q is obtained from P by the addition of one superior-subordinate pair. Let $(p_1,\ldots,p_n)$ and $(q_1,\ldots,p_n)$ be the cutset vectors of P and Q respectively. Can one characterize, in terms of the structures of P and Q, the situations where it is *not* the case that, for $1 \leq i \leq n, p_i \leq q_i$? Can one characterize the situations where, for $1 \leq i \leq n, p_i \geq q_i$?

Problem 2. Is the best z-ary poset with a single leader always a tree? If so, Campos, Chvátal, Devroye, and Taslakian 2007 have determined the structure of those trees.

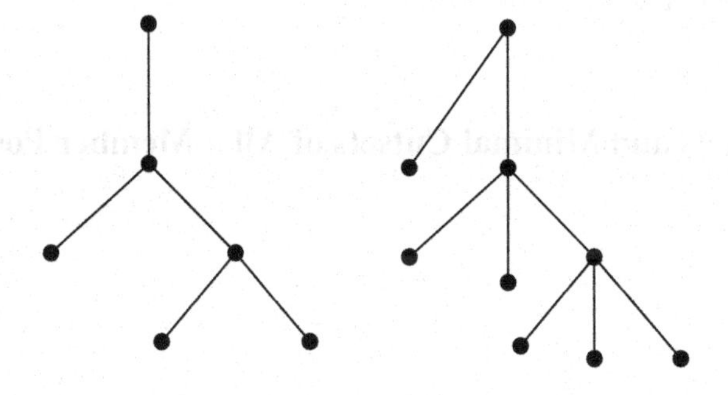

Figure 1(i). A binary tree. **Figure 1(ii).** A ternary tree that is not binary.

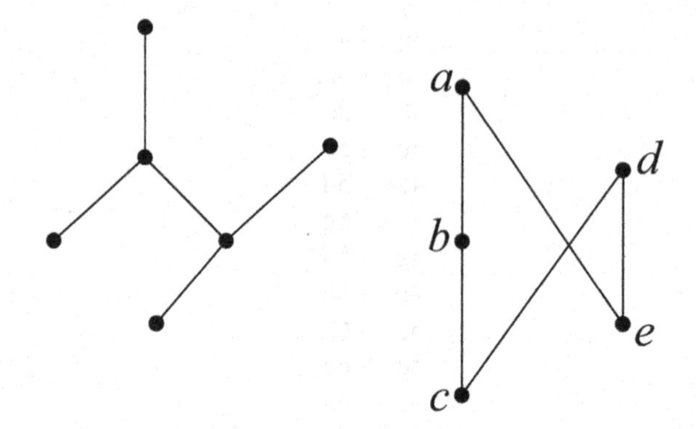

Figure 2. A poset that is not a tree. **Figure 3.** The 5-member poset 6g.

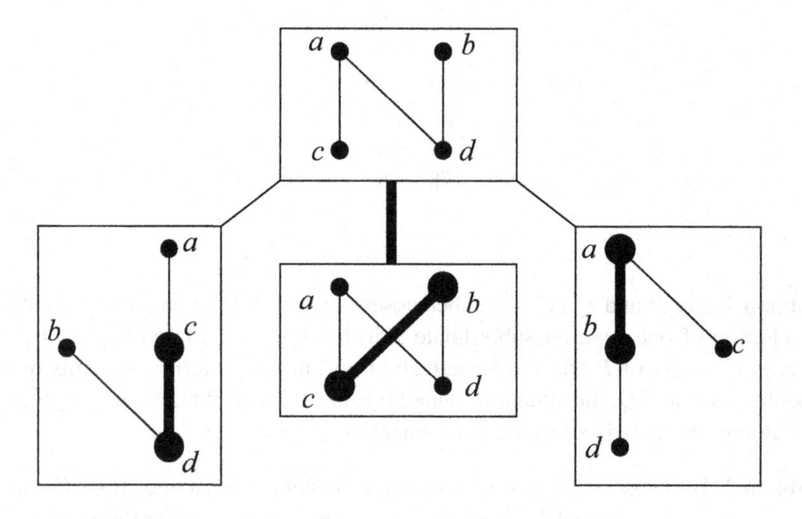

Figure 4. Adding superior-subordinate pairs to a poset.

Appendix: Cutsets and Minimal Cutsets of All n-Member Posets ($n \leq 5$)

$n = 1$

0a

Min Cutsets (1)
Cutsets (1)

$n = 2$

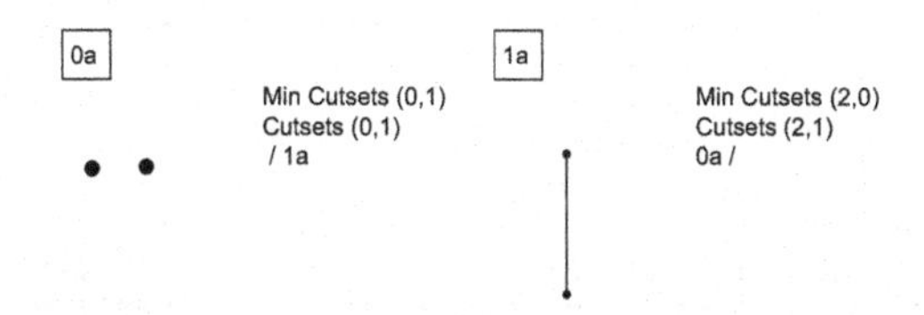
0a
Min Cutsets (0,1)
Cutsets (0,1)
/ 1a
1a
Min Cutsets (2,0)
Cutsets (2,1)
0a /

$n = 3$

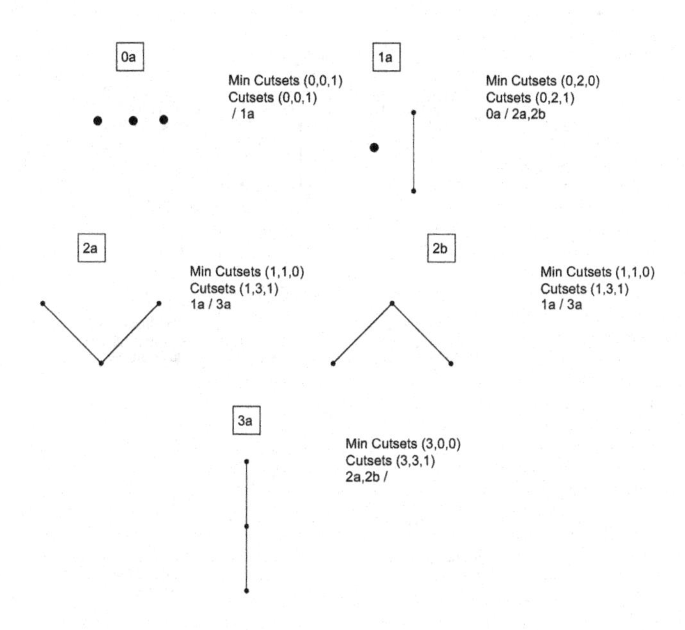
0a
Min Cutsets (0,0,1)
Cutsets (0,0,1)
/ 1a
1a
Min Cutsets (0,2,0)
Cutsets (0,2,1)
0a / 2a,2b
2a
Min Cutsets (1,1,0)
Cutsets (1,3,1)
1a / 3a
2b
Min Cutsets (1,1,0)
Cutsets (1,3,1)
1a / 3a
3a
Min Cutsets (3,0,0)
Cutsets (3,3,1)
2a,2b /

$n = 4$

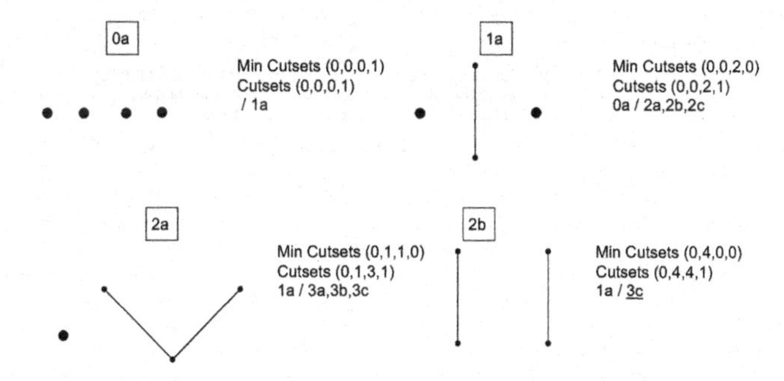
0a
Min Cutsets (0,0,0,1)
Cutsets (0,0,0,1)
/ 1a
1a
Min Cutsets (0,0,2,0)
Cutsets (0,0,2,1)
0a / 2a,2b,2c
2a
Min Cutsets (0,1,1,0)
Cutsets (0,1,3,1)
1a / 3a,3b,3c
2b
Min Cutsets (0,4,0,0)
Cutsets (0,4,4,1)
1a / 3c

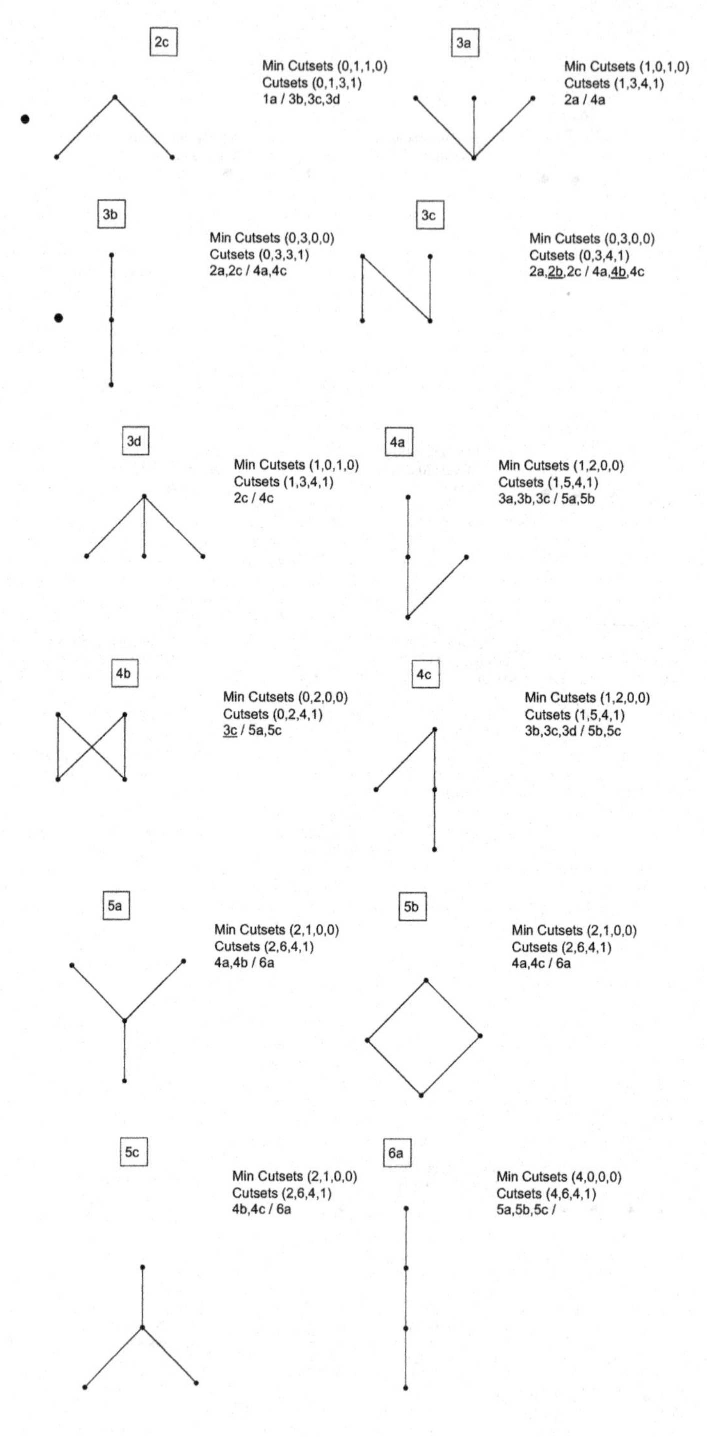
2c
Min Cutsets (0,1,1,0)
Cutsets (0,1,3,1)
1a / 3b,3c,3d
3a
Min Cutsets (1,0,1,0)
Cutsets (1,3,4,1)
2a / 4a
3b
Min Cutsets (0,3,0,0)
Cutsets (0,3,3,1)
2a,2c / 4a,4c
3c
Min Cutsets (0,3,0,0)
Cutsets (0,3,4,1)
2a,2b,2c / 4a,4b,4c
3d
Min Cutsets (1,0,1,0)
Cutsets (1,3,4,1)
2c / 4c
4a
Min Cutsets (1,2,0,0)
Cutsets (1,5,4,1)
3a,3b,3c / 5a,5b
4b
Min Cutsets (0,2,0,0)
Cutsets (0,2,4,1)
3c / 5a,5c
4c
Min Cutsets (1,2,0,0)
Cutsets (1,5,4,1)
3b,3c,3d / 5b,5c
5a
Min Cutsets (2,1,0,0)
Cutsets (2,6,4,1)
4a,4b / 6a
5b
Min Cutsets (2,1,0,0)
Cutsets (2,6,4,1)
4a,4c / 6a
5c
Min Cutsets (2,1,0,0)
Cutsets (2,6,4,1)
4b,4c / 6a
6a
Min Cutsets (4,0,0,0)
Cutsets (4,6,4,1)
5a,5b,5c /

$n = 5$

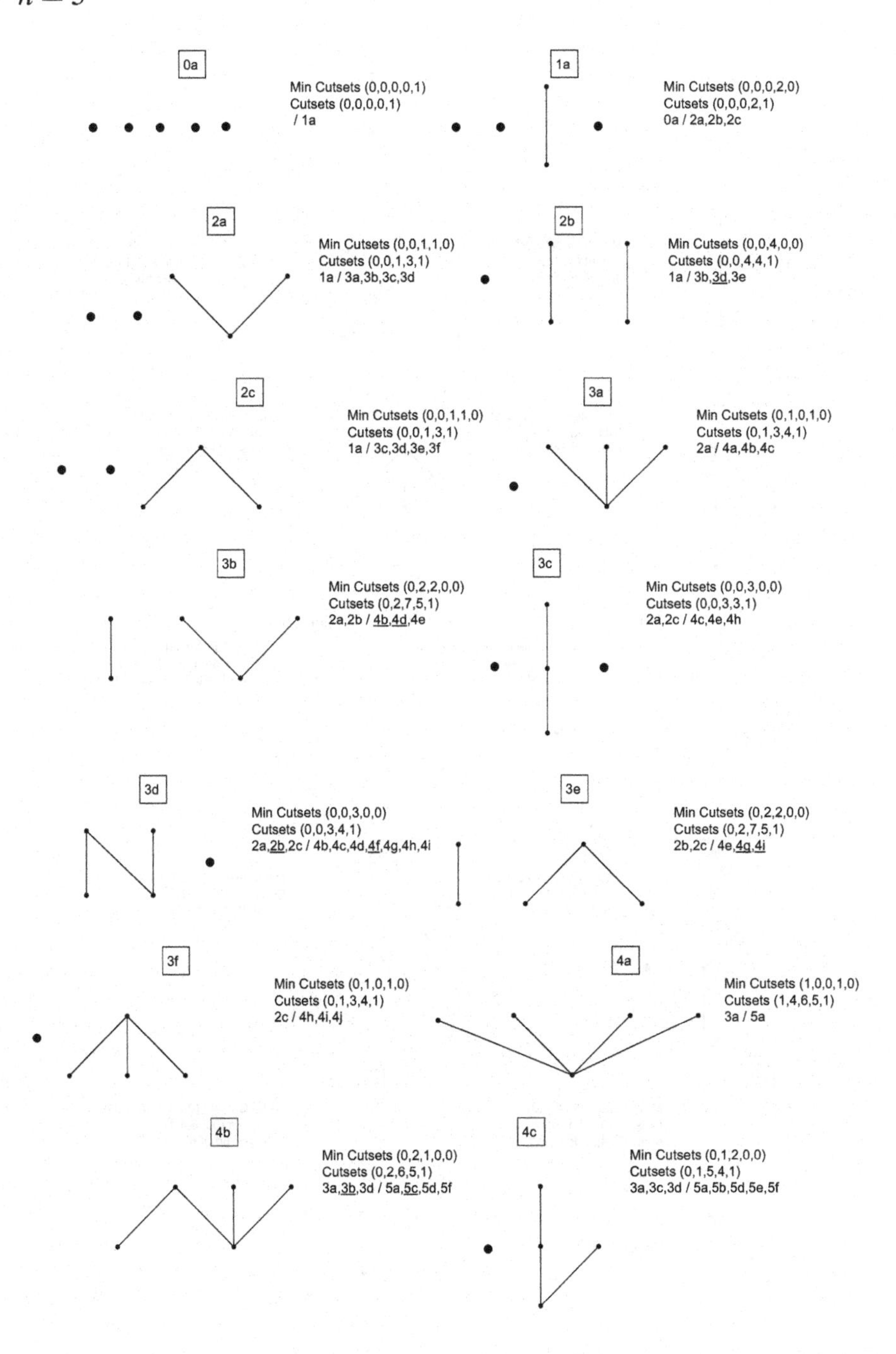
0a
Min Cutsets (0,0,0,0,1)
Cutsets (0,0,0,0,1)
/ 1a
1a
Min Cutsets (0,0,0,2,0)
Cutsets (0,0,0,2,1)
0a / 2a,2b,2c
2a
Min Cutsets (0,0,1,1,0)
Cutsets (0,0,1,3,1)
1a / 3a,3b,3c,3d
2b
Min Cutsets (0,0,4,0,0)
Cutsets (0,0,4,4,1)
1a / 3b,3d,3e
2c
Min Cutsets (0,0,1,1,0)
Cutsets (0,0,1,3,1)
1a / 3c,3d,3e,3f
3a
Min Cutsets (0,1,0,1,0)
Cutsets (0,1,3,4,1)
2a / 4a,4b,4c
3b
Min Cutsets (0,2,2,0,0)
Cutsets (0,2,7,5,1)
2a,2b / 4b,4d,4e
3c
Min Cutsets (0,0,3,0,0)
Cutsets (0,0,3,3,1)
2a,2c / 4c,4e,4h
3d
Min Cutsets (0,0,3,0,0)
Cutsets (0,0,3,4,1)
2a,2b,2c / 4b,4c,4d,4f,4g,4h,4i
3e
Min Cutsets (0,2,2,0,0)
Cutsets (0,2,7,5,1)
2b,2c / 4e,4g,4i
3f
Min Cutsets (0,1,0,1,0)
Cutsets (0,1,3,4,1)
2c / 4h,4i,4j
4a
Min Cutsets (1,0,0,1,0)
Cutsets (1,4,6,5,1)
3a / 5a
4b
Min Cutsets (0,2,1,0,0)
Cutsets (0,2,6,5,1)
3a,3b,3d / 5a,5c,5d,5f
4c
Min Cutsets (0,1,2,0,0)
Cutsets (0,1,5,4,1)
3a,3c,3d / 5a,5b,5d,5e,5f

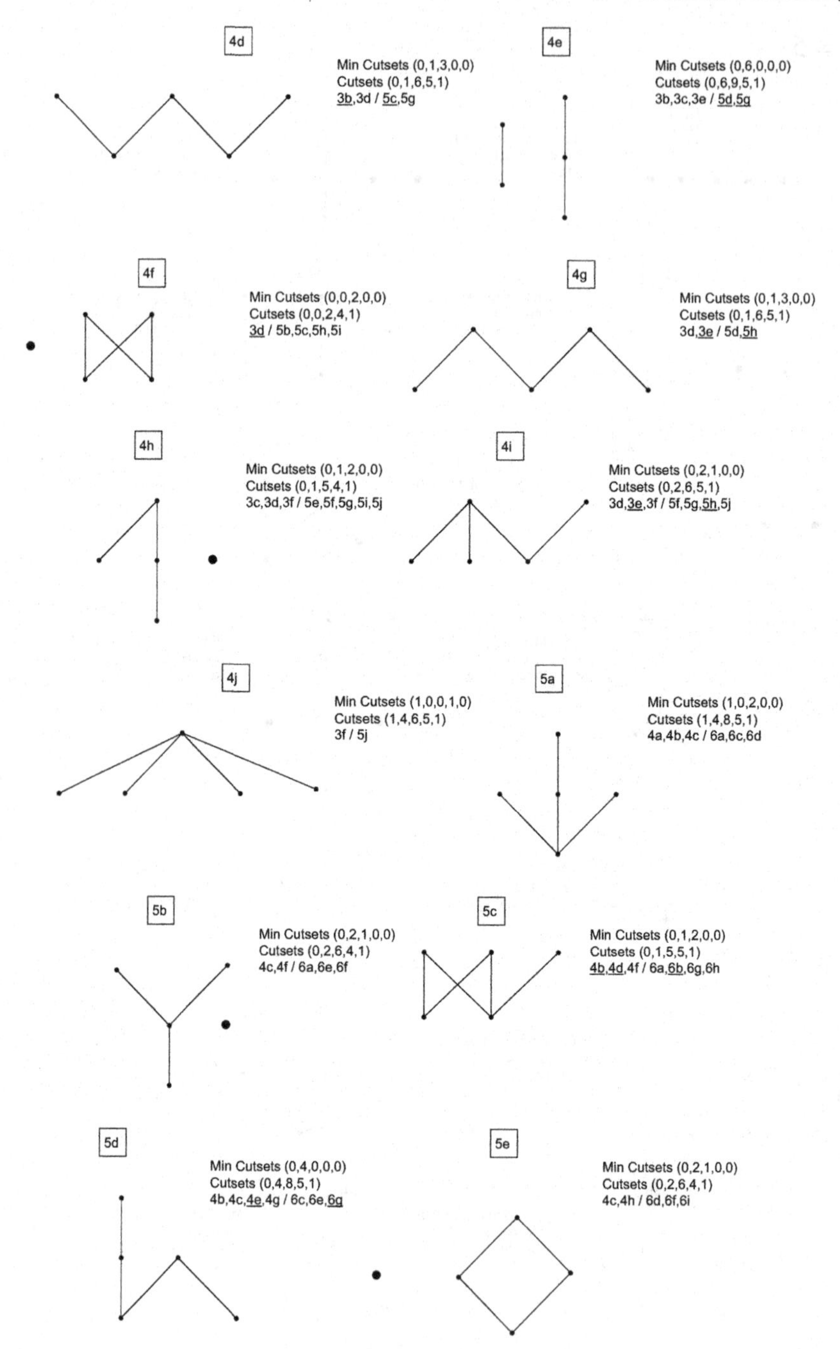
4d
Min Cutsets (0,1,3,0,0)
Cutsets (0,1,6,5,1)
3b,3d / 5c,5g
4e
Min Cutsets (0,6,0,0,0)
Cutsets (0,6,9,5,1)
3b,3c,3e / 5d,5g
4f
Min Cutsets (0,0,2,0,0)
Cutsets (0,0,2,4,1)
3d / 5b,5c,5h,5i
4g
Min Cutsets (0,1,3,0,0)
Cutsets (0,1,6,5,1)
3d,3e / 5d,5h
4h
Min Cutsets (0,1,2,0,0)
Cutsets (0,1,5,4,1)
3c,3d,3f / 5e,5f,5g,5i,5j
4i
Min Cutsets (0,2,1,0,0)
Cutsets (0,2,6,5,1)
3d,3e,3f / 5f,5g,5h,5j
4j
Min Cutsets (1,0,0,1,0)
Cutsets (1,4,6,5,1)
3f / 5j
5a
Min Cutsets (1,0,2,0,0)
Cutsets (1,4,8,5,1)
4a,4b,4c / 6a,6c,6d
5b
Min Cutsets (0,2,1,0,0)
Cutsets (0,2,6,4,1)
4c,4f / 6a,6e,6f
5c
Min Cutsets (0,1,2,0,0)
Cutsets (0,1,5,5,1)
4b,4d,4f / 6a,6b,6g,6h
5d
Min Cutsets (0,4,0,0,0)
Cutsets (0,4,8,5,1)
4b,4c,4e,4g / 6c,6e,6g
5e
Min Cutsets (0,2,1,0,0)
Cutsets (0,2,6,4,1)
4c,4h / 6d,6f,6i

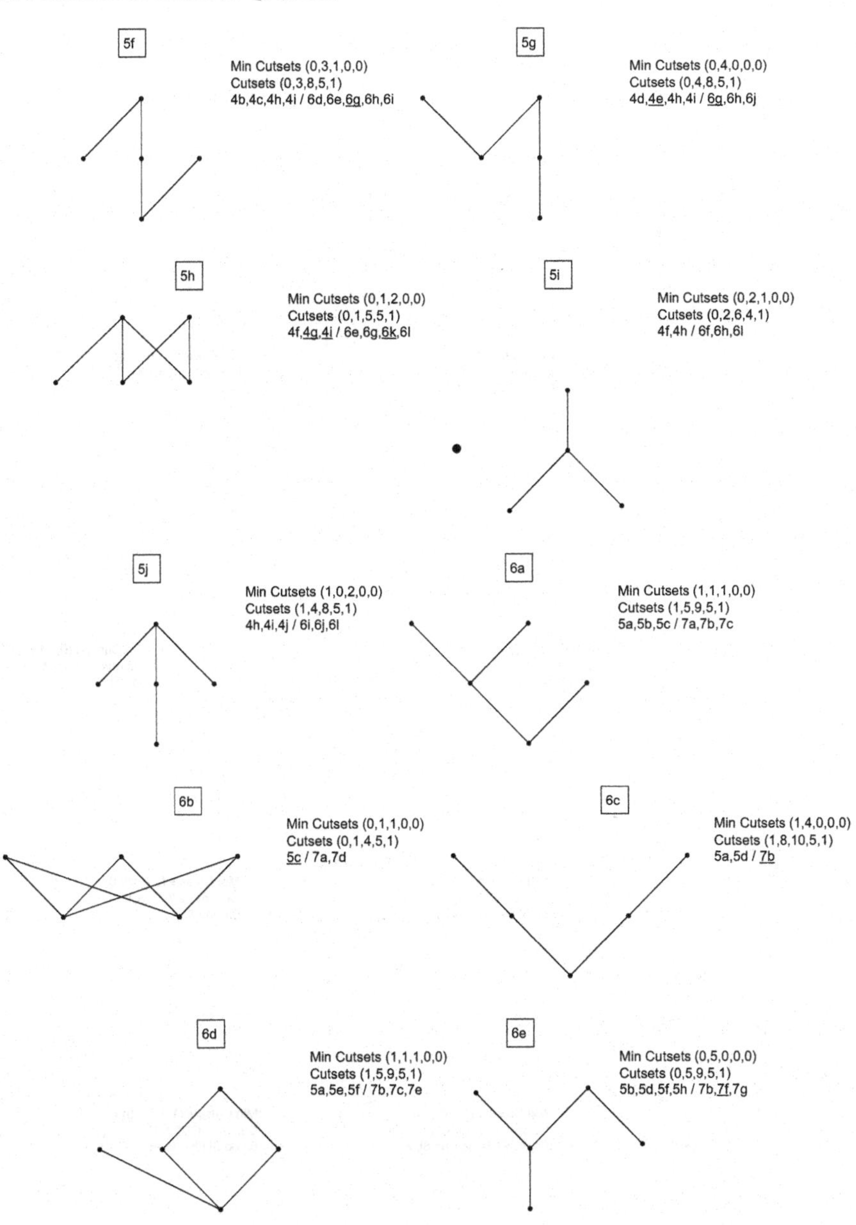
5f
Min Cutsets (0,3,1,0,0)
Cutsets (0,3,8,5,1)
4b,4c,4h,4i / 6d,6e,6g,6h,6i
5g
Min Cutsets (0,4,0,0,0)
Cutsets (0,4,8,5,1)
4d,4e,4h,4i / 6g,6h,6j
5h
Min Cutsets (0,1,2,0,0)
Cutsets (0,1,5,5,1)
4f,4g,4i / 6e,6g,6k,6l
5i
Min Cutsets (0,2,1,0,0)
Cutsets (0,2,6,4,1)
4f,4h / 6f,6h,6l
5j
Min Cutsets (1,0,2,0,0)
Cutsets (1,4,8,5,1)
4h,4i,4j / 6i,6j,6l
6a
Min Cutsets (1,1,1,0,0)
Cutsets (1,5,9,5,1)
5a,5b,5c / 7a,7b,7c
6b
Min Cutsets (0,1,1,0,0)
Cutsets (0,1,4,5,1)
5c / 7a,7d
6c
Min Cutsets (1,4,0,0,0)
Cutsets (1,8,10,5,1)
5a,5d / 7b
6d
Min Cutsets (1,1,1,0,0)
Cutsets (1,5,9,5,1)
5a,5e,5f / 7b,7c,7e
6e
Min Cutsets (0,5,0,0,0)
Cutsets (0,5,9,5,1)
5b,5d,5f,5h / 7b,7f,7g

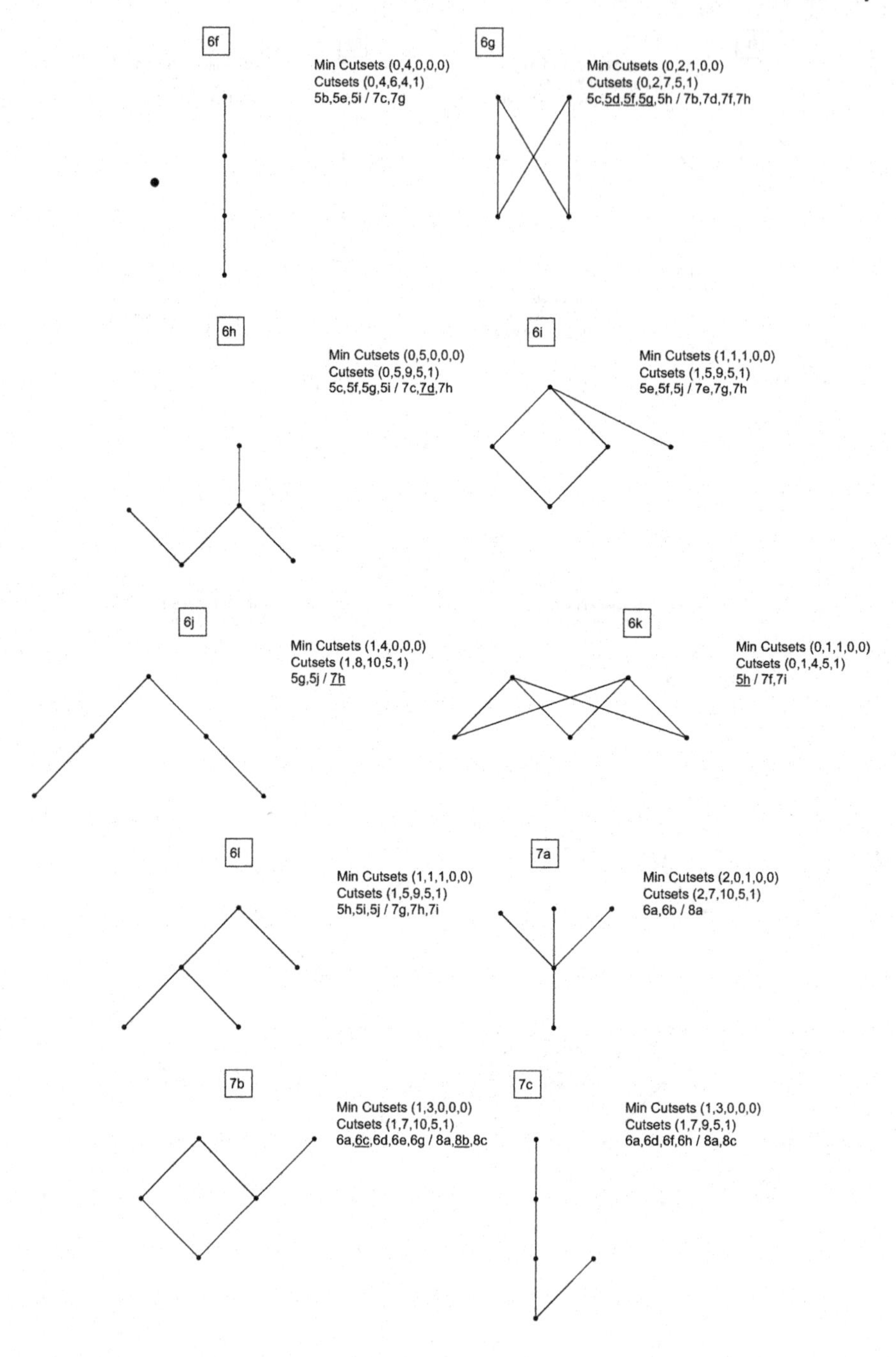
6f
Min Cutsets (0,4,0,0,0)
Cutsets (0,4,6,4,1)
5b,5e,5i / 7c,7g
6g
Min Cutsets (0,2,1,0,0)
Cutsets (0,2,7,5,1)
5c,5d,5f,5g,5h / 7b,7d,7f,7h
6h
Min Cutsets (0,5,0,0,0)
Cutsets (0,5,9,5,1)
5c,5f,5g,5i / 7c,7d,7h
6i
Min Cutsets (1,1,1,0,0)
Cutsets (1,5,9,5,1)
5e,5f,5j / 7e,7g,7h
6j
Min Cutsets (1,4,0,0,0)
Cutsets (1,8,10,5,1)
5g,5j / 7h
6k
Min Cutsets (0,1,1,0,0)
Cutsets (0,1,4,5,1)
5h / 7f,7i
6l
Min Cutsets (1,1,1,0,0)
Cutsets (1,5,9,5,1)
5h,5i,5j / 7g,7h,7i
7a
Min Cutsets (2,0,1,0,0)
Cutsets (2,7,10,5,1)
6a,6b / 8a
7b
Min Cutsets (1,3,0,0,0)
Cutsets (1,7,10,5,1)
6a,6c,6d,6e,6g / 8a,8b,8c
7c
Min Cutsets (1,3,0,0,0)
Cutsets (1,7,9,5,1)
6a,6d,6f,6h / 8a,8c

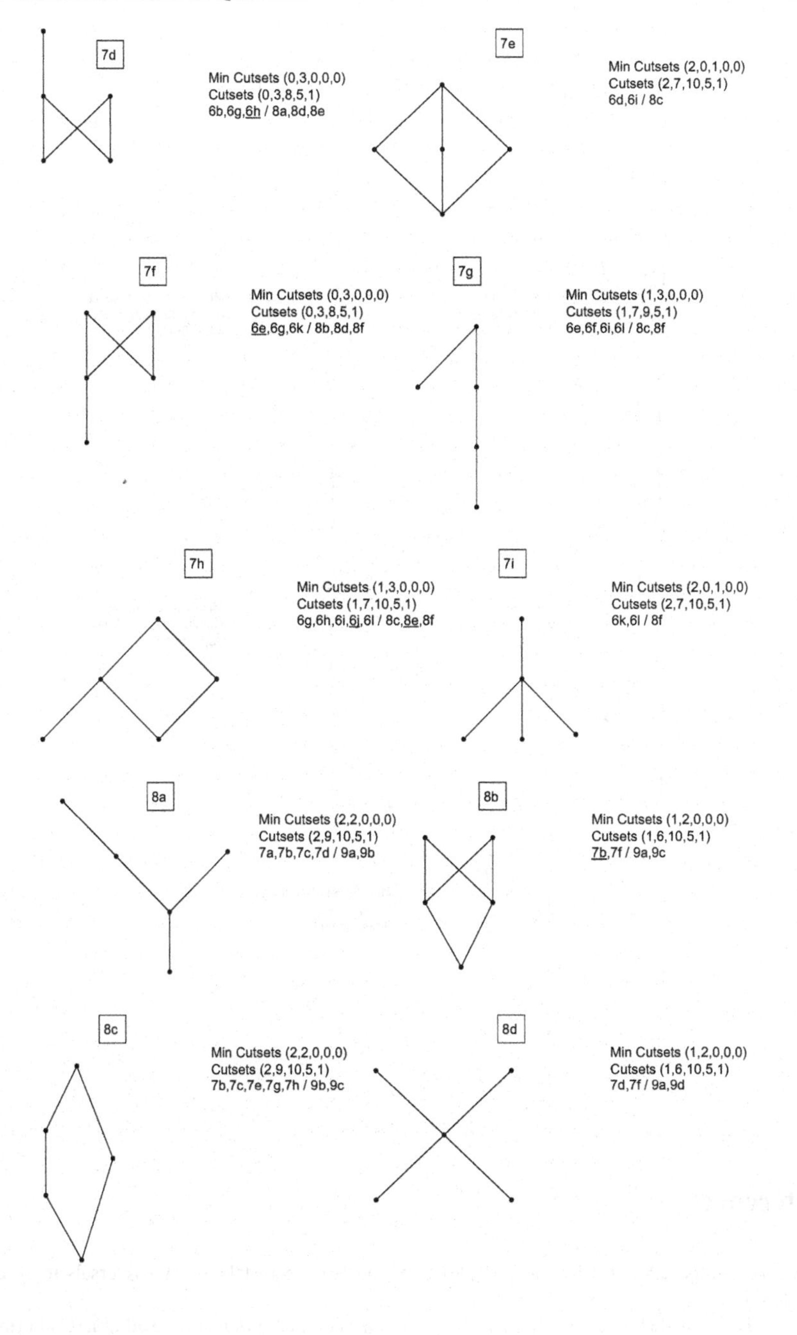
7d
Min Cutsets (0,3,0,0,0)
Cutsets (0,3,8,5,1)
6b,6g,6h / 8a,8d,8e
7e
Min Cutsets (2,0,1,0,0)
Cutsets (2,7,10,5,1)
6d,6i / 8c
7f
Min Cutsets (0,3,0,0,0)
Cutsets (0,3,8,5,1)
6e,6g,6k / 8b,8d,8f
7g
Min Cutsets (1,3,0,0,0)
Cutsets (1,7,9,5,1)
6e,6f,6i,6l / 8c,8f
7h
Min Cutsets (1,3,0,0,0)
Cutsets (1,7,10,5,1)
6g,6h,6i,6j,6l / 8c,8e,8f
7i
Min Cutsets (2,0,1,0,0)
Cutsets (2,7,10,5,1)
6k,6l / 8f
8a
Min Cutsets (2,2,0,0,0)
Cutsets (2,9,10,5,1)
7a,7b,7c,7d / 9a,9b
8b
Min Cutsets (1,2,0,0,0)
Cutsets (1,6,10,5,1)
7b,7f / 9a,9c
8c
Min Cutsets (2,2,0,0,0)
Cutsets (2,9,10,5,1)
7b,7c,7e,7g,7h / 9b,9c
8d
Min Cutsets (1,2,0,0,0)
Cutsets (1,6,10,5,1)
7d,7f / 9a,9d

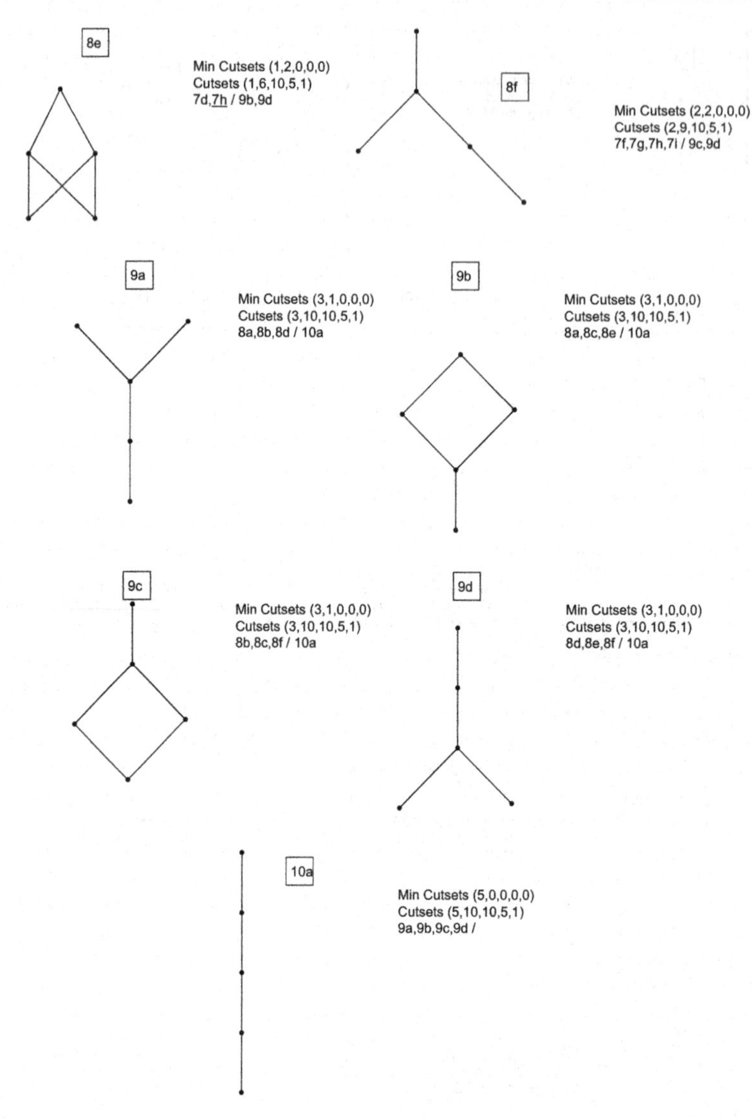

References

1. Victor Campos, Vašek Chvátal, Luc Devroye, and Perouz Taslakian, "Transversals in Trees" (2007).
2. B. A. Davey and H. A. Priestley, *Introduction to Lattices and Order*, second edition (Cambridge University Press, 2002).
3. Jonathan David Farley, "Breaking Al Qaeda Cells: A Mathematical Analysis of Counterterrorism Operations (A Guide for Risk Assessment and Decision Making)," *Studies in Conflict and Terrorism* **26** (2003), 399–411.

4. Jonathan David Farley, *Toward a Mathematical Theory of Counterterrorism: Building the Perfect Terrorist Cell* (The Proteus Monograph Series, 2007).

Part II
Forecasting

Understanding Terrorist Organizations with a Dynamic Model

Alexander Gutfraind

Abstract Terrorist organizations change over time because of processes such as recruitment and training as well as counter-terrorism (CT) measures, but the effects of these processes are typically studied qualitatively and in separation from each other. Seeking a more quantitative and integrated understanding, we constructed a simple dynamic model where equations describe how these processes change an organization's membership. Analysis of the model yields a number of intuitive as well as novel findings. Most importantly it becomes possible to predict whether counter-terrorism measures would be sufficient to defeat the organization. Furthermore, we can prove in general that an organization would collapse if its strength and its pool of foot soldiers decline simultaneously. In contrast, a simultaneous decline in its strength and its pool of leaders is often insufficient and short-termed. These results and other like them demonstrate the great potential of dynamic models for informing terrorism scholarship and counter-terrorism policy making.

1 Introduction

Our goal is to study terrorist organizations using a dynamic model. Generally speaking, in such a model a phenomenon is represented as a set of equations which describe it in simplified terms. The equations represent how the phenomenon changes in time or space, and cast our empirically-based knowledge in precise mathematical language. Once the model is constructed, it can be studied using powerful mathematical techniques to yield predictions, observations and insights that are difficult or impossible to collect empirically [1, 2]. For example, a dynamic model could be constructed for the various militant groups operating in Iraq and then used to predict

Alexander Gutfraind
Center for Applied Mathematics, 657 Rhodes Hall, Ithaca, NY, USA 14853, e-mail: ag362@cornell.edu

N. Memon et al. (eds.), *Mathematical Methods in Counterterrorism*,
DOI 10.1007/978-3-211-09442-6_7, © Springer-Verlag/Wien 2009

their strength a year in the future. Moreover, given the model, it would be possible to evaluate the efficacy of various counter-insurgency policies.

Mathematical models can help fill a large methodological void in terrorism research: the lack of model systems. Whereas biologists studying pathogens can do experiments *in vitro*, there are no such model systems in terrorism research, except for mathematical models. In this sense, the method developed below provides an *in vitro* form of terrorism, which can be investigated in ways not possible in its *in vivo* kind. Like all model systems, mathematical models are imperfect because they rely on large simplifications of the underlying political phenomena, and one can rightfully ask whether their predictions would be sufficiently accurate. Fortunately, complex phenomena in fields like biology have been studied very successfully with this mathematical technique [2]. Therefore, even phenomena as complex as found in terrorism research may, in some cases, be productively studied using mathematical models and indeed, previous models have brought considerable insights[1].

In the rest of the paper we describe a simple model of a terrorist organization. The model is new in its focus, methodology and audience: We focus on a single terrorist organization and model its processes of recruitment, its internal dynamics as well as the impact of counter-terrorism measures on it. As to methodology, with a few exceptions [4, 6, 8, 9] and perhaps a few others the powerful mathematical technique of differential equations has not been applied in terrorism research. Finally, the paper is specifically written to an audience of non-mathematicians: the main body of the paper uses non-technical language to explain the terminology and to describe the equations and assumptions used in the model, while the technical analysis is exposed in the appendix.

The model described below was built following two design principles. First, it was desired to have a model of broad applicability across organizations and conflicts. Indeed, the model is so general that it can be applied to insurgencies or even to some non-terrorist organizations. As we shall see, despite this generality it makes non-trivial observations and more importantly it specifies sufficient conditions for victory over the organization (see subsection 4.2). Second, it was desired to build a simple model so as to facilitate interpretation, analysis and further development. It was hoped that the model would establish a methodological prototype that could be easily extended and modified to fit specific cases.

The organization of the paper is as follows. Section 2 describes the model - its variables, parameters and relations between them. Section 3 graphically illustrates the model's predictions about terrorist organizations. Sections 4 and 5 discuss the insights gleaned from the model, and the implications to counter-terrorism policies. The conclusions are in Section 6. Finally, all of the technical arguments are gathered in the appendix.

[1] E. g. dynamic models: [3, 4, 5, 6, 7, 8, 9], rational choice models: [10, 11, 12, 13], agent-based models: [14, 15].

2 A Mathematical Model

There are many ways of describing a terrorist organization, such as its ideology or political platform, its operational patterns, or its methods of recruitment. Here we consider it from the "human resources" point of view. Namely, we are interested in examining how the numbers of "leaders" and "foot soldiers" in the organization change with time. The former includes experienced managers, weapon experts, financiers and even politicians and religious leaders who help the organization, while the latter are the more numerous rank-and-file. These two quantities arguably give the most important information about the strength of the organization. The precise characteristics of the two groups and their relative sizes would depend on the organization under consideration. Nevertheless, this distinction remains relevant even in the very decentralized organizations like the post-Afghanistan al-Qaeda movement, because we can identify the "leaders" as the experienced terrorists, as compared to the new recruits (see discussions in [16, 17]). The division between those two groups is also important in practice because decision makers often need to choose which of the groups to target [18, 19, Ch.5]: while the leaders represent more valuable targets, they are also harder to reach. Later on in section 5 we actually compare the policy alternatives.

Therefore, let us represent a terrorist organization as two time-varying quantities, L and F, corresponding to the number of leaders and foot soldiers, respectively. Also, L and F determine the overall "strength" S of the organization. Because leaders possess valuable skills and experience, they contribute more to the strength than an equivalent number of foot soldiers. Hence, strength S is taken to be a weighted sum of the two variables, with more weight ($m > 1$) given to leaders:

$$S = mL + F$$

We now identify a set of processes that are fundamental in changing the numbers of leaders and foot soldiers. These processes constitute the mathematical model. While some of them are self-evident, others could benefit from quantitative comparison with data. The latter task is non-trivial given the scarcity of time-series data on membership in terrorist organizations and hence we leave it out for future work.

The histories of al-Qaeda and other terrorist organizations e. g., [20, 21, 22] suggest that the pool of terrorist leaders and experts grows primarily when foot soldiers acquire battle experience or receive training (internally, or in terrorist-supporting states, [23]). Consequently, the pool of leaders (L) is provisioned with new leaders at a rate proportional to the number of foot soldiers (F). We call this process "promotion" and label the parameter of proportionality p. This growth is opposed by internal personnel loss due to demotivation, fatigue, desertion as well as in-fighting and splintering [24, Ch.6]. This phenomenon is modeled as a loss of a fraction (d) of the pool of leaders per unit time. An additional important influence on the organization are the counter-terrorism (CT) measures targeted specifically at the leadership, including arrests, assassinations as well as efforts to disrupt communications and to force the leaders into long-term inactivity. Such measures could be modeled as the removal of a certain number (b) of people per unit time from the pool of leaders. CT is modeled as a constant rate of removal rather than as a quantity that depends on the size of the organization because the goal is to see how a fixed resource allocation towards CT would impact the organization. Presumably, the human resources and funds available to fight the given terrorist organization lead, on average, to the capture or elimination of a fixed number of operatives. In sum, we assume that on average, at every interval of time the pool of leaders is nourished through promotion, and drained because of internal losses and CT (see appendix, equation (1).)

The dynamics of the pool of foot soldiers (F) are somewhat similar to the dynamics of leaders. Like in the case of leaders, some internal losses are expected. This is modeled as the removal of a fraction (d) of the pool of operatives per unit time where for simplicity the rate d is the same as the rate for leaders (the complex case is discussed in subsection 5.2.) Much like in the case of leaders above, counter-terrorism measures are assumed to remove a fixed number (k) of foot soldiers per unit time. Finally and most importantly, we need to consider how and why new recruits join a terrorist organization. Arguably, in many organizations growth in the ranks is in proportion to the strength of the organization, for multiple reasons: Strength determines the ability to carry out successful operations, which increase interest in the organization and its mission. Moreover, strength gives the organization the manpower to publicize its attacks, as well as to man and fund recruitment activities. By assuming that recruitment is proportional to strength, we capture the often-seen cycle where successful attacks lead to greater recruitment, which leads to greater strength and more attacks. Overall, the pool of foot soldiers is nourished through recruitment, and drained because of internal losses and CT (see appendix, equation (2))[2,3].

The numerical values of all of the above parameters (p,d,b,r,m,k) are dependent on the particular organization under consideration, and likely vary somewhat with time[4]. Fortunately, it is possible to draw many general conclusions from the model without knowing the parameter values, and we shall do so shortly. Finally, it should be noted that counter-terrorism need not be restricted to the parameters b,k (removal of leaders and foot soldiers, respectively), and measures such as public advocacy, attacks on terrorist bases, disruption of communication and others can weaken the organization by reducing its capabilities as expressed through the other parameters.

---

[2] A minor assumption in our model is that once a foot soldier is promoted a new foot soldier is recruited as a replacement. It is shown in the appendix that if in some organizations such recruitment is not automatic, then the current model is still valid for these organizations as long as $p < r$. In any case the drain due to promotion is marginal because foot soldiers are far more numerous than leaders even in relatively "top heavy" organizations.

[3] This model is similar to structured population models in biology, where the foot soldiers are the "larvae" and the leaders are the "adults". However, an interesting difference is that whereas larvae growth is a function of the adult population alone, in a terrorist organization the pool of foot soldiers contributes to its own growth.

[4] The simplest approach to estimating them would be to estimate the number and leaders and foot soldiers at some point in time, and then find the parameter values by doing least-squares fitting of the model to the data on the terrorist attacks, where we consider the terrorist attacks to be a proxy of strength. However, this approach has some limitations.

In the above description, we assumed that counter-terrorism measures are parameters that can be changed without affecting recruitment. This is a significant simplification because in practice it may be difficult to respond to terrorist attacks without engendering a backlash that actually promotes recruitment e. g., [25, 26]. Nevertheless, the advantages of this simplification outweigh the disadvantages: Firstly, it is clear that any model that would consider such an effect would be much more complicated than the current model and consequently much harder to analyze or use. Secondly, the current model can be easily extended to incorporate such an effect if desired. Thirdly, the strength of this effect is difficult to describe in general because it depends extensively on factors such as the specific CT measures being used, the terrorist actions and the political environment. Indeed, [9] who incorporated this effect, constructed their model based on observations of a specific context within the current conflict in Iraq.

The model includes additional implicit assumptions. First, it assumes a state of stable gradual change, such that the effect of one terrorist or counter-terrorism operation is smoothed. This should be acceptable in all cases where the terrorist organization is not very small and thus changes are not very stochastic. Second, the model assumes that an organization's growth is constrained only by the available manpower, and factors such as money or weapons do not impose an independent constraint. Third, it is assumed that the growth in foot soldiers is not constrained by the availability of potential recruits - and it is probably true in the case of al-Qaeda because willing recruits are plentiful (for the case of England, see [27]). We discuss this point further in subsection 4.3.

3 Analysis of the Model

Having written down the governing equations, the task of studying a terrorist organization is reduced to the standard problem of studying solutions to the equations. Because the equations indicate rates of change in time, the solutions would be two functions, $L(t)$ and $F(t)$, giving the number of leaders and foot soldiers, respectively, at each time. Let us suppose that currently (time 0) the organization has a certain number of leaders and foot soldiers, L_0 and F_0, respectively and is subject to certain CT measures, quantified by b and k. We want to see whether the CT measures are adequate to defeat the organization. Mathematically, this corresponds to the question of whether at some future time both L and F would reach zero. Intuitively, we expect that the organization would be eliminated if it is incapable of recovering from the losses inflicted on it by CT. In turn, this would depend on its current capabilities as well as the parameters p, d, r, m which characterize the organization.

Mathematical analysis of the model (see the appendix) shows that most terrorist organizations[5] evolve in time like the organizations whose "orbits" are displayed in Fig. 1(a,b). In Fig. 1(a) we plotted eight different organizations with different starting conditions. Another perspective can be seen in Fig. 1(b) which graphically illustrates the dynamical equations via arrows: the direction of each arrow and its length indicate how an organization at the tail of the arrow would be changing and it what rate. By picking any starting location (L_0, F_0) and connecting the arrows, it is possible to visually predict the evolution into the future. Another illustration is found in Fig. 2, which shows how two example organizations change with time.

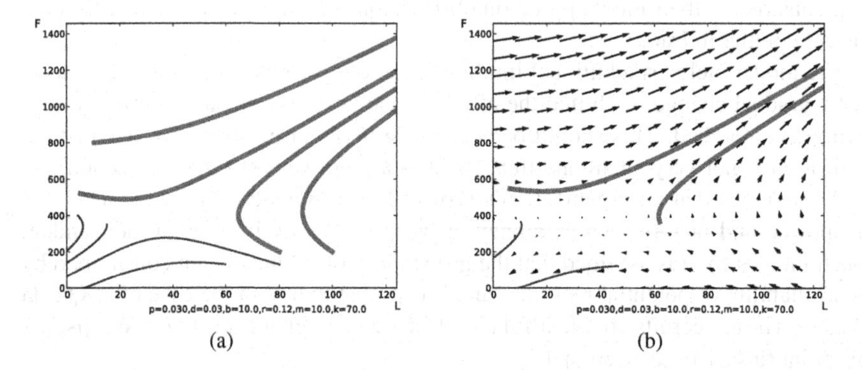

Fig. 1 (a) Typical solution curves of the equations coded by ultimate fate: thin blue for successfully neutralized organizations and thick red for those remaining operational and growing. The parameters were set to representative values, but as was said earlier, all realistic organizations are qualitatively similar and resemble these. (b) "Vector field" of L and F. At each value of L, F the direction and length of the arrow give the rate of change in L and F.

In general, it is found that the dynamics of the organization is dependent upon the position of the organization with respect to a threshold line, which can be termed the "sink line": an organization will be neutralized if and only if its capabilities are below the sink line. In other words, the current CT measures are sufficient if and only if the organization lies below that threshold (thick red line on Fig. 3). The threshold is impassable: an organization above it will grow, and one below it is sure to collapse. This threshold is also very sharp: two organizations may lie close to the line, but the one above it would grow, while the one below it would shrink even if the differences in initial capabilities are small. In addition to the sink line, the model also predicts that all successful organizations would tend towards a particular trajectory. This "trend line" (a dashed black line on Fig. 3) is discussed further in subsection 4.1.

[5] That is, those with realistically low rates of desertion: $d < \frac{1}{2}(r + r\sqrt{1+\frac{mp}{r}})$. A higher rate of desertion d always causes the organization to collapse and is not as interesting from a policy perspective (see subsection 5.2 for a discussion of desertion).

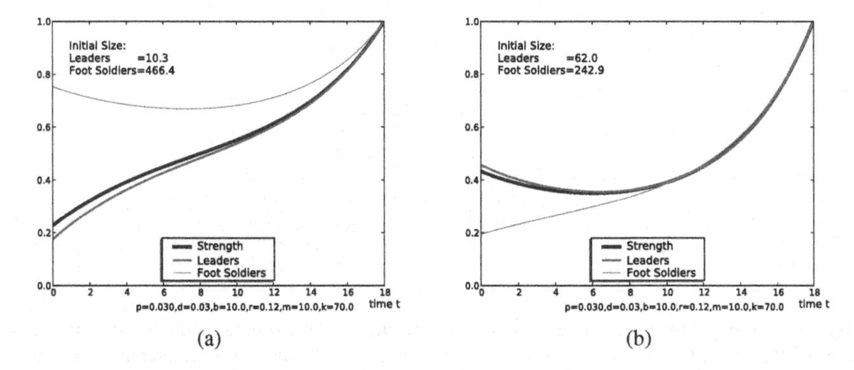

(a) (b)

Fig. 2 Evolution of strength, leaders and foot soldiers (S, L, F, respectively) in two terrorist organizations as a function of time. In (a), due to CT, F falls initially but eventually the organization recovers through promotion. In (b), L and S fall initially but eventually the organization recovers through recruitment. The vertical axis has been rescaled by dividing each quantity by the maximum it attains during the time evolution. This makes it possible to represent all three quantities on the same plot. The units of time are unspecified since they do not affect the analysis. Of course, in a more complex model it would be desirable to consider periodic events like election cycles or generational changes.

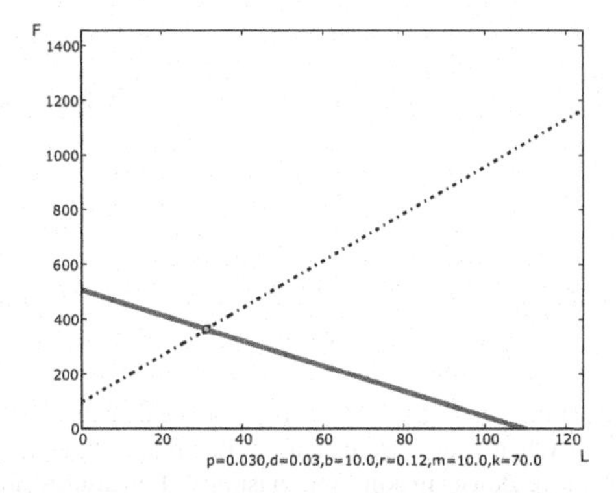

Fig. 3 Plot of the sink (thick red) and trend lines (thin dashed black). The two lines intersect at a "saddle point".

Suppose now that the model predicts that the given organization is expected to grow further despite the current CT measures, and therefore increased CT measures would be needed to defeat it. To see the effect of additional CT measures, we need to examine how the dynamical system changes in response to increases in the values of the parameters, in particular, the parameters b and k which express the CT measures directed at leaders and foot soldiers, respectively (Fig. 4).

It is also possible to affect the fate of the organization by influencing the values of other parameters affecting its evolution, such as recruitment and promotion (Fig. 5). In general, to bring the terrorist organization under control it is necessary

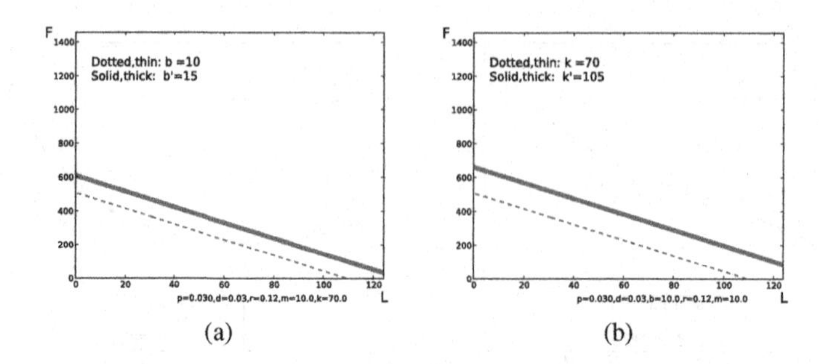

Fig. 4 The effects of the parameters b and k on the dynamical system, (a) and (b) respectively, as seen through the effect on the sink line. In each case, as the CT measures are increased, the sink line moves up confining below it additional terrorist organizations.

to change the parameters individually or simultaneously so that the organization's current state, (L,F), is trapped under the sink line. An interesting finding in this domain is that both b and k are equivalent in the sense that both shift the sink link up in parallel (Fig. 4).

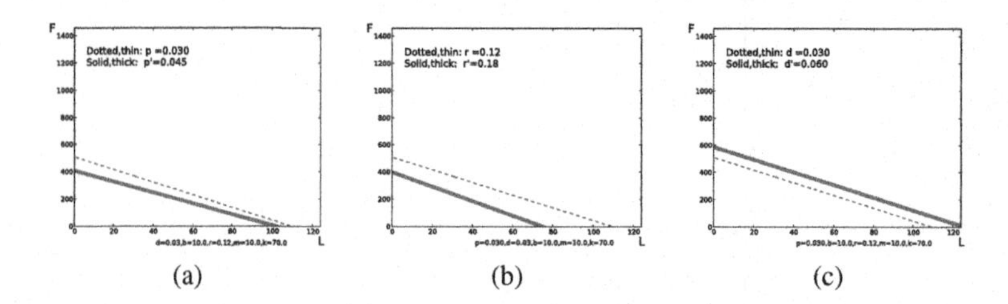

Fig. 5 The effects of the parameters p (a), r (b) and d (c) on the dynamical system as seen through the effect on the sink line. When p or r are increased the organizations are able to grow faster, causing the sink line to move down, making the existing CT measures no longer sufficient to neutralize some terrorist organizations. In contrast, when d is increased, the sink line moves up because the organization is forced to replace more internal loses to survive.

4 Discussion

4.1 Nascent terrorist organizations

Recall that the sink line (Fig. 3) distinguishes two classes of terrorist organizations - those destined to be neutralized and those that will continue growing indefinitely.

Within the latter group, another distinction is introduced by the trend line - a distinction with significance to counter-terrorism efforts: organizations lying to the left of it have different initial growth patterns compared to those lying to the right (Fig. 1). The former start with a large base of foot soldiers and a relatively small core of leaders. In these organizations, F may initially decline because of CT, but the emergence of competent leaders would then start organizational growth (e. g. Fig. 2(a)). In contrast, the latter type of organizations start with a large pool of leaders but comparatively few recruits. CT could decimate their leadership, but they would develop a wide pool of foot soldiers, recover and grow (e. g. Fig. 2(b)). Thus, all successful terrorist organizations may be classified as either "p-types" (to the left of the trend line) or "r-types" (to the right of the trend line) in reference to the parameters p of promotion and r of recruitment. In p-type organizations early growth occurs mainly through promotion of their foot soldiers to leaders, while in the r-types mainly through recruitment of new foot soldiers.

This classification could be applied to many actual organizations. For example, popular insurgencies are clearly p-type, while al-Qaeda's history since the late 1990s closely follows the profile of an r-type: Al-Qaeda may be said to have evolved through three stages: First, a core of followers moved with bin Laden to Afghanistan. They were well-trained but the organization had few followers in the wider world (for a history see [28]). Then the attacks and counter-attacks in the Fall of 2001 reduced the organization's presence in Afghanistan leaving its operatives outside the country with few leaders or skills. Finally the organization cultivated a wide international network of foot soldiers but they were ill-trained as compared to their predecessors. This description closely matches the profiles in Fig. 1 where r-type organizations start from a small well-trained core, move toward a smaller ratio of leaders to foot soldiers and then grow through recruitment.

As was noted, nascent organizations tend towards the trend line, regardless of how they started (Fig. 1). The slope of this line is $\frac{r+\sqrt{r^2+4rmp}}{2p}$, and this number is the long-term ratio between the number of foot soldiers and the number of leaders. Notice that this formula implies that ratio is dependent on just the parameters of growth - r, m, p - and does not depend on either d or the CT measures k, b. This ratio is generally not found in failing organizations, but is predicted to be ubiquitous in successful organizations. It may be possible to estimate it by capturing a division of an organization and it can help calculate the model's parameters. However, it is important to note that L includes not just commanding officers, but also any individuals with substantially superior skills and experience. The existence of the ratio is a prediction of the model, and if the other parameters are known, it could be compared to empirical findings.

4.2 Conditions for Victory

Recall, that the model indicates that all terrorist organizations belong to one of three classes: r-types, p-types and organizations that will be defeated. Each class exhibits characteristic changes in its leaders, foot soldiers and strength (L, F and S resp.) over time. This makes it possible to determine whether any given organization belongs to the third class, i. e., to predict whether it would be defeated.

One finding is that if a terrorist organization weakens, i. e. shows a decline in its strength S, it does not follow that it would be defeated. Indeed, in some r-type organizations it is possible to observe a temporary weakening of the organization and yet unless counter-terrorism (CT) measures are increased, the organization would recover and grow out of control (see Fig. 2(b)). Even a decline in the leadership is not by itself sufficient to guarantee victory. The underlying reason for this effect is out-of-control growth in F, which would ultimately create a new generation of terrorist leaders. Similarly, it is possible for an organization to experience a decline in its pool of foot soldiers and yet recover. These cases indicate that it is easy during a CT campaign to incorrectly perceive apparent progress in reducing the organization as a sign of imminent victory.

Fortunately, under the model it is possible to identify reliable conditions for victory over the organization (see the appendix for the proof):

1. *For a p-type organization, it is impossible to have a decline in strength S. If such a decline is made to happen, the organization would be defeated.*
2. *For an r-type organization, it is impossible to have a decline in foot soldiers F. If such a decline is made to happen, the organization would be defeated.*

Consequently:

> *A terrorist organization would collapse if counter-terrorism measures produce both: (1) a decline in its strength S and (2) a decline in its foot soldiers F.*

In a notable contrast, declines in strength and the *leaders* are *not* sufficient in all cases (see Fig. 2(b)). To apply the theorem to an organization of an unknown type, one needs merely to estimate whether the organization's pool of foot soldiers and strength are declining. The latter could be found indirectly by looking at the quantity and quality of terrorist operations. It is not necessary to know the model's parameters or changes in the pool of leaders - the latter could even be increasing. Furthermore, while it may take some time to determine whether S and F are indeed declining, this time could be much shorter compared to the lifetime of the organization. Therefore, the theorem suggests the following two-step approach:

1. Estimate the scale of CT measures believed to be necessary to defeat the organization.
2. Measure the effect on S and F. If they both declined, then sustain the scale of operations (i. e. do not reduce b, k); Otherwise an increase in CT measures would be necessary.

The theorem and findings above give sufficient conditions for victory but they do not characterize the only possible victory scenario. For example, it is possible for an organization to see an increase in its pool of foot soldiers F yet ultimately collapse:

these are organizations that lie to the right of the trend line and just slightly under the sink line. More generally, it should be remembered that to prove the theorem it was necessary to use a simplified model of a terrorist organization, as described in section 2. Nevertheless, it is likely that some form of the theorem would remain valid in complicated models because the model is built on fundamental forces that are likely to be retained in these models.

4.3 Stable Equilibria

Recall that the model does not have a stable equilibrium (Fig. 3). Yet, in many practical cases, terrorist organization seem to reach a stable equilibrium in terms of their structure and capabilities. It is plausible that such stability is the result of a dynamic balance between the growing terrorist organization and increasing CT measures directed against it. Indeed, rather than staying constant numbers like b,k, CT may actually grow when the organization presents more of a threat[6]. Aside from CT, stability may be the result of organizations reaching an external limit on their growth - a limit imposed by constraints such as funding, training facilities or availability of recruits. The case of funding could be modeled by assuming that the growth of the organization slows as the organization approaches a maximum point, (L_{max}, F_{max}). Alternatively, it is quite possible and consistent with the model that there would be a perception of stasis because the organization is changing only slowly.

5 Counter-Terrorism Strategies

Recall that the general counter-terrorism (CT) strategy in this model is based on the location of the sink line, which we want to place above the terrorist organization (in Fig. 1). To implement this strategy, it is necessary first to calculate the model's parameters for a given organization (p,r,m,d), and second, to determine the efficacy of the current counter-terrorism measures (b,k). Then, it remains "just" to find the most efficient way of changing those parameters so as to move the sink line into the desired location. Let us now consider several strategic options.

5.1 Targeting the leaders

An important "counter terrorist dilemma" [19] is whom to target primarily - the leaders or the foot soldiers. Foot soldiers are an inviting target: not only do they do the vital grunt work of terrorism, they also form the pool of potential leaders, and

[6] It would be a straightforward task to modify the model to incorporate such a control-theoretic interaction, but the task is more properly the subject of a follow-up study.

thus their elimination does quiet but important damage to the future of the organization. Moreover, in subsection 4.1 we saw that while an organization can recover from a decline in both its strength and leadership pool, it cannot recover from declines in both its strength and its foot soldiers pool. That finding does not say that attacking leaders is unlikely to bring victory - indeed, they form an important part of the organization's overall strength, but it does suggest that a sustained campaign against an organization is more likely to be successful when it includes an offensive against its low-level personnel. Yet, it seems that the neutralization of a terrorist leader would be more profitable since the leader is more valuable to the organization than a foot soldier, and his or her loss would inevitably result in command and control difficulties that may even disrupt terrorist attacks.

When we use the model to address the problem quantitatively, we find that the optimal strategy is actually dependent upon the organization, that is to say the parameters p,d,r,m (but not on b,k). For example, for the parameter values used in the figures above, an increase in b gives a greater rise in the sink line than an equal increase in k. Specifically, for those parameter values every two units of b are equivalent to about ten units of k. In general, when m,r,d are high but p is low then attack on the leadership is favored, while attack on the foot soldiers is best when p is high but m,r,d are low - in agreement with intuition[7]. In the first parameter range, foot soldiers are recruited so rapidly that attacking them is futile, while in the second set leaders are produced quickly so the only strategy is to fight the foot soldiers to prevent them from becoming leaders. In any case, policy prescriptions of this kind must be applied with consideration of counter-terrorism capabilities and policy costs. Thus, while on paper a particular strategy is better, the other strategy could be more feasible.

It is often argued that counter-terrorism policies have considerable side effects. For instance, there is evidence that targeted assassinations of leaders have led terrorist organizations to escalate, in what has been called the "boomerang effect" [29, p.125]. Fortunately, the model suggests that the policy maker has useful substitutes, with possibly fewer policy side effects. As Fig. 5 shows, making recruitment (r) lower has an effect similar to increasing k. Likewise, decreasing the rate of promotion to leadership (p) can substitute for increasing b. This agrees with intuition: for example, in the case of the foot soldiers, growth can be contained either actively through e. g. arrests or proactively by slowing the recruitment of new operatives (through e. g. attacks on recruitment facilities or advocacy).

[7] Mathematically to obtain this result we first compute the derivatives of the fixed point with respect to both b and k, then project them to the orthogonal to the sink link and then use an optimization solver to find the parameter values of the model which maximize (and minimize) the ratio between the lengths of the projections.

5.2 Encouraging desertion

Fatigue and attrition of personnel have been empirically found to be an important effect in the evolution of terrorist organizations. In interviews with captured or retired terrorists, they often complained about the psychological stress of their past work, its moral contradictions, and the isolation from relatives and friends [24, Ch.6]. This is part of the reason why terrorist organizations cannot remain inactive (as in a cease fire) for very long without experiencing irreplaceable loss of personnel due to loss of motivation, and many organizations even resort to coercion against desertion. Therefore, encouraging operatives to leave through advocacy or amnesties may be an effective counter-terrorism strategy.

The model introduced here brings theoretical insight into this phenomenon. One prediction of the model is that even if such desertion exceeds recruitment (i. e. $d > r$) the organization would still sustain itself as long as it has a sufficiently large rate of promotion (p) or leaders of sufficiently high caliber (m). However, if d is even greater, namely, exceeds $d = \frac{1}{2}(r + r\sqrt{1 + \frac{mp}{r}})$, then the model predicts that the organization would be destroyed regardless of starting conditions, or counter-terrorism efforts (b, k).

Organizations with lower d are, of course, also effected by desertion. Earlier, in Fig. 5 we saw how increasing d raises up the sink line. To see the phenomenon in more detail, we replaced d by two (not necessarily equal) parameters d_L and d_F for the desertion of L and F, respectively. The two parameters change the slope of the sink line: increasing d_L flattens it, while increasing d_F makes it more steep (Fig. 6). Therefore, increasing d_L could be a particularly effective strategy against nascent r-type organizations, while increasing d_F could be effective against the nascent p-types.

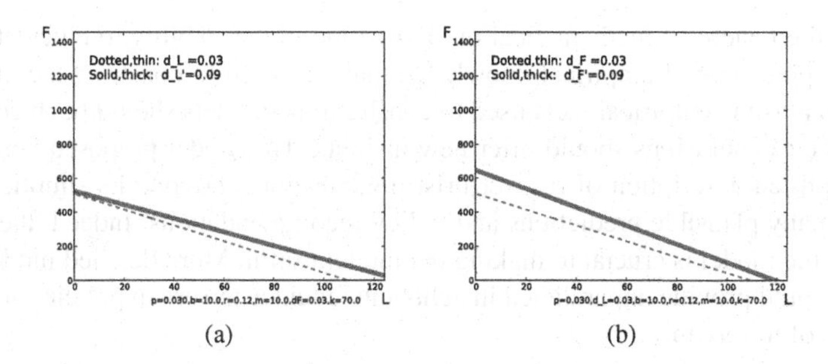

Fig. 6 The effects of d_L (a) and d_F (b) on the dynamical system, as seen through the effect on the sink line. As the desertion rates increase, the sink line moves up and its slope changes, thus trapping additional terrorist organizations.

5.3 Minimization of Strength S

Counter terrorism (CT) is often the problem of resource allocation among competing strategies. Therefore, suppose that resources have become available towards a CT operation against the terrorist organization. Namely, suppose we can remove leaders and operatives in a single blow (unlike the parameters b,k in the model which take a gradual toll). A reasonable approach to allocating these resources efficiently would be to divide them between operations targeting the leadership and those targeting the foot soldiers in such a way that the terrorist organization's strength S is minimized[8]. However, by some simplified economic analysis, it is possible to show (see appendix) that this counter-terrorism strategy is in general *suboptimal.* Instead, for a truly effective resource allocation, it is necessary to consider the dynamics of the organization being targeted and the true optimum may be considerably different. For example, when the ratio of promotion to recruitment is relatively large (i. e. $\frac{p}{r} \gg 0$), then the optimum shifts increasingly towards attacking the foot soldiers since they become much harder to replace than leaders.

On an intuitive level, the reason why the strategy is suboptimal is because often, the losses we can inflict most effectively on the organization are precisely those losses that the organization can restore most easily. Hence, in the long-term a strategy targeting strength S would be ineffective. Instead, when making a CT strategy it would be valuable to understand the target organization's dynamics, and in particular, to build a dynamical model. Such a model would help because it can identify an organization's unique set of vulnerabilities due to the composition of its human capital and its properties as a dynamical system.

6 Conclusions

Much of the benefit of mathematical models is due to their ability to elucidate the logical implications of empirical knowledge that was used to construct the model. Thus, whereas the empirical facts used to construct the models should be uncontroversial, their conclusions should offer new insights. The model proposed here is a very simplified description of real terrorist organizations. Despite its simplicity, it leads to many plausible predictions and policy recommendations. Indeed, the simplicity of the model is crucial to making the model useful. More detailed models of this kind could provide unparalleled insights into counter-terrorism policies and the dynamics of terrorism.

Acknowledgments

I would like to thank Steven Strogatz, Michael Genkin, and Peter Katzenstein for help and commentary, Richard Durrett for unconditional support, Rose Alaimo for infinite patience and my colleagues at the Center for Applied Mathematics for

[8] Mathematically, this would be two variable minimization of S constrained by a budget.

encouragement.

Note An earlier shorter version of this paper appeared in *Studies in Conflict and Terrorism* [30].

7 Appendix

The original differential equations are:

$$\frac{dL}{dt} = pF - dL - b \tag{1}$$

$$\frac{dF}{dt} = r\underbrace{(mL+F)}_{S} - dF - k \tag{2}$$

If we wished to incorporate the drain of the foot soldiers due to promotion ($-pF$) in Eqn.2, then we could adjust the original parameters by the transformation $r \to r-p$ and $m \to \frac{rm}{r-p}$. However, this would affect some of the analysis below, because for $r < p$ it would not longer be the case that $r > 0$, even though $rm > 0$ would still hold true. Alternatively, we could change the internal losses parameter for foot soldiers : $d_F \to d_F + p$ and break the condition $d_F = d_L$.

The linearity of the system of differential equations makes it possible to analyze the solutions in great detail by purely analytic means. The fixed point is at:

$$L_* = \frac{kp - b(r-d)}{d(r-d)+rmp} \quad F_* = \frac{kd+rmb}{d(r-d)+rmp} \tag{3}$$

The eigenvalues at the fixed point are

$$\lambda_{1,2} = \frac{r - 2d \pm \sqrt{(r-2d)^2 + 4(rmp + d(r-d))}}{2} \tag{4}$$

From Eqn.(4), the fixed point is a saddle when $rmp+d(r-d) > 0$, i. e. $\frac{r-\sqrt{r^2+4rmp}}{2} < d < \frac{r+\sqrt{r^2+4rmp}}{2}$ (physically, the lower bound on d is 0). The saddle becomes a sink if $r < 2d$ and $rmp + d(r-d) < 0$. By Eqn.(3), this automatically gives $F_* < 0$, i. e. the organization is destroyed[9]. It is impossible to obtain either a source because it requires $r-2d > 0$ and $rmp+d(r-d) < 0$, but the latter implies $d > r$, and so $r-2d > 0$ is impossible; or any type of spiral because $(r-2d)^2 + 4(rmp+d(r-d)) < 0$ is algebraically impossible[10]. It is also interesting to find the eigenvectors because they

[9] Of course, the dynamical system is unrealistic once either F or L fall through zero. However, by the logic of the model, once F reaches zero, the organization is doomed because it lacks a pool of foot soldiers from which to rebuild inevitable losses in its leaders.

[10] The degenerate case of $\lambda = 0$ has probability zero, and is not discussed.

give the directions of the sink and trend lines:

$$e_{1,2} = \left(\frac{2p}{r \pm \sqrt{r^2 + 4rmp}} \right) \tag{5}$$

We see that the slope of e_2, which is also the slope of the sink line - the stable manifold - is negative. Therefore, we conclude that the stable manifold encloses, together with the axes, the region of neutralized organizations. Concurrently, the slope of e_1 - the trend line i. e. the unstable manifold - is positive. Thus, the top half of the stable separatrix would point away from the axes, and gives the growth trend of all non-neutralized organizations ($\frac{\Delta F}{\Delta L} = \frac{r + \sqrt{r^2 + 4rmp}}{2p}$).

7.1 Proof of the theorem

Recall, we wish to show that a terrorist organization that experiences both a decline in its strength and a decline in the number of its foot soldiers will be destroyed. The proof rests on two claims: First, a p-type organization cannot experience a decline in strength, and second, an r-type organization cannot experience a decrease in F (for a graphic illustration see Fig. 7 below). Thus, both a decline in strength and a decline in the number of foot soldiers cannot both occur in an r-type organization nor can they both occur in a p-type organization. Hence, such a situation can only occur in the region of defeated organizations.

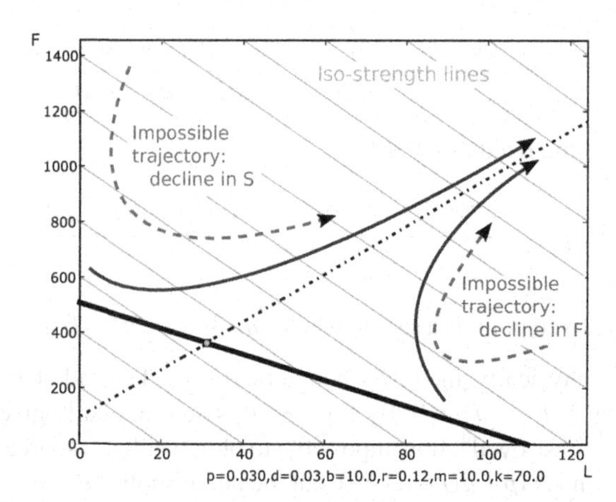

Fig. 7 The phase plane with possible (solid blue) and impossible (dashed red) trajectories, and lines of equal organization strength (green). Because orbits of the p-type must experience an increase in strength S, the left red line cannot be an orbit. Also, r-type orbits must experience an increase in F, and so the right red line cannot be an orbit either.

As to the first claim, we begin by showing that the slope of the sink line is always greater than the slope of the iso-strength lines ($= -m$). By Eqn. (5) the slope is $\frac{r-\sqrt{r^2+4rmp}}{2p} = -m\frac{2}{1+\sqrt{1+4\frac{mp}{r}}} > -m$. Therefore, the flow *down* the sink line has $\frac{dS}{dt} > 0$ (Down is the left-to-right flow in the figure). Now, we will show that in a p-type organization, the flow must experience an even greater increase in strength. Let A be the matrix of the dynamical system about the equilibrium point and let the state of the terrorist organization be $(L, F) = d_1 e_1 + d_2 e_2$ where e_1, e_2 are the distinct eigenvectors corresponding to the eigenvalues λ_1, λ_2. Consideration of the directions of the vectors (Eqn.(5)) shows that for a p-type organization, $d_1 > 0$ and $d_2 < 0$. The direction of flow is therefore $d_1\lambda_1 e_1 + d_2\lambda_2 e_2$. Notice that $\lambda_1 > 0, \lambda_2 < 0$, and so the flow has a positive component ($= d_1\lambda_1$) in the e_1 direction (i. e. up the trend line). Since the flow along e_1 experiences an increase in both L and F, it must experience an increase in strength. Consequently, a p-type organization must have $\frac{dS}{dt}$ which is even more positive than the flow along the sink line (where $d_1 = 0$). Thus, $\frac{dS}{dt} > 0$ for p-types.

As to the second claim, note that r-type organizations have $d_1 > 0$ and $d_2 > 0$. Moreover, in an r-type organization, the flow $d_1\lambda_1 e_1 + d_2\lambda_2 e_2$ has $\frac{dF}{dt}$ greater than for the flow *up* the right side of the sink line (right-to-left in the figure): the reason is that e_1 points in the direction of increasing F and while in an r-type $d_1 > 0$, along the sink line $d_1 = 0$. The flow up the sink line has $\frac{dF}{dt} > 0$, and so $\frac{dF}{dt} > 0$ in an r-type organization. In sum, $\frac{dS}{dt} < 0$ simultaneously with $\frac{dF}{dt} < 0$ can only occur in the region $d_1 < 0$ - the region of defeated organizations. QED.

7.2 Concrete Example of Strength Minimization

In subsection 5.3 we claimed that the task of minimizing S is different from the optimal counter-terrorism strategy. Here is a concrete example that quantitatively illustrates this point. Suppose a resource budget B is to be allocated between fighting the leadership and fighting the foot soldiers, and furthermore, that the cost of removing l leaders and f foot soldiers, respectively, is a typical convex function: $c_1 l^\sigma + c_2 f^\sigma$ (c_1 and c_2 are some positive constants and $\sigma > 1$)[11]. Notice that whereas uppercase letters L, F indicate the number of leaders and foot soldiers, respectively, we use lowercase l, f to indicate the number to be *removed*. The optimal values of l and f can be easily found graphically using the standard procedure in constrained optimization: the optimum is the point of tangency between the curve $B = c_1 l^\sigma + c_2 f^\sigma$ and the lines of constant difference in strength: $\Delta S = ml + f = constant$ (Fig.8(a)). However, as illustrated in Fig. 8(b), if such a strategy is followed, the terrorist organization may still remain out of control. It is preferable to choose a different strategy - in the example it is the strategy that focuses more on attacking the foot soldiers

[11] $\sigma > 1$ because e. g. once the first say 20 easy targets are neutralized, it becomes harder to find and neutralize the next 20 (the law of diminishing returns.) In any case the discussion makes clear that for most cost functions the suggested optimum would be different from the true optimum.

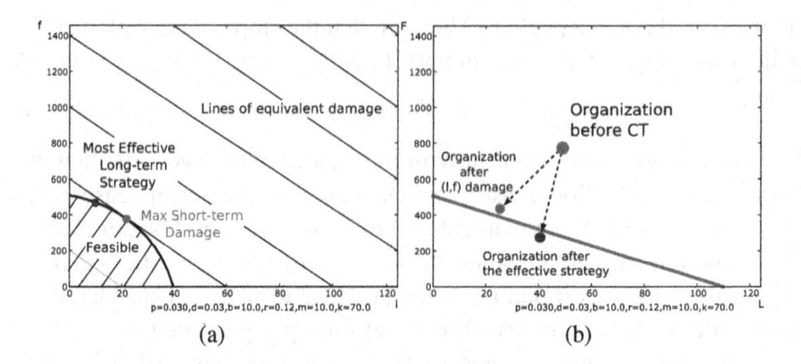

(a) (b)

Fig. 8 Graphical calculation of optimal budget allocation (a) and contrast between minimization of S and the actual optimum (b). In (a), the optimal choice of (l, f) is given by the point of tangency between the feasible region and the lines of constant S. In (b), the red line is the sink line. The minimization of S through the removal of about 20 leaders and 400 foot soldiers would not bring the organization under the sink line, but a different (still feasible) strategy would.

and thus brings the organization under the sink line (red line), even though the ΔS is not as large. In general, the difference between the strategies is represented by the difference between the slope of the sink line and the slopes of the lines of equivalent damage to strength. The latter always have slope $-m$ while the former becomes arbitrarily flat as $\frac{p}{r} \to \infty$.

References

1. Aris, R.: Mathematical Modelling Techniques. Dover Publications, Mineola, NY (1995)
2. Ellner, S.P., Guckenheimer, J.: Dynamic Models in Biology. Princeton University Press, Princeton, NJ (2006)
3. Allanach, J., Tu, H., Singh, S., Willett, P., Pattipati, K.: Detecting, tracking, and counteracting terrorist networks via Hidden Markov Models. In Profet, R., Woerner, D., Wright, R., eds.: IEEE Aerospace Conference Proceedings. (2004) 3246–3257
4. Chamberlain, T.: Systems dynamics model of Al-Qa'ida and United States "Competition". J. Homeland Security and Emergency Management **4**(3:14) (2007) 1–23
5. Farley, J.D.: Evolutionary dynamics of the insurgency in Iraq: A mathematical model of the battle for hearts and minds. Studies in Conflict and Terrorism **30** (2007) 947–962
6. Feichtinger, G., Hartl, R.F., Kort, P.M., Novak, A.J.: Terrorism control in the tourism industry. J. Optimization Theory and Applications **108**(2) (2001) 283–296
7. Johnson, N.F., Spagat, M., Restrepo, J.A., Becerra, O., Bohorquez, J.C., Suarez, N., Restrepo, E.M., Zarama, R.: Universal patterns underlying ongoing wars and terrorism. arxiv.org **physics/0506213** (2006)
8. Stauffer, D., Sahimi, M.: Discrete simulation of the dynamics of spread of extreme opinions in a society. Physica A **364** (2006) 537–543
9. Udwadia, F., Leitmann, G., Lambertini, L.: A dynamical model of terrorism. Discrete Dynamics in Nature and Society **2006**(85653) (2006) 32 pages.
10. Anderton, C.H., Carter, J.R.: On rational choice theory and the study of terrorism. Defence and Peace Economics **16**(4) (2005) 275–282

11. Sandler, T., Tschirhart, J.T., Cauley, J.: A theoretical analysis of transnational terrorism. American Political Science Rev **77**(1) (1983) 36–54
12. Sandler, T.: Collective action and transnational terrorism. The World Economy **26**(6) (2003) 779–802
13. Wintrobe, R.: Rational Extremism: The Political Economy of Radicalism. Cambridge University Press, Cambridge, UK (2006)
14. MacKerrow, E.P.: Understanding why - dissecting radical Islamic terrorism with agent-based simulation. Los Alamos Science **28** (2003) 184–191
15. Tsvetovat, M., Carley, K.M.: On effectiveness of wiretap programs in mapping social networks. Computional Mathematical Organization Theory **13**(1) (2007) 63–87
16. Hoffman, B.: Al–Qaeda, trends in terrorism, and future potentialities: An assessment. Studies in Conflict and Terrorism **26**(6) (2003) 429–442
17. Sageman, M.: Understanding Terror Networks. University of Pennsylvania Press, PA (2004)
18. Wolf, J.B.: Antiterrorist Initiatives. Plenum Press, New York (1989)
19. Ganor, B.: The Counter-Terrorism Puzzle: A Guide for Decision Makers. Transaction Publishers, Piscataway, NJ (2005)
20. Laqueur, W.: A History of Terrorism. Transaction Publishers, Piscataway, NJ (2001)
21. Harmon, C.C.: Terrorism Today. 1 edn. Routledge, Oxford, UK (2000)
22. Hoffman, B.: Inside Terrorism. Columbia University Press, USA (2006)
23. Jongman, A.J., Schmid, A.P.: Political Terrorism. Transaction Publishers, Piscataway, NJ (2005)
24. Horgan, J.: The Psychology of Terrorism. Routledge, New York, NY (2005)
25. Ganor, B.: Terrorist organization typologies and the probability of a boomerang effect. Studies in Conflict and Terrorism **31** (2008) 269–283
26. Hanson, M.A., Schmidt, M.B.: The Impact of Coalition Offensive Operations on the Iraqi Insurgency. SSRN eLibrary (2007) http://ideas.repec.org/p/cwm/wpaper/56.html.
27. Manningham-Buller, E.: The international terrorist threat to the UK. MI5 website (2006) http://www.mi5.gov.uk/output/the-international-terrorist-threat-to-the-uk-1.html.
28. Wright, L.: The Looming Tower. Allen Lane, London (2006)
29. Crenshaw, M.: The counter-terrorism and terrorism dynamic. In Thompson, A., ed.: Terrorism and the 2000 Olympics, Defence Studies Center, Canberra, Australia (1996) 125
30. Gutfraind, A.: Understanding terrorist organizations with a dynamic model. Studies in Conflict and Terrorism **32**(1) (2009) 45–59

Inference Approaches to Constructing Covert Social Network Topologies

Christopher J. Rhodes

Abstract Social network analysis techniques are being increasingly employed in counter-terrorism and counter-insurgency operations to develop an understanding of the organisation, capabilities and vulnerabilities of adversary groups. However, the covert nature of these groups makes the construction of social network topologies very challenging. An additional constraint is that such constructions often have to be made on a fast time-scale using data that has a limited shelf-life. Consequently, developing effective processes for constructing network representations from incomplete and limited data of variable quality is a topic of much current interest. Here we show how Bayesian inference techniques can be used to construct candidate network topologies and predict missing links in two different analysis scenarios. The techniques are illustrated by application to data from open-source publications.

1 Introduction

Today the predominant threats that occupy the attention of defence and security agencies around the world have their origins in terrorist groups, insurgency movements, militia organisations and trans-national organised crime. However, much of the defence and intelligence resources that governments use to counter these threats have, until recently, been habituated to confront and overcome conventional military capabilities fielded by other nation states. Consequently governments often talk of the "asymmetric threat", a term that consciously draws attention to the mismatch between conventional military force structures, hardware and doctrines (which have evolved over many decades) and the fast-moving adaptive threats posed to largely civil populations by ideologically motivated opponents who have little recognisable

---

Christopher J. Rhodes
Institute of Mathematical Sciences, Imperial College London, 53 Prince's Gate, Exhibition Road, South Kensington, London, SW7 2PG, UK, e-mail: c.rhodes@imperial.ac.uk

N. Memon et al. (eds.), *Mathematical Methods in Counterterrorism*,
DOI 10.1007/978-3-211-09442-6_8,

war-fighting ability but nevertheless possess a great capacity for violence, destruction and destabilisation.

Historically mathematics has always been used to conceptualise adversary threats and their capabilities. Before the last century basic arithmetic was often all that was required; usually the side that assembled the numerically larger force defeated its opponent. Over recent decades, however, with the development of increasingly advanced weapon systems and delivery capabilities a more sophisticated analysis has been developed, much of which now depends on simulation, systems analysis and mathematical modelling. This activity is termed "operational analysis", and it is widely used by defence and security agencies to inform procurement decisions as well as to refine command structures and operational procedures.

In the era of the asymmetric threat it has become apparent that conventional operational analysis is struggling to engage with an opposition that is constituted from ideologically unified groups of individuals capable of deadly acts using minimal technology. Consequently, new ways of conceptualising current threats are needed so that the old question of how these threats can be disrupted, diminished or overcome can be answered.

Because the most significant assets that a terrorist or insurgency group possesses are people, attempts to develop quantitative insights into their structure, organisation and vulnerabilities have drawn heavily on the field of social network research. Network structures are appealing as an organising paradigm because they immediately reveal an underlying pattern to the relationships between individuals. Additionally, they suggest the possibility of gaining insight into the function, purpose and possible future evolution of the entities that are represented by these patterns. Consequently, social network analysis is now emerging as one of the cornerstones of a new analysis geared for the challenges of conflict in the 21st Century. The first difficulty, however, is constructing the network structure on a rapid time-scale, often using limited and/or poor quality data that has a limited shelf-life. Due to the covert nature of terrorist groups, reliable data that relates to them is generally sparse, fragmentary and sometimes contradictory. The purpose here is to introduce a statistical inference method to utilise such data in order to derive candidate social network topologies and to predict the presence of missing links in a sample of network data. Using data from open source publications two different covert social network analysis scenarios are discussed.

2 Network Analysis

Network structures are an efficient and convenient way of representing the complexities of systems that are comprised of many interacting elements [1]. Social scientists have long been accustomed to using network representations as a means of capturing the strength, duration and nature of the interactions between members of defined groups [2], [3] and software is freely available to determine network properties [4]. However, data collection for social network analysis can be labour-intensive and time-consuming. With inevitably finite collection resources it is always necessary to make decisions on how these resources are deployed. Also, some groups take active steps to conceal the links and relationships between individuals, so generating candidate network topologies can be particularly challenging; this is often true of organised crime and terrorist groups [5], [6], [7], [8], [9]. As well as shedding light on the structure of groups, the ability to infer likely network topologies should also assist in deciding how to prioritise future intelligence gathering efforts, and where to deploy surveillance assets.

Most social network analysis with a counter-terrorism or counter-insurgency focus falls into the categories of either:

Case 1: detailed data-gathering on a defined group of individuals (usually < 10 individuals)

or,

Case 2: a broader and less detailed investigation of the links between a much larger selection of individuals

In Case 1 the objective might be to go on to use the detailed information about the defined small group to "grow" the network, that is, predict network changes as new individuals peripheral to the initial analysis emerge. In this way it is possible to infer where they are likely to fit into the existing network. The challenge is to grow the network when there are few data about the additional individuals or suspects. In Case 2, by contrast, it may be desirable to infer the presence of links that have not been reported in the initial data-gathering campaign on a defined large group of people. Because proportionately less effort can be expended on gathering data on each individual in this larger population it is likely that some links will be missed, so it is desirable to be able to infer the presence of links in these "sampled" or "partially observed" networks. Sometimes it is also desirable to provide guidance on the possible existence of dormant but nevertheless important ties in a network [7].

In both analysis scenarios the objective is to exploit data that of itself may be thought to be only weakly indicative of a link but, when aggregated with other data, gives more or less confidence that there is an association between individuals. Also the methods we discuss are able to accommodate the absence of data; specifically, if some data relating to individuals is missing (or unobtainable over the timescale of the analysis) then the performance of the inference degrades gracefully rather than failing catastrophically. As we shall see, this approach to constructing networks suggests that any incidental data relating to individuals can be of potential benefit in constructing network topologies. It is the aggregation of data within a Bayesian framework that gives this approach its utility.

3 A Bayesian Inference Approach

The process of inferring links depends on assessing evidence for interaction against a known sample of positive (and negative) links in the network. This is done using

Bayes' Theorem [10], and our approach follows that of Jansen et al. [11] who investigated protein interactomes. Positive links are those links that connect any two individuals in the population whereas negative links are simply the absence of a link. Each individual in the network has a number of possible independent attributes assigned to it. In [12], several attributes such as the role played in the organization, which faction they belonged to and the resources they controlled were exploited to infer network structure.

In a Bayesian approach (following from a standard definition of statistical odds), for a given probability of a link $P(pos)$, the "prior" odds of finding a positive link is given by

$$O_{prior} = \frac{P(pos)}{1 - P(pos)} \tag{1}$$

which can be written as

$$O_{prior} = \frac{P(pos)}{P(neg)} \tag{2}$$

By contrast, the "posterior" odds is the odds of finding a positive link *after* we have considered N pieces of evidence (in our case the attributes) with values $A_1 \ldots A_N$, and is given by

$$O_{post} = \frac{P(pos|A_1 \ldots A_N)}{P(neg|A_1 \ldots A_N)} \tag{3}$$

According to Bayes' rule the prior and posterior odds are related by

$$O_{post} = L(A_1 \ldots A_N) O_{prior} \tag{4}$$

where the likelihood ratio L is given by

$$L(A_1 \ldots A_N) = \frac{P(A_1 \ldots A_N|pos)}{P(A_1 \ldots A_N|neg)} \tag{5}$$

When the pieces of evidence under consideration are conditionally independent the likelihood ratio factorises into a product of individual likelihoods, i. e.,

$$L(A_1 \ldots A_N) = \prod_{i=1}^{N} \frac{P(A_i|pos)}{P(A_i|neg)} \tag{6}$$

The probabilities of positive $P(A_i|pos)$ links given an attribute A and negative $P(A_i|neg)$ link given an attribute A are calculated from the sampled network data and the likelihood is simply the product of the ratios for each attribute.

Estimating the prior odds requires some assumption to be made about the number of positive links one would expect to see in the network. Using such an estimate, and by evaluating the likelihood ratio in equation 6, it is possible to calculate the posterior odds of there being a link, i. e., the probability that individuals are linked following consideration of the available evidence.

In what follows we show how the Bayesian inference method can be used to undertake both the Case 1 and Case 2 analysis scenarios described above.

4 Case 1 Analysis

The objective of this analysis is to predict the structure of a covert network given detailed knowledge of a restricted portion of the network. We presume that an initial intelligence gathering campaign has yielded knowledge of the links between individuals in a small subset of the network. The aim is to exploit this knowledge to then predict the links between other less well-known peripheral figures in order to attempt to establish an estimate of the topology of the wider network. We then compare those predicted links with the ones that are believed to exist.

The method is dependent upon identifying a "gold-standard" reference data set of known positive and negative links, then assessing further sources of evidence against it. In this case negative links are declared where (in a finite period of surveillance time) no evidence for a link has been observed. To illustrate this approach we use open source data from the believed-defunct Greek terrorist group November 17 (N17). Details of the data sources and historical background information are provided in [12] as well as in [13] and [14], where more conventional static network analyses are presented. The open source reporting permitted the construction of a network for N17. In what follows we exploit our detailed knowledge of this network to test the inference approach.

For the purposes of this study the gold-standard data set consists of six individuals: Alexandros Giotopoulos, Nikitas, Savas Xiros, Dimitris Koufontinas, Pavlos Serifis and Constantinos Karatsolis. The social network formed by these individuals is shown in Fig. 1. The reason that these individuals were chosen is that from a visual inspection of the network they have the appearance of being well connected and individuals, and therefore likely to be representative of the sort of group member that would be subject to an initial surveillance gathering campaign. However, this choice is essentially arbitrary, and these individuals are no better characterized in terms of the data that relates to them than the others. The gold-standard network consists of 9 positive links (grey lines) and 6 negative links (note, the total possible number of links is given by $6^2 - 6/2 = 15$.

The individuals in the network each have certain "attributes" associated with them. In this data set the attributes that were used were *role* played in the organisation (i. e., a leadership figure or an operational functionary), the *faction* they belong to (broadly, three separate factions encompassed most of the membership) and *resources* controlled (be this money, weapons, safe houses etc). These attributes are the basis of the data sets that are sources of evidence of interaction in the Bayesian analysis, i. e., the A_N in equation 3. These attributes were used in this instance because this was the kind of information relating to each individual in the open source reporting on N17 [12]. Table 1 lists the data attribute data for N17. In other situations, when analysing other groups, it would be necessary to make a judgement

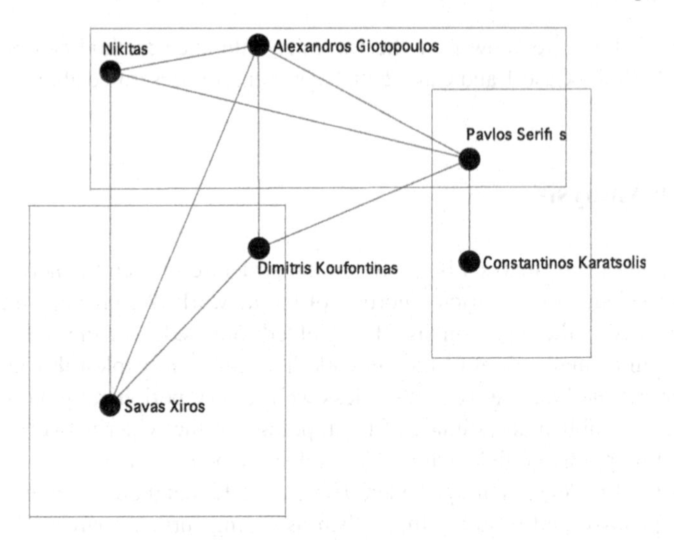

Fig. 1 This network represents the result of a detailed initial surveillance campaign on six individuals of interest. The three boxes in this and subsequent figures are demarcations of three factions in N17.

about how the available intelligence could be organised into evidence. Once the available evidence is organised, the likelihood ratios for all possible pairs in the network are then calculated for all the attributes.

The Bayesian framework then allows us to multiply these ratios together using equation 6 and, in turn, to predict that a pair is linked if the combined likelihood ratio exceeds a cut-off. This procedure generates true-positive (TP) pairs (those pairs who are predicted to be linked and are actually linked) and false-positive (FP) pairs (those pairs who are predicted to be linked but are not actually linked). The ratio of true positive to false positive (TP/FP) can be used as a measure of the effectiveness of the prediction.

Applying the inference approach to Fig. 1 we find that there are 21 *de-novo* predictions, of which 13 (True Positives) are represented in the original N17 adjacency matrix. This gives a prediction accuracy of 62 %, *assuming* that the data we had on the original N17 network represents a complete and accurate representation of the links within N17. (Alternatively, TP/FP $= 13/8 = 1.6$). The resulting predicted network is shown in Fig. 2.

The links shown in black lines are those links that are predicted correctly (i. e., a link is predicted where one is thought to exist from the open source reporting).The dashed grey lines are links that are predicted but are not in the original reported network structure. That does not mean that those links are necessarily incorrect, rather, the open source reporting did not indicate that the links existed. It can be seen that an additional four individuals (black nodes) have been integrated into the network. Taking the gold-standard links and the correctly predicted links ($9+13 = 22$) together, it turns out that there are only 3 links out of the total of 25 links that are not accounted between the 10 individuals in Fig. 2, so we now have 88 % of the

Table 1 Attribute data for N17

	money	weapons	safe houses	attacks	drugs	human trafficking	weapons smuggeling	cash acquisition	weapon acquisition	**Role**	**Faction**
Alexandros Giotopoulos	1	0	0	1	0	0	0	1	1	L	G
Anna	0	0	0	1	0	0	0	0	0	L	G
Christodoulos Xiros	0	1	1	1	0	0	0	1	1	L	K
Constantinos Karatsolis	0	1	1	1	1	1	1	0	0	O	S
Constantinos Telios	0	1	1	1	0	0	0	1	1	O	K
Dimitris Koufontinas	1	1	1	1	0	0	0	1	1	L	K
Dionysis Georgiadis	0	1	1	1	0	0	0	1	1	O	K
Elias Gaglias	0	1	1	1	0	0	0	1	1	O	K
Iraklis Kostaris	0	1	1	1	1	1	1	0	0	O	S
Nikitas	0	0	0	1	0	0	0	0	0	L	G
Ojurk Hanuz	0	1	1	1	0	0	0	1	1	O	–
Patroclos Tselentis	0	1	1	1	1	1	1	0	0	O	S
Pavlos Serifis	0	0	0	1	1	1	1	0	0	L	S
Sardanopoulos	1	0	0	1	1	1	1	0	0	L	S
Savas Xiros	0	1	1	1	0	0	0	1	1	O	K
Sotirios Kondylis	0	1	1	1	0	0	0	1	1	O	–
Fotis	0	0	0	1	0	0	0	0	0	L	G
Thomas Serifis	0	1	1	1	1	1	1	0	0	O	S
Vassilis Tzortzatos	0	1	1	1	0	0	0	1	1	O	K
Vassilis Xiros	0	1	1	1	0	0	0	1	1	O	K
Yiannis	0	1	1	1	0	0	0	1	1	O	–
Yiannis Skandalis	0	1	1	1	0	0	0	1	1	O	–

links correctly accounted for. The details of how these calculations are performed using attribute data are presented in [12].

Consequently it can be seen that the inferential approach is a robust way to integrate peripheral figures into an existing network. This is often useful because as new individuals or suspects become the focus of attention it is desirable to use what little information may be available to work out where they might fit in to the network. Also, the inference approach permits other "what if" questions to be addressed. For example, if it was desirable to integrate an individual into an adversary network it is possible to predict where (given the attributes of the infiltrator) where they would best fit into the network. Or, conversely, it suggests what attributes an infiltrator should adopt if they wish to maximise their chances of occupying a given location in the network. Finally, using this approach it should be possible to self-consistently re-calculate how a network structure re-configures following the removal of an individual from the network.

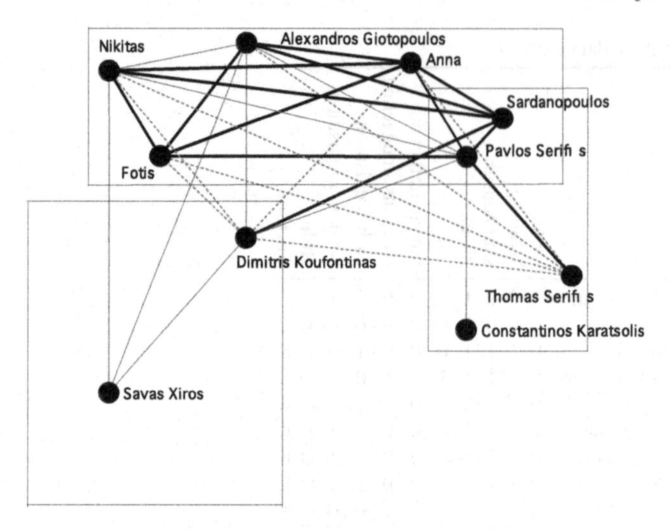

Fig. 2 The resulting network assuming a 0.5 probability that the predicted links are true links. Correctly predicted links are shown in black, false positives are grey dotted lines. In this way, additional individuals can be integrated into the original network.

5 Case 2 Analysis

In the Case 2 analysis here, it is assumed that an intelligence collection effort for a fixed period of time has yielded data on a large number of individuals. In doing so it has generated a *sample* of network data from the underlying social network. The objective is to then infer the presence of links that may have been missed by the data collection process. The sampling of a social network can be simulated using the full data set on the twenty-two known members of N17 as shown in [12]. Figure 3 shows what is believed to be the social network structure of the full organisation. The links in Fig. 3 indicate that open source reporting has demonstrated some connection between the two individuals at some point in the past. As in the Case 1 analysis, there is no attempt here to gauge the specific nature or strength of the links as is sometimes done in network analysis. Open source social networks of this size, with detailed associated information on each individual, are rare.

To simulate an intelligence collection campaign on a network of this size for a fixed period of time we randomly delete thirty-two of the sixty-four (50 %) extant links. Fig. 4 shows a sample of the full network following the deletion of randomly selected N17 links. This results in a network consisting of seventeen individuals. Five individuals with low numbers of links to others in the full network (black nodes) have not survived the sampling process and so we assume that they are not detected in the initial data-gathering operation which the sampling process aims to simulate.

Table 2 shows the link (degree) centrality measure for the original network and the sampled network. The centrality ranking of individuals is not wholly preserved

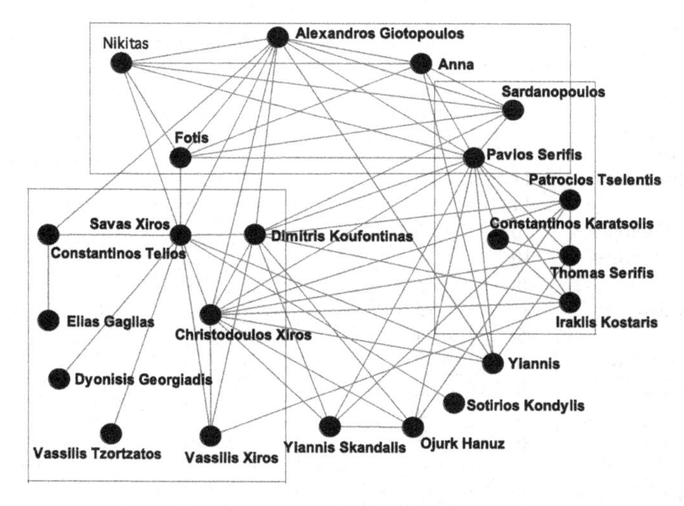

Fig. 3 The full November 17 network from open source reporting.

in the sampling process, but the general pattern is preserved and is in accord with similar observations by Borgatti *et al.* [15]. The sampling scheme is a simple one based on deletion of a given proportion of links in the whole network, and corresponds to each individual being given an equal weighting in the sampling process. In practice, however, it might be desirable to weight the intelligence collection activity with a bias towards certain individuals, so other sampling schemes would be appropriate. We do not pursue that possibility here and just work with the basic uniform link deletion method.

We now assume that the data in Fig. 4 is the raw data set derived from a data-gathering campaign/surveillance and the objective is to use it to infer the presence of the missing (i. e., the deleted) links. In this analysis the social network remains bounded by the seventeen sampled N17 membership and we do not consider the addition of any further individuals. It is the possibility of ties between members of this specified group that are of interest here.

Using the Bayesian inference approach described above it is possible to use it to infer the presence of the missing links in the sampled data. As before, it is the attribute data for each individual that is central to this analysis. However, here the attribute list has now been extended by the addition of a "link centrality" attribute. This additional attribute is a measure of how many links each individual has in the sampled network and should reflect the tendency of individuals with relatively high and low links to be connected.

The likelihood ratios for all possible pairs in the sampled network are calculated for the attributes based on the data collected. The Bayesian framework then allows us to multiply these ratios together and, in turn, to predict that a pair is linked if the combined likelihood ratio exceeds a cut-off.

It is now possible to use the information captured in the likelihood estimates, L_i, to predict the existence of links between individuals that were not detected during

Table 2 Link centrality measures for the full N17 network and the 50 % sampled network

	L. C. Full n/w	L. C. Sampled n/w
Pavlos Serifis	14	9
Christodoulos Xiros	11	7
Savas Xiros	11	3
Alexandros Giotopulos	10	6
Dimitris Koufontinas	10	5
Anna	6	3
Iraklis Kostaris	6	3
Nikitas	6	4
Patroclos Tselentis	6	2
Sardanopoulos	6	2
Fotis	6	4
Yiannis	6	3
Thomas Serifis	6	4
Ojurk Hanuz	5	2
Yiannis Skandalis	5	1
Vassilis Xiros	4	3
Constantinos Karatsolis	3	3
Constantinos Telios	3	0
Dionysis Georgiadis	1	0
Elias Gaglias	1	0
Sotirios Kondylis	1	0
Vassilis Tzortzatos	1	0

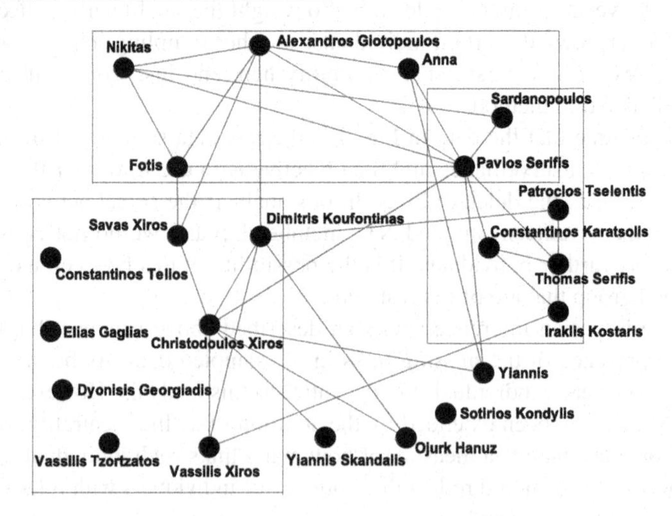

Fig. 4 A sample of the full network generated by removing a random selection of 50 % of the links.

the initial network sampling campaign. To do this it is necessary to propose a cut-off likelihood value (L_{cut}) and then establish which pairs exceed this cut-off. In attempting to make this estimate there is limited information and it is at this point the "subjective" aspects of Bayesian analysis that are so often referred to emerge. At this stage, what is unknown is the expected number of links in a network of this size, that is, an estimate of O_{prior}. The original (full) network has a link density (number of links/maximum possible number of links) of $64/231 \sim 0.28$, whereas the sampled network has a link density of $32/136 \sim 0.24$. By comparison, a fully connected network in which each individual is connected to every other has a link density of unity. So, having sampled the network we realize the link density of an inferred network must lie somewhere between 0.24 and 1. The only available guide is the empirically observed link density in other social networks of this kind. Because our purpose here is to determine how well the algorithm performs we exploit the knowledge of the original network link density, but it must be remembered that this parameter may not be accurately known for a given network.

Therefore, from equation 1, the prior odds estimate of there being a link between two individuals is determined by $\frac{64}{231}/1 - \frac{64}{231} = 0.28$. Consequently (using equation 4) to impose posterior odds $O_{post} > 1$ it is necessary for $L_{cut} > 2.6$. This posterior odds gives a probability of at least 0.5 that any predicted link is a true link. By contrast, setting $L_{cut} > 10.5$ gives a probability of at least 0.8 that there is a true link.

Using equation 6, the attribute likelihoods are multiplied together. For a given L_{cut} those pairs of individuals that exceed the cut-off are taken to be linked. Taking the pairs with likelihood $L_{cut} > 10.5$ the network shown in Fig. 5 is produced. At this level of prediction we get 4 *de-novo* predictions (black links), all of which are represented in the original N17 network. No false positive links are predicted, but the remaining 22 links are missed. Therefore, 18 % of the links missed by the sampling process are correctly inferred. Recall that there are five individuals who are not part of the sampled network so the six links that lead to/from them cannot be predicted in this analysis. Hence, there are $(32 - 6 =)$ 26 links in the sampled network that we are aiming to correctly predict.

Reducing L_{cut} lowers the probability that there is a link between two individuals. Taking pairs with likelihood $L_{cut} > 2.6$ yields the network shown in Fig. 6. The links shown black are those links that are predicted correctly. The dashed grey lines are links that are predicted but are not in the original network structure. Of the 23 *de-novo* predictions 9 are true positives and 14 are false positives. This gives a true positive/false positive = 0.64; we would expect a TP/FP ratio of ~ 1 at this level of likelihood. Therefore, the precision of this prediction (50 % correctly predicted links) broadly matches the expected precision of prediction. Of the 26 links that could in principle be positively identified 35 % are correctly inferred. Reducing L_{cut} increases the number of predicted links, but at the cost of lowering confidence in those links. In this example we know which of the links are true and false positives (from Fig. 1); in practice that knowledge would not be available. Were the 24 predicted links scattered randomly across the network it would be expected that there would only be 4 correctly identified links.

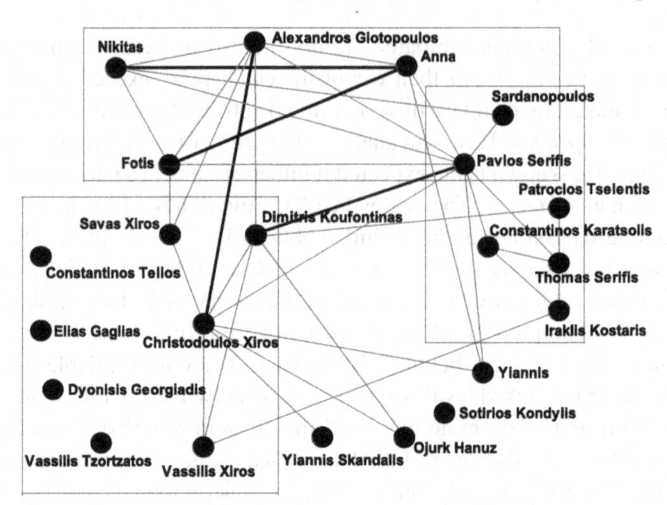

Fig. 5 The predicted missing links (black) assuming that the probability of a predicted link being correct is 0.8.

The network generated in Fig. 4 is, or course, just one realisation of a sampling of the full network. Repeating this calculation for other samples produces similar results. Therefore, using the inference approach it is possible to indicate the presence of links that might have been missed by the original network data collection campaign.

6 Conclusions

The era of asymmetric conflict has generating the need for new paradigms for threat analysis. Effort is being made on a number of different fronts, but it is characteristic of the new situation that many of the most promising research initiatives involve a fusion of quantitative methods with approaches from the social and political sciences. Here we have discussed network analysis and recent work that attempts to move beyond simply providing static pictorial representations of social networks. Network analysis does appear to offer valuable insight into the structure, organisation and capabilities of covert social groups. However, constructing social network structures for such groups (particularly for large groups) is a very labour-intensive process involving data-gathering, validation, data analysis and network construction. The need remains to find methods that are able to extract additional value and insight from that data which is known and the work described here aims to do that.

It is expected that the methods presented here will be useful in assisting the planning of social network data collection resources. They should assist decisions into how the inevitably resource-constrained intelligence collection effort is to be deployed. The results discussed here suggest that inference methods are a promising

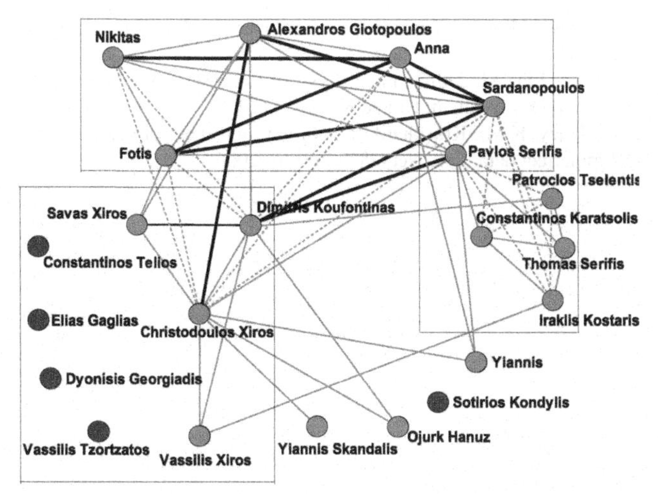

Fig. 6 The predicted missing links (black) assuming that the probability of a predicted link being correct is 0.5. False positives are the dotted lines.

way to address analysis of covert networks, but there is a clear need to apply them to other data sets in order to reach a more comprehensive assessment of their merits. In this way, social network analysis of covert organisations should continue to mature into a useful and reliable analysis method for counter-terrorism applications.

References

1. Newman, M.E.J.: The structure and function of complex networks. SIAM Rev. 45(2), 167–256 (2003)
2. Wasserman, S., Faust, K.: Social Network Analysis: Methods and Applications. Cambridge University Press, Cambridge (1994)
3. Scott, J.: Social Network Analysis: A Handbook. Sage Publications, London (2000)
4. Borgatti, S.P., Everett, M.G., Freeman, L.C.: UCINET 6.29 for Windows: Software for Social Network Analysis. Analytic Technologies, Harvard (2002)
5. Carley, K.M., Lee, J., Krackhardt, D.: Destabilising Networks. Connections 24(3), 31–34 2001
6. Farley, J.D.: Breaking Al Qaeda: A mathematical analysis of counter-terrorism operations. Studies in Conf. and Terr. 26, 399–411 (2003)
7. Fellman, P.V., Wright, R.: Complex systems at the mid-range. http://www.psych.lse.ac.uk/complexity/Conference/FellmanWright.pdf
8. Krebs, V.: Uncloaking Terrorist Networks. First Monday 7(4) (2002)
9. Sparrow, M.K.: The application of network analysis to criminal intelligence: an assessment of the prospects. Soc. Net. 13(3), 251–274 (1991)
10. Sivia, D.S.: Data Analysis: A Bayesian Tutorial. Oxford University Press, Oxford (2004)
11. Jansen, R., Yu, H., Greenbaum, D., Kluger, Y., Krogan, N.J., Chung, S., Emili, A., Snyder, M., Greenblatt, J.F., Gerstein, M.: A Bayesian networks approach to predicting protein-protein interactions from genomic data. Science 302, 449–451 (2003)
12. Rhodes, C.J., Keefe, E.M.J.: Social network topology: a Bayesian approach. J. Op. Res. Soc. 58(12), 1605–1611.

13. Irwin, C., Roberts, C., Mee, N.: Counter Terrorism Overseas. Dstl Report, Dstl/CD053271/1.1 (2002)
14. Abram, P.J., Smith, J.D.: Modelling and analysis of terrorist network disruption. MSc thesis, Cranfield University, Shrivenham, United Kingdom (2004)
15. Borgatti, S.P., Carley, K.M., Krackhardt, D. On the robustness of centrality measures under conditions of imperfect data. Soc. Net. 28, 124–136 (2006)

A Mathematical Analysis of Short-term Responses to Threats of Terrorism

Edieal J. Pinker

Abstract A terror threat information model capturing the uncertainty in timing and location of terror attacks is developed to create a mathematical framework for analyzing counterterrorism decision making. Using this framework two important defensive mechanisms, warnings and the deployment of physical resources are studied. Warnings are relatively inexpensive to issue but their effectiveness suffers from false alarms. Physical deployments of trained security personnel can directly thwart attacks but are expensive and need to be targeted to specific locations. By structuring the tradeoffs faced by decision makers in a formal way we try to shed light on an important public policy problem. We show that the interaction between the use of warnings and physical defenses is complex and significant.

1 Introduction

Keohane and Zeckhauser [8] define four mechanisms by which a government may combat terrorism: reducing the stock of terrorists, limiting the flow of resources to terrorists, taking averting actions, and taking ameliorating steps. Counterterrorism methods also differ with regard to the time horizon in which they are applied. They could be long-term strategic moves or they could be short-term tactical actions. In this paper we do not consider stocks and flows of terrorism but focus on two defensive measures, deployment of physical resources, an example of an averting action, and issuance of warnings, an example of amelioration. We also limit our study to short-term counterterrorism activities. The purpose of this paper is to develop a mathematical framework for analyzing short-term counterterrorism decisionmaking situations. By developing such a framework we create an opportunity for mathematical scientists to bring their vast array of tools to bear on an important set of public policy issues. At the heart of any mathematical analysis of counterterrorism one

Edieal J. Pinker Simon School of Business, University of Rochester, Rochester, NY 14627, USA, e-mail: ed.pinker@simon.rochester.edu

N. Memon et al. (eds.), *Mathematical Methods in Counterterrorism*,
DOI 10.1007/978-3-211-09442-6_9, © Springer-Verlag/Wien 2009

must have a model of the information available to the decision maker. This will be the starting point for this paper. We formulate a model that represents the likelihood of an attack, the possible targets, and the potential damage. We then demonstrate how such a model can be used as a basis for analysis of decisionmaking by studying the possible responses of a government or security service to such information, focusing on the optimal deployment of physical resources and issuance of warnings. By developing a theoretical model of the interplay between resource deployment, terror warnings, and the information available to decision makers we create a structured way to think about these defensive mechanisms and identify when and how they can be used most effectively to provide for public security.

We also distinguish between long-term and short-term defensive measures. Long-term measures can be defined as permanent or semi-permanent shifts in policies, procedures and allocations of resources in response to the threat of terrorism. For example the creation of the Department of Homeland Security (DHS), its budget allocation, standards for airline security, and the creation of the Transportation Safety Administration (TSA) can all be viewed as long-term defensive measures. The creation of emergency reaction plans at different levels of government, the training and equipping of first-responders, and the hardening of physical facilities are also examples of long-term defensive measures. Short-term defensive measures can be defined as immediate responses to the analysis of real-time intelligence. Temporarily boosting security at an embassy, by placing more guards, or issuing a travel advisory are examples of short-term defensive measures. Spot checks of trucks entering Manhattan or reviewing emergency procedures with staff are also examples of short-term responses. To better structure our thinking about terrorism defenses we begin by considering a single potential target and then expand our discussion to a spectrum of multiple targets.

When analyzing the long-term risk of terrorism to a particular target several factors are important: the probability of an attack, the difficulty of doing substantial damage to the target (its hardness), and the economic and political value of the target. In Figure 1 we use an influence diagram to depict how these factors work together to determine the cost of terrorism from the perspective of a single target.

The total cost of terrorism is comprised of damage from attacks and the cost of defensive measures taken. Defensive measures are in addition to the inherent hardness of the target. For example, a shopping mall, designed for open access to the general public is not as difficult to damage as a nuclear power plant that has restricted access and many safeguards to make it robust. A shopping mall can be hardened by installing concrete barriers that prevent vehicle bombs from getting too close or by installing metal detectors in its entrance ways. Such measures combined with the inherent hardness of the target will influence the damage an attack causes. The damage caused by an attack is also influenced by the value of the target. Damage caused to a high value target is by definition more costly to society than damage caused to a low value target. The value of a target is determined by its economic significance as well as its political or symbolic significance. Loss of life has both economic and political significance. Finally, attack damage is influenced by the type of attack, where "no attack" is a possible type of attack. The type of attack is of

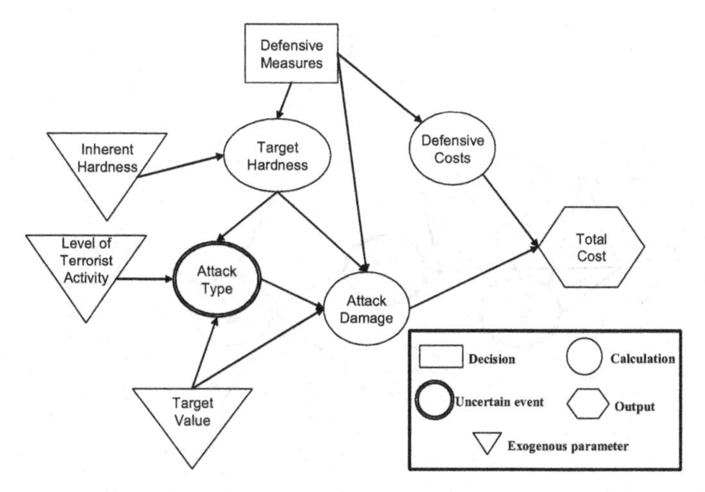

Fig. 1 Influence diagram of long-term risk of damage from terrorism to a single target

course uncertain but is itself influenced by the hardness of the target and its value, because we assume that the terrorists themselves perform a cost benefit analysis when selecting a target. Terrorists want to cause significant damage but want a high probability of success.

The analysis outlined above, when applied to all potential targets in a geographic area will yield what we can term a risk profile for that space. Several insurance companies are currently developing such risk profiles to determine insurance coverage criterion and parameters (see Risk Management Solutions [20]) The purpose of the risk profile is to assess the expected damage costs of terrorism to each target over the lifetime of the insurance policy. A short-term perspective assumes that the outcome of the long-term risk analysis outlined above has been conducted and operationalized and that its outcome is known by the decision maker and to a large extent the terrorist groups as well. Powell [18] [19] has used a game theoretic approach to show how the long-term risk profile may evolve over time as terrorists respond to defensive measures when choosing targets. In Figure 2 we use an influence diagram to depict how intelligence information influences decisions about short-term defensive measures.

Given the long-term risk profile there is a current state of the world that influences the short-term likelihood of attack and type of attack. Intelligence data gives the decision maker imperfect knowledge of this state that influences his decision making. For example, a long-term risk profile, as developed by insurance companies, may indicate that the set of high-rise office buildings in Manhattan are high risk targets for a truck bomb. However, security surveillance tapes may indicate suspicious activity by small groups of men filming specific buildings in the financial district. Such information merged with the long-term risk profile would yield an updated probability of an attack on high-rise office buildings in the financial district. Looking at Figure 2 this means that the probability of an attack on a specific tar-

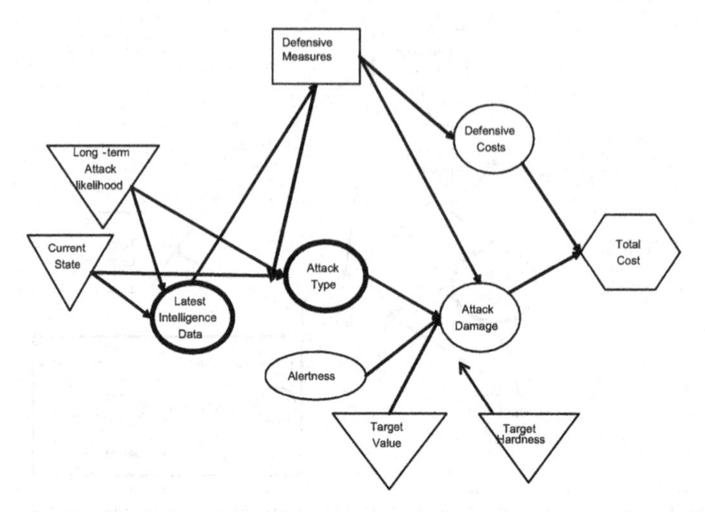

Fig. 2 Influence diagram of short-term risk of damage from terrorism to a single target

get is influenced by the long-term risk for that target and the current intelligence data. Target value and target hardness have already influenced the probability of attack through the long-term risk analysis. In the short-term perspective of Figure 2 these factors only influence the damage from an attack. In the short-term setting, the decision maker must decide what defensive measures to take to potentially reduce damage from attacks. Defensive measures can reduce the damage from attacks in multiple ways. They can influence the attack mechanism, the probability of attack success, and the expected damage from an attack.

The structure provided by the influence diagrams allow us to illustrate several important observations about defending against terrorism. First, there is a distinction between long-term and short-term decision making that closely resembles planning for natural disasters. For example, there is a historical record of hurricanes that informs our expectations about where and when they will hit and the damage they will cause. Such expectations influence insurance assessments, disaster planning, and building codes. However, when real time intelligence is available that a hurricane has, for example, formed in the Caribbean, the government updates its assessment of where it may strike, issues warnings to the public, deploys emergency services, and prepares shelters for storm refugees all in an effort to mitigate the potential damage from the hurricane. Second, the influence diagrams indicate a key difference between terrorism and natural disasters, namely terrorists are agents that decide when and where to attack. Long-term measures to harden the defenses of some targets will influence terrorist's choices of targets since they perform their own cost benefit analysis of targets. The insurance industry is in fact using this perspective on terrorist behavior to inform their risk modeling [11]. Short-term intelligence gives the decision maker an indication of the outcome of the terrorists' target analysis. Third, we note that both warnings and physical deployments work together to influence the outcomes of terrorist attacks but through different mechanisms. As a result decision making processes for these two defense modalities must be integrated.

In Section 2 we develop an information model of terrorism risk that gives mathematical structure to the concept of short-term counterterrorism intelligence. In Sections 3 and 4 we develop and analyze mathematical models of defensive mechanisms. Following with a numerical investigation in Section 5. We conclude in Section 6.

2 Information Model

In any time period decision makers face the following situation. They have a risk profile for a wide range of potential targets that defines the expected damage to these targets from terrorism. They also have current intelligence on some of the activities of known terrorist groups. The short-term or real-time intelligence can be of many forms, for example: communications intercepts, informants, observations of suspects and/or suspicious behavior, or public threats made by terrorist leaders. Recent political events may also be take into consideration. For example, elections may influence terrorist activities [9]. Whatever the source or content of the intelligence data its analysis yields the following: an assessment of the likelihood of an attack during that time period and a revised risk profile for the possible target locations. Given these data the decision maker must decide upon defensive measures. Notation is summarized in Table 1.

We assume that, in any time period, there are discrete possible threat states of the world $s \in [0,k]$ and the decision maker knows which state the world is in. For each state, s, there is a probability p_s that a terrorist attack will be attempted in that time period. Without loss of generality we assume that $p_s > p_{s'}$ for $s > s'$. The probability of being in state s is given by π_s. The vectors p_s and π_s in combination define both the decision maker's expectations about the frequency of attacks and his ability to collect intelligence to anticipate the timing of attacks.

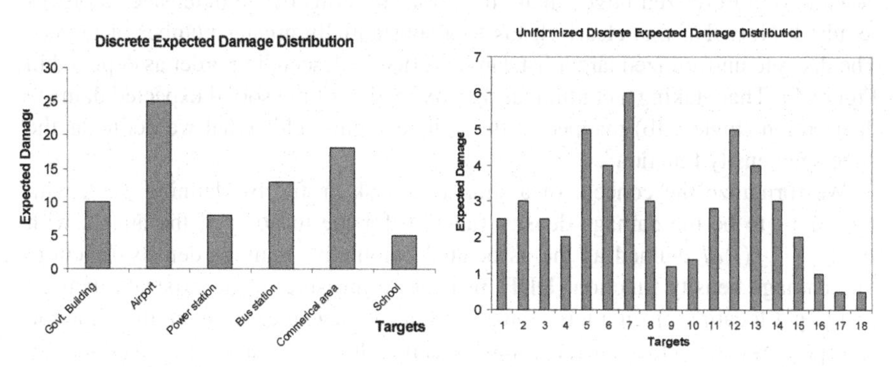

Fig. 3 3a: Expected damage distribution to discrete target list, 3b: Uniformized expected damage distribution

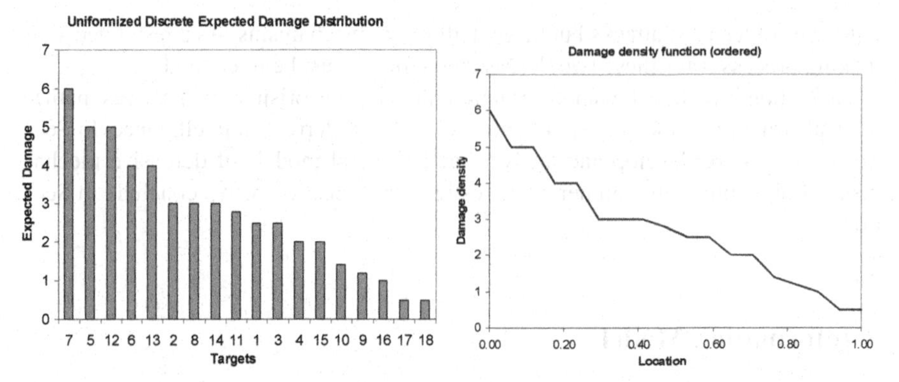

Fig. 4 4a: Sorted uniformized expected damage distribution, 4b: Continuous approximation of damage density function.

We define the set of targets as a set of locations that are uniform in terms of the resources required to defend them. For example, while some would designate both Los Angeles International Airport (LAX) and the Seattle Space Needle as potential terrorist targets it makes little sense to view them as equivalent from a defensive standpoint. On the other hand we can split these into sub targets like, the northern, southern, eastern and western borders of the LAX grounds, the airspace over LAX, the domestic terminal at LAX, the grounds of the Space Needle park, the airspace over the Space Needle etc. These sub targets forms a new set of targets that are less dissimilar in terms of the resources required to defend them. To simplify our analysis we map this discrete, uniformized list of targets to the continuous unit segment $[0,1]$. Underlying this simplification is the assumption that there are a sufficient number of targets to make a continuous model a reasonable approximation of what is inherently a discrete space. The process of uniformizing the target list is depicted in Figures 3a and 3b. Figure 3a shows a set of target types and represents the expected damage combining the likelihood that a target is actually attacked and the potential for damage on some scale. Figure 3b is the same after target types have been broken down into uniformized targets that are similar in terms of the defensive resources required. Note that this clearly leads to a substantially greater number of targets. The discrete uniformized target list can be sorted in descending order as depicted in Figure 4a. Then, taking a continuous approximation of the sorted expected damage distribution (Figure 4b) mapped to the unit segment yields what we define as the "damage density function".

We formalize the concept of a short-term risk profile by defining $f(x)$, with $x \in [0,1]$, to be the damage density function for the location of the attack, with $F(x) = \int_0^x f(l)dl$ defined as the associated cumulative damage density function. The damage density function (DDF) provides a measure of the expected damage to a target location given a terror attack occurs somewhere. As such, this function incorporates data on the value, hardness, and likelihood of attack for each target. We define $f_l(x)$ to be the DDF of target locations after they have been sorted from highest expected damage to least expected damage and note that it is by definition a decreasing function. We define $F_l(x)$ as the corresponding cumulative DDF. Given that the likely modalities of terrorist attacks, e.g. bombings, hijackings, CBRN (chemical biological, radiological, and nuclear), are not changing rapidly and that the values and hardness of potential targets are not changing rapidly either, it is reasonable to assume that the long-term damage density function is constant over time. For the purposes of this paper we further assume that the ranked short-term damage density function is time invariant. The implication is that while intelligence data may boost the likelihood of attack at some locations in one period and different locations in a later period the overall ranked damage density function won't change significantly. If the short-term damage density function changed over time it would mean that the capabilities of terrorists and or security forces were changing over time as well. While entirely likely, and of interest, modeling such shifts is beyond the scope of this paper and involves analysis of longer time horizons than we consider here. That said, the concept of a damage density function as defined here could be used in more dynamic environments as well.

We define $C = \int_0^1 f_l(l)dl$ as the total expected damage from a terrorist attack given that it occurred. By scaling $f_l(x)$ by C, we can form a normalized damage density function defined on $[0,1]$, with $F_l^*(x)$ the associated cumulative damage density function. The current state, s and normalized DDF $f_l^*(x)$ are outputs of the intelligence gathering process. The state information indicates what is known about the timing of an attack and the DDF captures what is known about its location and damage potential. We now have a mathematical representation of the information available to a decision maker considering various counterterrorism actions. In the next section we discuss two such actions and how they relate to one another.

Table 1 Notation

Symbol	Description
q	decision variable, vector of the range of deployment of defensive resources in different threat states.
w	decision variable, vector of the warnings given in each threat state.
W	set of threat states for which a private warning is issued.
s	threat state of the world.
π_s	probability of being in state s.
p_s	probability of an attempted attack given in state s.
$f(x)$	damage density function.
$f_l(x)$	damage density function sorted highest to lowest.
$f_l^*(x)$	normalized sorted damage density function.
C	total expected damage cost if a terrorist attack occurs.
R	cost, relative to C, of deploying physical resources per unit of territory covered.
g	reduction in expected damage from attack for a target with extra physical defensive resources.
$\gamma(\phi)$	fraction of damage caused by successful attack when warning is given.

3 Defensive Measures

Given intelligence on short-term terrorist threats a government has essentially two broad classes of defensive mechanisms to choose from. Those that involve the physical deployment of security forces and those that involve warnings to increase the effectiveness of the existing security infrastructure. Physically deploying security forces in the field can have several benefits. If the deployment is very visible it can serve as a sign that the government is "doing something" to protect the public and is a visible sign that perhaps the government is anticipating an attack. Similarly these deployments may deter terrorists from striking. Clearly the physical presence of security forces can also prevent or limit the extent of a terrorist attack. Physical resources are also expensive, and their cost effectiveness has much to do with the intelligence available on likely threats. Placing a battalion of the New York State National Guard around the Statue of Liberty will not be very effective if terrorists are planning an attack on Los Angeles International Airport. Likewise, if the US government believes there is an imminent attack planned on a fast food restaurant in the continental US deploying a squad of armed troops at every McDonalds and Burger King in the country will be a very expensive exercise. Physical deployments may also reveal defensive procedures if terrorists are probing defenses with feints.

Since March 2002, the United States has adopted a system of color coded terror warning levels. At any point in time the Department of Homeland Security (DHS) states the current assessment of the likelihood of a terrorist attack on the USA. These assessments are based on information and analyses from the various intelligence agencies of the US government. In Israel, the security services regularly report on the number of active terror threats they are tracking (see for example [3]). In Israel it is also not uncommon to hear news reports that refer to recent security operations as related to some previous warnings [13]. Warnings can serve many of the same purposes as physical deployments. Public warnings can create the perception that government authorities are active and informed in a struggle that is often conducted in a secretive way. Public warnings can also deter terror attacks. If a terrorist cell, in the midst of planning an attack, comes to believe (as a result of the warning) their plans have been partially revealed they may abort them. Warnings can also serve as a mechanism to directly aid in the prevention of terrorist attacks and the reduction in the damage caused by terrorist attacks. The first goal is achieved by increasing the alertness of security personnel at possible targets. The second goal is clearly expressed by the DHS in its description of its alert system. In its literature, the DHS describes the steps that all government agencies should take to prepare their facilities and employees at different levels of threat, with similar guidelines for the general public. Raising preparedness means that if important infrastructure is damaged, government and civil society can continue to function thereby minimizing the damage of the attack. It can also mean canceling or altering activities so that fewer people are present at likely targets, again minimizing the damage from an attack.

Compared to physical deployments, warnings have an important cost advantage. Given modern communications technologies, disseminating a warning across a broad geographic area is relatively inexpensive. However, the effectiveness of warnings is limited by their credibility. Pate-Cornell ([15] and [16]) models the optimization of warning systems. In particular, she provides a structure for understanding how warning systems work and can be evaluated. She characterizes the costs and benefits of a warning system as being a function of "(1) of the quality of the signals issued and (2) of the response of the people to whom warnings are directed." These two factors determine the trade-off between failures to issue alarms when something bad happens and the issuance of false alarms. If the public does not believe the warning it will not take the precautions dictated by the raised alert. This holds true for security personnel as well whose work is typically quite monotonous. Therefore, while warnings may be relatively cost effective they cannot be used indiscriminately or their credibility and utility will be diminished.

In this paper we analyze two types of warnings, private and public. Private warnings are defined as warnings issued to security forces to boost their alertness and preparedness for possible attacks. Public warnings are defined as warnings issued to the general public to similarly boost their alertness and preparedness for possible attacks. Being public, these warnings also serve as warnings to security forces and are observed by terrorists and as such may affect terrorist behavior. The economic analysis of terrorism, and responses to it, has found that substitution is an important phenomenon in terrorist behavior [4]. Defensive actions that make one type of target more difficult to attack lead terrorists to different targets. Terrorists are also forced to make intertemporal substitutions shifting attacks from one time period to another in response to attacks on them by security forces. [5] gives an overview of the development of this literature. Therefore, in this paper, when we model public warnings we include the possibility that terrorists will shift their attacks to periods in which there is no public warning.

We define R as the cost of defending all targets with physical resources to an acceptable standard. Here R is not measured in dollars but relative to C. If $q \in [0,1]$ is the range of targets defended by physical resources, then the defensive cost generated is Rq. The expected damage to any target that is defended by additional physical resources is assumed to be reduced by a constant factor g. That is, if the interval of targets defended by additional resources is $[0,q]$ the expected damage from an attempted terrorist attack is $gF_l^*(q)+(1-F_l^*(q))$. We first formulate the decision problem assuming the only defensive mechanism available to the decision maker is deploying physical resources. In this case the problem is:

$$P1 : \min_{q} \sum_{s=0}^{k} \pi_s[p_s[gF_l^*(q_s)+(1-F_l^*(q_s))]+q_sR \tag{1}$$

Subject to:$q_s \in [0,1]$. In each state the decision maker must select the deployment range that minimizes total costs. We will refer to the region receiving extra defensive resources (locations with positions $x < q_s$) as the defended region and the rest of the space as the *undefended region*[1] . We have assumed that the benefit of deploying

[1] In this paper the terms undefended and defended are defined with respect to the extra resources being deployed. All locations may have some basic level of security.

physical resources is the same g reduction in expected damage regardless of location and that the level of resources expended per target defended R is an exogenously defined constant. The implication of these assumptions is that the decision maker cannot choose to defend one target more than another within the defended region and that the effect of defensive measures is proportional to the expected damage from the attack. Making g and R constant greatly simplifies the optimization and relaxing this assumption is an avenue for extending the model in future research. Having a proportional effect is motivated by the observation that there is a difference between preventing an attack on a specific target and preventing damage from an attack by a terrorist cell that was targeting a specific location. For example, a terrorist cell may be planning a truck bomb attack on a landmark building. The planning involves preparing the explosives and the truck, mapping a route to the target building that best avoids security forces, identifying the optimal points to detonate the bomb, and conducting rehearsals. In theory, enough physical resources could be deployed in such a way that even the most carefully planned truck bombing attack would be thwarted from destroying the targeted landmark. However, there are many ways the terrorist could still cause damage in the attempted attack. First, the terrorists could cause less than catastrophic damage to the target by exploding the bomb farther away from the building than planned. Second, the terrorists could kill and injure security forces and bystanders when they were blocked from attacking the target directly. Third, the terrorists might at the last minute attack a nearby, less defended, less significant building when they observe very tight security at their planned target. Fourth an attack, even if unsuccessful at its main objective, causes fear in society with significant economic and political impact. Because the destructive capability of a terrorist cell is probably proportional to the damage they are attempting to cause by attacking a particular target, it is reasonable to assume that the impact of the defensive resources is proportional to the expected damage to the target.

We have also assumed that the deployment does not affect the damage density function by shifting the likelihood of attack to a different location. First we note that as discussed above, part of the justification for a proportional defensive effect, g, is that terrorists may shift targets at the last minute in response to defensive deployment. Second, we note that even though deployment of physical resources tends to be observable, terrorists do not always have the ability to observe these resources in time to change their plans. For example, surprise roadblocks can catch unsuspecting terrorists on their way to an attack, and raids of suspected terrorist hideouts can thwart attacks in planning stages. Also, an increased presence of undercover security is very difficult to detect. For example, a hijacker will have difficulty detecting additional air marshals placed on a flight that is believed to be at high risk by security forces.

Private warnings are another mechanism for defending against terror attacks. We model private warnings as having zero cost and reducing the probability of a successful terror attack in any location by a multiplicative factor $\gamma \in [0,1]$. Define w_s as an indicator variable that takes value 1 if a private warning is issued when in state s and 0 otherwise. The decision maker must decide in which threat states he will issue warnings. Call this set W, i. e., $W \equiv \{s|w_s = 1\}$. Given the vectors w,π, and

p there is a long run rate ϕ at which false alarms are issued[2]. The greater the false alarm rate the less warnings will be heeded and therefore the less of a benefit that will come from the warnings, and thus $\frac{d}{d\phi}\gamma(\phi) > 0$. We reformulate the problem with warnings as follows.

$$P2: \min_{q,\pi} \sum_{s\in W} \pi_s[p_s\gamma(\phi)[gF_l^*(q_s) + (1 - F_l^*(q_s))] + q_sR$$

$$+ \sum_{s\notin W} \pi_s[p_s[gF_l^*(q_s) + (1 - F_l^*(q_s))] + q_sR \tag{2}$$

Subject to: $\phi = \sum_{s\in W} \pi_s(1 - p_s)$ and $q_s \in [0,1]$, with w_s defined as 0 or 1.

In problem P2 we are putting a formal structure on the situation faced by a decision maker facing potential terrorist attacks. While the formulation somewhat simplifies reality we believe that structurally it does capture the inherent trade-offs in defending against terror attacks. In our model, deploying defensive resources has a direct cost that is directly related to the size of the space defended. The deployment of physical resources in one period has no effect on future periods. For example, the fact that a roadblock at an intersection is up on a Monday does not increase the ability to discover a car bomb at that intersection on Friday. Warnings are less costly than physical resources but do have an effect across periods. For example, a false alarm on Monday can reduce the impact of a warning on Friday.

In theory, warnings could have levels of severity, e.g. low, high, etc. Each threat state could receive a different level of warning instead of the warning/no warning dichotomy we model. We do not believe that differing warning levels are practical. It is difficult for individuals to calibrate themselves to different levels of alertness. In fact this has been one of the criticisms of the U.S. DHS color-coded scheme (NY Times 2004). It is also worth noting that since its inception soon after 9/11/2001 the warning system has only taken on two settings, Yellow and Orange [14]. These have formed a de facto two level system, normal and heightened state of alert.

We do not capture the fact that threat probabilities may change over time. However, as long as the impact of false alarms on warning effectiveness operates on a time scale that is significantly shorter than the timescale in which there are significant shifts in the threat probabilities our model should be applicable. We also do not model the effect that defensive actions have on the behavior of terrorists. It is possible that terrorists who are aware of a higher level of security will shift their planned attack to a later time. If we view warnings as being given to security forces only, then it is unlikely that terrorists will be aware of them. In [17] an extension to public warnings is studied in which terrorist reactions to warnings are explicitly considered.

[2] We are interested in the rate at which false alarms are issued and not the fraction of warnings that are false alarms. We model false alarms this way because it is believed [12] that when false alarms occur more frequently, recipients will be more skeptical even if the fraction of warnings that are false alarms is the same.

4 Analysis

In this section we investigate two sets of questions about short term defensive measures. First, we study how the use of physical resources and warnings interact with one another. Second, we study how the information available to the decision maker regarding the timing and location of attacks influences the optimal use of defensive measures. We then illustrate our observations with a set of numerical examples.

4.1 Interaction between warnings and physical deployments

Proposition 1. *Given a warning vector* **w** *the optimal defensive deployment in threat state* $s, q_s^*(w)$ *is 0, 1 or is given by:*

$$q_s^*(w) = \begin{cases} f_l^{*-1}(\frac{R}{p_s(1-g)}) & for\ w_s = 0 \\ f_l^{*-1}(\frac{R}{\gamma(\phi)p_s(1-g)}) & for\ w_s = 1 \end{cases} \tag{3}$$

Proof: When no warning is given in state s and defensive deployment is q_s, *the expected cost is given by:*

$$V(q_s) = p_s[gF_l^*(q_s) + (1 - F_l^*(q_s)] + q_s R$$

$$\frac{dV}{dq_s} = p_s[(g-1)f_l^*(q_s)R$$

$$\frac{d^2V}{dq_s^2} = p_s[(g-1)\frac{df_l^*(q_s)}{dq_s}$$

By definition $\frac{df_l^*(q_s)}{dq_s} \geq 0$ *because we rank location in decreasing order of likelihood of attack and* $g < 1$. *We can then apply the first order condition and have that if the optimum deployment* q_s^* *is an interior solution then it satisfies:*

$$\frac{R}{p_s(1-g)} = f_l^*(q^*s)$$

The same logic applies to the case when a warning is given in state s.

The numerators in Equation (3) are the marginal costs of broader deployment of defensive resources while the denominators are the difference between the expected damage in a defended region versus an undefended region with warnings and without. Because f_l^* is by definition decreasing it means that greater deployment cost reduces the scope of the deployment while greater expected damage increases the scope of deployment.

¿From Equation (3) we can also see that there is a complex relationship between the warning threshold and the deployment range. When a warning is given, the deployment of physical resources is smaller than when no warning is given since

$\gamma(\phi) \in [0,1]$. At the same time, the more frequently warnings are given, i. e., the larger the set W, the less effective the warning in any particular threat state, and therefore the greater the optimal deployment. We can also observe that a warning given in a particular state reduces costs for that state regardless of the false alarm rate. This means that it is always optimal to issue a warning in at least one of the states. We state this formally in Proposition 2.

Proposition 2. *In the optimal warning policy* ***w****:*

$$\sum_s w_s^* \geq 1 \tag{4}$$

Proof: Follows from discussion above.

Problem P2 integrates decision making about warnings and the deployment of physical resources. In practice, the organizational structure of a country's security services may cause these decisions to be made independently leading to inferior performance (see [17]).

We expect that the states in which we have a high terror threat are typically the least likely or least frequent, that is we expect that π_s is decreasing in s. Given this observation, we see that it makes sense to deploy resources more widely if we are in a high threat state because they are most likely to be useful as shown in Proposition 1 and the amount of time we will have them deployed is relatively short. It also makes sense to issue a warning to boost alertness and effectiveness of the defenses because in high threat states there will be fewer false alarms. In fact, as per Proposition 1, the warnings make it possible to use fewer physical resources by deploying less widely. On the other hand deploying physical resources during the more frequent low threat states is more costly and it is in these states that the low cost of warnings is attractive. This cost trade-off can lead to a somewhat counterintuitive result. Considering warnings independently of physical deployments one would expect that as the probability of an attack increases warnings would become more attractive. However, higher frequency states give warnings a larger cost advantage over physical resources. For example let's say there are three states of the world 1, 2 and 3 with π_s = (.8, .15, .05) and p_s = (.05, .23, .5). The fraction of attacks that occur in each state i is given by,

$$A_i = \frac{p_i \pi_i}{\sum_j p_j \pi_j} \tag{5}$$

In the example, **A** = (.4,.35,.25). Nearly half the attacks actually occur in the lowest threat state (state 1) so holding the false alarm rate constant, the most benefit from a warning comes in the lowest threat state. Therefore there may be situations in which it is optimal to issue a warning in low threat level states, not to issue a warning in an intermediate threat level, and to issue a warning in a high threat level state, a warning vector **w** = (1,0,1). Such a strategy will lead to more false alarms than **w**= (0,1,1) but may, in total, reduce costs if warning effectiveness is not very sensitive to false alarms.

The interaction between warnings and physical deployment has another effect. If the overall likelihood of an attack increases, one would expect defensive deployments to increase as well and total costs to increase. However, more frequent attacks reduce the false alarm rate of warnings making them more effective and reducing the need to rely as much on physical resources. The net result is that it may be the case that societies that are at a greater state of alertness because of more frequent attacks may be more efficient at defending themselves against terrorism than societies that face less frequent attacks. The complex interactions between physical deployments and warnings described here suggest that there is a potential for significant gaps between the performance of integrated and segmented defenses. Clearly the size of this gap will depend on parameter values. If warnings are not effective and/or resource deployments are not costly the gap will be small.

4.2 Effect of intelligence on defensive measures

Our model of the information available to the decision maker posits that the long term risk profile is updated with real time intelligence data that, over the short-term, determines the likelihood of an attack, i. e., identifies the threat state, and indicates the most likely and costly targets via the damage density function. Having more or less precise intelligence says something about the shape of the damage density function, $f_l^*(x)$, and the threat state structure. We expect that more precise intelligence leads to lower total costs from terrorism.

In an ideal world the decision maker would know exactly when an attack would occur. This is equivalent to there being two threat states, one in which an attack always occurs in the current period, and one in which an attack never occurs. In such a situation warnings would be given whenever there was a certainty of an attack and there would be no false alarms. The least informative situation would be one in which the threat states were indistinguishable, i. e., $p_s = p$ for all s. In such a situation the decision maker cannot know if an attack is more likely in the current period versus later periods. According to Proposition 2, in this situation it is optimal to issue warnings sometimes just to get some benefit from the increased alertness. The rationale is that without specific attack timing information, randomly issuing warnings will occasionally thwart an attack and are better than never issuing a warning. The closer the decision maker is to the ideal world case the more effective his use of warnings. Information theory provides a mechanism for determining how close the threat state structure is to the ideal case. We can adapt the concept of conditional information [2] to give a measure of how much uncertainty in the occurrence of an attack is resolved by knowledge of the threat state. For our purposes we calculate this quantity, I_π as follows:

$$I_\pi = 1 + \frac{1}{\ln(2)} \sum_s \pi_s [p_s \ln(p_s) + (1 - p_s) \ln(1 - p_s)] \tag{6}$$

In the ideal world $I_\pi = 1$. When threat states are indistinguishable $I_\pi = 1 + (p\ln(p) + (1-p)\ln(1-p)) \setminus \ln(2)$. If $p = .5$, $I_\pi = 0$, it is at its minimum because an attack is equally likely to no attack. In the numerical examples we show that costs increase when I_π decreases.

In an ideal world the decision maker would also know exactly where an attack was going to occur and its damage potential. In our model this corresponds to having intelligence data that makes the probability of an attack, and therefore the damage density, zero everywhere except one point. In this case $f_l^*(x)$ would be an impulse function at $x = 0$. Defensive resources would be deployed only at $x = 0$ and therefore deployment costs would be zero. The least informative situation would be one in which the damage density function was uniform, i. e., $f_l^*(x) = 1$ for all x. In this case no one target or set of targets stands out as a priority for defensive resources. The result is that either all targets are covered at great expense, or none are worth defending. From Equation (3) we see that if $\frac{R}{(p_s(1-g))} > f_l^*(0)$ then $q^*(w) = 0$, and if $\frac{R}{(p_s(1-g))} > f_l^*(1)$ then $q^*(w) = 1$. The situation in which the decision maker is least informed about where an attack will occur is the case of $f_l^*(x) = 1$ for all $x \in [0,1]$, i. e., a uniform distribution. In all other cases $f_l^*(0) > 1$ and $f_l^*(1) < 1$, because $f_l^*(x)$ integrates to one and is decreasing by definition. As a result when the damage density function is uniform the optimal deployment is an all or nothing strategy, $q^*(w) = 0$ or $q^*(w) = 1$ depending upon the value of $\frac{R}{(p_s(1-g))}$. We generalize this observation to say that the less informed the decision maker is regarding where an attack will occur; the more likely the optimal deployment is all or nothing. One interpretation for the scenario with a uniform damage density function is that over a long period of time a terror risk homeostasis is achieved. All high value targets are hardened enough that from the terrorist's perspective there is little distinction between targets. For example passenger planes have historically been popular targets for terrorists. Enormous investments in airline security have made it increasingly difficult for terrorists to commandeer or destroy a passenger jet. On the other hand customers in check-in desk queues at most airports are vulnerable to attacks by suicide bombers. However, any place many civilians congregate, movie theaters, buses, train platforms, etc. are equally vulnerable to suicide bombers. As a result all these targets become equally likely choices for the terrorists with similar damage potential.

With risk homeostasis the only reason the damage density functions could be non-uniform, or in other words the only reason one location becomes a more likely target than another, is if short-term intelligence information indicates terrorist planning for a particular target and mode of attack. The implication of this observation is that while long-term defensive efforts may be effective in causing a general reduction in damage from terror attacks, the prevention of specific attacks will require short-term intelligence gathering and analysis capabilities. Such capabilities would apply to predicting both the location, mode of attack, and timing of attacks. In other words, a uniform damage density function is a symptom of intelligence failure. This failure could be any combination of failure to gather, analyze or disseminate intelligence effectively.

5 Illustrative numerical experiments

In this section we construct illustrative numerical examples to quantify and make more concrete the qualitative observations we have made above. Quantifying the behavior of the model is very challenging because there is great uncertainty surrounding the key parameters. Some of these parameters, such as those related to costs, are frequently estimated by risk analysts while others such as the effect of warnings are not well studied. For example [1]; analyzes the threat of a nuclear attack on the United States via sea port freight transport, estimating that a successful nuclear attack on a major seaport would cause from hundreds of billions to several trillion dollars of costs, including the value of statistical lives lost, direct property damage, the cost of trade disruption, and other indirect costs. The range of possible damage is quite large varying by a factor of at least ten and would correspond to the parameter C. The study also estimates the costs and benefits of several recommended security measures. They estimate that by reducing the probability of a successful attack the expected attack cost would be reduced by a factor of ten, corresponding to $g = .1$ in our model. They also estimate an annual cost of $10 billion to implement the security measures. This corresponds to a defensive deployment cost R relative to C ranging from .005 to .1. This analysis, while perhaps accurate for seaports, greatly underestimates the costs of defending against a terrorist nuclear attack. A successful nuclear attack on a major U.S. city whether a sea port or not would probably lead to damage similar to those estimated in [1] however sea freight is only one method of conducting a nuclear attack. Achieving a factor of ten reduction in risk from all possible modes of nuclear attack would cost far more than $10 billion per year.

Assessing the costs and benefits of defense against a potential attack is a meaningless exercise without an assessment of the likelihood or frequency of attack. An example of an attempt to address this question is a survey of experts conducted by Senator Richard Lugar the chairman of the senate foreign relations committee [10]. The survey finds that on average these experts believe that the probability of a nuclear attack on the United States in the next decade is 29.2 %, of a radiological attack 40 %, of a biological attack 32.6 %, and of a chemical attack 30.5 %. 62 % of the respondents thought the probability of a nuclear attack was between 10 % and 50 %. The high end of this range would justify a five fold increase in defensive spending over the low end of the range. We can conclude from these examples that even in areas that risk analysts and policy makers are experienced in, there is considerable uncertainty about parameter values and when we layer on the less well understood aspects of the problem the uncertainties increase. As a result one should use great care in interpreting the results of any model and focus on identifying general insights that are not very sensitive to particular parameter values. In the following we use parameter values chosen for illustrative purposes but consistent with the studies reported above.

For the numerical examples we consider an environment in which the overall probability of a terrorist attack in any period is $\sum \pi_s p_s = .1$, under two threat state scenarios each with three states. In the "No information" scenario the probability of an attack in a state s, $p_s = .1$ for all s. This means that the decision maker does not have the ability to detect activity leading to an attack and by default $s = 1/3$ for all s. In the "Informative states" scenario p_s varies with s, we set $p_s = [.05, .1, .8]$ and $s = [.7, .25, .05]$. We also consider two normalized damage density functions, one representing the case of no information about attack location and one representing the case in which the decision maker has information that makes some set of targets much more likely. The first case we model as a uniform distribution and the second as a beta distribution $\beta(1, 7.5)$ with 80 % of the potential damage concentrated in 20 % of the possible targets.

For each pair of threat state and damage density function we determine the optimal defensive strategy for three values of the deployment cost, R, $[.01C, .1C, .5C]$. The effect of defensive resource deployment is assumed to be $g = .1$. These values of R capture a wide range of deployment costs relative to terror attack damage. As a reference point we compare the cost of deploying over the entire target set in all states with the reduction in expected damage from an attack that the deployment provides, assuming no warnings. With an overall attack probability of .1 the benefit of a full deployment is $(.1)(1-g)C = .09C$. $R = .01C$ represents the case where deployment costs are significantly lower than the benefits they provide. In this case, terror damage costs dominate the objective function and a full deployment will be close to optimal. When $R = .1C$ deployment costs are of a similar order of magnitude to the benefits they provide thus the decision maker must tradeoff terror damage risk and deployment costs. The objective function is relatively flat so small deviations from the optimal deployment will not greatly increase expected costs. When $R = .5C$ deployment costs are high and thus optimization of deployments is most critical.

The warning response function can be characterized in two ways: the effectiveness of warnings at reducing damage from terror attacks and the sensitivity of the response to false alarms. Neither of these is well understood so for the purpose of the numerical examples presented here we consider two cases, A and B defined in Table 2. In Case A warnings tend to be less effective and their effectiveness is degraded significantly at a lower false alarm rate than Case B. We use a discrete representation of the warning response function because it is more realistic to assume that over ranges of false alarm rates there will be little change in people's responses to warnings. Given the lack of research on people's responses to terror warnings and resulting impact on terror attack damage the functional forms chosen here are purely speculative.

Table 2 Warning response functions $\gamma(\phi)$ used in numerical examples

False Alarm Rate ϕ	Warning Effect Case A	Warning Effect Case B
$\phi \leq .2$	.3	.1
$.2 < \phi \leq .5$	.5	.35
$.5 < \phi \leq .85$	.8	.45
$.85 < \phi$	1	1

Table 4 presents the optimal defenses for four different information scenarios with warning response Case A. Within each information scenario (see Table 3) we display the optimal deployment in each state as a triplet on a (0, 1) scale, and the total expected cost for three different deployment cost levels. Information scenario I represents the case in which the least information is available, the damage density function (DDF) is uniform and attack probabilities are constant across all threat states. In scenario II treat states are informative. In scenario III the DDF is more concentrated, i. e., more informative but threat states are not. In scenario IV both the threat states and DDF are informative. The optimal warning strategy for each state is displayed as a triplet of zeroes and ones representing respectively no warning and private warnings. Table 5 displays the same information but with more effective warnings, warning response Case B.

In Table 4 we see that less informative scenarios have higher costs than more informative ones. More precision regarding the location and/or timing of an attack makes more efficient resource allocation possible. When the damage density function is uniform we see that the optimal deployment is full or nothing. The results for the uninformative threat states indicate that even if the decision maker can not distinguish one threat state from another it still makes sense to issue warnings some of the time, but deployments are lower when warnings are issued than when warnings are not issued.

Scenario I in Tables 4 and 5 can be interpreted as a situation in which an intelligence management failure has occurred. Comparing the results of scenario I with scenario II indicates the value of conveying threat state information to those decision makers that control physical deployments. The comparison also indicates how better threat state intelligence can lead to more effective and selective use of warnings. Comparing the results of scenarios I and II with scenarios III and IV we can see how an inability to turn intelligence into operational assessments of types and locations of attacks severely limits the use of physical deployments. For example, in the case of R = .1C it is optimal to use more physical resources when the DDF is concentrated then when it is uniform. The result is that when the DDF is uniform a higher proportion of the costs incurred are from terror than when the DDF is more informative. Comparing scenarios II and IV in Table 5 we see that as the DDF changes the physical deployments can change dramatically but so does the optimal warning policy. These observations highlight the fact that warnings cannot be managed independently of physical deployments.

6 Summary

We have found that the intelligence available to decision makers strongly influences the optimal defensive measures taken. Our analysis also highlights some important political challenges in defending against terrorism. Issuing a warning because it is better than not issuing a warning, even if the government doesn't know the threat is any greater, seems dishonest. If deployment when threat states are high are not sig-

Table 3 Information Scenarios

	Information Scenario I	Information Scenario II	Information Scenario III	Information Scenario IV
Damage density function	Uninformative Uniform(0,1)	Uninformative Uniform(0,1)	Informative Beta(1,7.5)	Informative Beta(1,7.5)
Threat states	Uninformative (I =.53)	Informative (I =.65)	Uninformative (I =.53)	Informative (I =.65)

Table 4 Optimal defenses under different information scenarios (Case A)

	Information Scenario I	Information Scenario II	Information Scenario III	Information Scenario IV
R	Deploy	Deploy	Deploy	Deploy
.01 C	(1.0,1.0,1.0)	(1.0,1.0,1.0)	(.42,.48,.48)	(.42,.42,.58)
.1 C	(0,0,0)	(0,0,1.0)	(.17,.26,.26)	(.07,.17,.40)
.5 C	(0,0,0)	(0,0,0)	(0,.05,.05)	(0,0,.23)
R	Warn	Warn	Warn	Warn
.01 C	(1,0,0)	(0,1,1)	(1,0,0)	(0,1,1)
.1 C	(1,0,0)	(0,1,1)	(1,0,0)	(0,1,1)
.5 C	(1,0,0)	(0,1,1)	(1,0,0)	(0,1,1)
R	Cost	Cost	Cost	Cost
.01 C	.0183C	.0168C	.0136C	.0118C
.1 C	.0833C	.0545C	.0412C	.0354C
.5 C	.0833C	.0675C	.0808C	.0578C

Table 5 Optimal defenses under different information scenarios (Case B)

	Information Scenario I	Information Scenario II	Information Scenario III	Information Scenario IV
R	Deploy	Deploy	Deploy	Deploy
.01 C	(1.0,1.0,1.0)	(1.0,1.0,1.0)	(.48,.41,.41)	(.34,.48,.57)
.1 C	(0,0,0)	(0,0,1.0)	(.26,.6,.16)	(.06,.26,.39)
.5 C	(0,0,0)	(0,0,0)	(.05,0,0)	(0,.05,.22)
R	Warn	Warn	Warn	Warn
.01 C	(0,1,1)	(0,1,1)	(0,1,1)	(1,0,1)
.1 C	(0,1,1)	(1,0,1)	(0,1,1)	(1,0,1)
.5 C	(0,1,1)	(0,1,1)	(0,1,1)	(1,0,1)
R	Cost	Cost	Cost	Cost
.01 C	.0163C	.0158C	.0114C	.0106C
.1 C	.0633C	.0476C	.0361C	.0302C
.5 C	.0633C	.0577C	.0621C	.0496C

nificantly different than when threat states are low, because warnings have boosted alertness, and if this is revealed to the public, then the public may feel the government is not doing enough to defend them. Not deploying any additional physical resources because the attack location uncertainty is large is also politically risky. More importantly the framework developed here can be used as a foundation for investigation of many aspects of counterterrorism.

References

1. Abt, C. C. (2003) The Economic Impact of Nuclear Terrorist Attacks on Freight Transport Systems in an Age of Seaport Vulnerability – Executive Summary. Abt Associates, Cambridge MA.
2. Brillouin, L. (1962) Science and Information Theory. 2nd Edition. Academic Press, Inc. New York.
3. Dudkevitch, M. (2004). Official: Israel facing 57 daily terror attack warnings. The Jerusalem Post, July, 22.
4. Enders, W. and T. Sandler. (1993). The effectiveness of antiterrorism policies: A vector-autoregression intervention analysis. American Political Science Review. 87 829–844.
5. Enders, W. and T. Sandler. (2004). What do we know about the substitution effect in transanational terrorism?. In A. Silke and G. Illardi (eds.) Researching Terrorism: Trends , Achievements, Failures. Frank Cass publishers.
6. Green, K. C. and Armstrong, J. S. (2004). "Value of expertise for forecasting decisions in conflicts" – Monash University Econometrics and Business Statistics Working Paper 27/04
7. Green, K. C. (2002). Forecasting decisions in conflict situations: A comparison of game theory, role-playing, and unaided judgement. International Journal of Forecasting, 18, 321–344.
8. Keohane, N. O., and R. J. Zeckhauser. (2003). The ecology of terror defense. Journal of Risk and Uncertainty. 26 (2) pp. 201–29.
9. Klor, E. and Berrebi, C. (2006). On Terrorism and Electoral Outcomes: Theory and Evidence from the Israeli-Palestinian Conflict. Journal of Conflict Resolution. 50(6) 899–925.
10. Lugar, R. (2005). The Lugar Survey on Proliferation Threats and Responses. U.S. Senate Foreign Relations Committee.
11. Markram, B. (2002) An insoluble Problem? Reactions, pp. 24–30 July.
12. Mileti, D. S., and L. Peek. (2000). The social psychology of public response to warnings of a nuclear power plant accident. Journal of Hazardous Materials. 75 181–194.
13. Myre, G. (2004). With about 50 warnings daily, Israel handles most quietly. The New York Times, August 6.
14. NY Times, (2004). Editorial: The Terror Alerts. The New York Times, August 5.
15. Pate-Cornell, M. E. (1986). Warning Systems in Risk Management. Risk Management. 6 (2) 223–234.
16. Pate-Cornell, M. E.and C. P. Benito-Claudio. (1984). Warning Systems: Response Models and Optimization. Proceedings of the Society for Risk Analysis International Workshop on Uncertainty in Risk Assessment, Risk Management, and Decision Making. pp. 457–468.
17. Pinker, E. (2007). An Analysis of Short-term Responses to Threats of Terrorism. Management Science 53(6) 865–880.
18. Powell, R. (2007). Defending Against Terrorist Attacks with Limited Resources. American Political Science Review (forthcoming).
19. Powell, R. (2007). Defending Against Terrorist Attacks with Private Information about Vulnerability. American Journal of Political Science (forthcoming).
20. Risk Management Solutions. (2005). Managing Terrorism Risk. http://www.rms.com/publications/terrorism_risk_modeling.pdf.

Network Detection Theory

James P. Ferry, Darren Lo, Stephen T. Ahearn, and Aaron M. Phillips

Abstract Despite the breadth of modern network theory, it can be difficult to apply its results to the task of uncovering terrorist networks: the most useful network analyses are often low-tech, link-following approaches. In the traditional military domain, detection theory has a long history of finding stealthy targets such as submarines. We demonstrate how the detection theory framework leads to a variety of network analysis questions. Some solutions to these leverage existing theory; others require novel techniques – but in each case the solutions contribute to a principled methodology for solving network detection problems. This endeavor is difficult, and the work here represents only a beginning. However, the required mathematics is interesting, being the synthesis of two fields with little common history.

1 Introduction

Network theory is a broad field which includes deep mathematics; popular literature; researchers in sociology, computer science, molecular biology, and epidemiology; and data-deluged intelligence analysts. Network theory is particularly relevant to counterterrorism because in the asymmetric war against Al-Qaeda and other terrorist organizations, the enemy's key assets are not concrete military hardware, but rather the covert human network of terrorists and their sympathizers. Naturally, the broad,

James P. Ferry
Metron, Inc., 1818 Library St., Suite 600 Reston, VA 20190, USA, e-mail: ferry@metsci.com

Darren Lo
Metron, Inc., 1818 Library St., Suite 600 Reston, VA 20190, USA, e-mail: lo@metsci.com

Stephen T. Ahearn
Metron, Inc., 1818 Library St., Suite 600 Reston, VA 20190, USA, e-mail: ahearn@metsci.com

Aaron M. Phillips
Metron, Inc., 1818 Library St., Suite 600 Reston, VA 20190, USA, e-mail: phillips@metsci.com

N. Memon et al. (eds.), *Mathematical Methods in Counterterrorism*,
DOI 10.1007/978-3-211-09442-6_10,

vibrant field of modern network theory should be leveraged to expose and exploit these hidden terrorist networks. But how? What specifically does network theory have to offer?

The mathematical study of networks is often traced back to the work on random graphs in the late 1950s, when the Erdős–Rényi random graph model was developed in a series of papers culminating in [1]. This work introduced the idea of *sharp thresholds*: i. e., for various graph properties, a small change in a structural parameter of the graph model can lead to the property going from being almost certain to occur to being almost certain not to. Over the course of 50 years, this work has grown in exciting and unexpected ways. The 1975 Szemerédi regularity lemma [2], for example, made a surprising connection between the structure of large, dense graphs and arithmetic combinatorics. A recent outgrowth of this connection is the proof that there are arbitrarily long arithmetic sequences of primes [3], a result which helped Tao win the 2006 Fields Medal. However, much of this mathematical work, although deep, is still focused only on the Erdős–Rényi model [4, 5], which is a poor model of real-world networks; and the results, correspondingly, are often of interest only to other mathematicians. As interesting as arithmetic sequences of primes may be, they are unlikely to have applications to counterterrorism.

Social network analysis began with the study of networks of human relationships, applying statistical techniques to the field of sociology. A famous early result is the 1967 *small world* experiment of Stanley Milgram [6], which was later popularized as "six degrees of separation." Wasserman and Faust [7] provide a comprehensive summary of the state of the field in 1994. Since then, the field has become more mathematical and rigorous due to an influx of physicists, and has widened in scope to encompass networks arising in other contexts (e. g., genomics and the World Wide Web). Newman's 2003 review article [8] summarizes the state of what had become known as *network science*. A major development during this period was the study of scale-free random graph models, which exhibit the power-law degree distributions that some have claimed are a property of a variety of real-world networks. The first and most famous of these is the preferential attachment model of Barabási and Albert [9]. By this time, the body of interesting insights had been popularized by several authors (e. g., [10, 11]). The activity of the physicists had also gained the attention of mathematicians such as Bollobás [12], who have crossed over from the study of classical Erdős–Rényi random graphs, seeking to raise the level of discourse further still by providing precise definitions of network models and rigorous proofs.

How, then, does network theory help expose hidden terrorist networks? The work of Valdis Krebs [13] provides a good example of what network theory does, and does not, provide. Krebs demonstrated how the 9/11 terrorist network might have been discovered beforehand from available data. He applies standard social network analysis techniques (à la [7]) to study the characteristics of the networks used to plan and execute the 9/11 attacks, noting that they are surprisingly sparse due to the need for secrecy. The later work of the physicists and mathematicians was not needed for this analysis. Indeed, even the social network analysis itself was largely incidental to the problem of *finding* the terrorist network. Krebs used open source material and judicious consideration of what information to use to construct the links in the network. In other words, the key element in detecting the 9/11 network was simply good, classical intelligence analysis, not the vast body of research on networks.

This chapter proposes the following approach to leveraging network analysis to detect terrorist networks: instead of focusing on the word "network," we focus on "detect." We rely on the general framework of *detection theory* to tell us *how* to leverage network theory to detect terrorist networks. This approach was pioneered by Mifflin *et al.* in 2004 [14], and leads to a variety of interesting, difficult mathematical problems. We will give a brief overview, referring the reader to [15] for further details.

Signal detection theory [16] is a staple of traditional military operations: it is employed whenever one needs to determine from raw radar or sonar data whether a target is present. The Neyman–Pearson lemma [17] states that the most powerful test of fixed size for discrimination between two point hypotheses is the likelihood ratio Λ, which in our context is defined by

$$\Lambda(\text{evidence}) = \frac{\Pr(\text{evidence} \mid \text{target present})}{\Pr(\text{evidence} \mid \text{target not present})}. \tag{1}$$

That is, suppose there two processes ("target present" and "target absent") that may have generated some given, observed dataset. The Neyman–Pearson lemma states that there is no procedure better than the following to determine which process generated the observed data. Pick a threshold Λ_0, and declare that the target was present when $\Lambda \geq \Lambda_0$. Choosing Λ_0 involves a trade-off between hit rate (the probability of calling "target present" then the target is indeed present) and false alarm rate (the probability of calling "target present" when it is, in fact, absent). However, if Λ_0 is tuned to some given false alarm rate, the best possible hit rate is achieved by (1).

We now consider a simple problem which illustrates how detection theory may be applied to the network domain. Suppose we are given a graph on n vertices and told that one of two processes generated it. The first process ("target absent") is just the Erdős–Rényi process $\mathscr{G}(n,p)$. An instance of $\mathscr{G}(n,p)$ is generated by instantiating each possible edge (on a set of n vertices) independently and with probability p. The second process ("target present") is the union of an instance of $\mathscr{G}(n,p)$ with the insertion of an isomorphic copy of some graph H – we denote this process $\mathscr{G}_H(n,p)$. The number of possible locations of H on n vertices is denoted $X_H(K_n)$, and we assume that the insertion is equally likely to occur at each location. Mifflin *et al.* prove the following formula for Λ in this case [14]:

$$\Lambda_H(G) \doteq \frac{\Pr(G \mid G \sim \mathscr{G}_H(n,p))}{\Pr(G \mid G \sim \mathscr{G}(n,p))} = \frac{X_H(G)}{\mathbb{E}[X_H(\mathscr{G}(n,p))]}, \tag{2}$$

where $X_H(G)$ is the number of subgraphs of G isomorphic to H, and, for any random graph model $\mathscr{R}$, $\mathbb{E}[X_H(\mathscr{R})]$ is the expected value of $X_H(G)$ given $G \sim \mathscr{R}$. This result has a simple, intuitive appeal. The likelihood ratio is simply the number of copies of H in the evidence graph G divided by the number one would expect from pure noise.

The expected value $\mathbb{E}[X_H(\mathscr{G}(n,p))]$ is simple to calculate. Letting $\mathscr{R}$ be any symmetric random graph process (i. e., one for which isomorphic instances have equal probabilities), we have

$$\mathbb{E}[X_H(\mathscr{R})] = X_H(K_n)\Pr(\mathscr{R} \supseteq H), \tag{3}$$

where $\Pr(\mathscr{R} \supseteq H)$ is the probability that an instance of $\mathscr{R}$ contains any specific isomorphic copy of H [18]. For the Erdős–Rényi case, $\Pr(\mathscr{G}(n,p) \supseteq H) = p^{e(H)}$ because each of the $e(H)$ edges necessary is instantiated independently. The formula for $X_H(K_n)$ is a simple exercise [5], so we have

$$\mathbb{E}[X_H(\mathscr{G}(n,p))] = \binom{n}{v(H)} \frac{v(H)!}{|\operatorname{Aut}(H)|} p^{e(H)} \tag{4}$$

where $|\operatorname{Aut}(H)|$ is the size of the automorphism group of H.

This chapter continues the argument made in [15] for developing a detection theory for networks, extending the results in various ways, and pointing toward future challenges. Although the results (2) and (4) are cited in [15], the insufficiency of the Erdős–Rényi model is recognized there, and more appropriate network models are surveyed. Section 2 generalizes (4) to one such model (random intersection graphs). Section 3 returns to the simple Erdős–Rényi case, but addresses a more difficult issue: getting information about the distribution of the likelihood ratio Λ_H in order to characterize the performance of this optimal detection statistic. These first two sections give an indication of the kind of mathematics that arises in addressing the questions raised by the detection theory methodology. In each case, there is a body of prior literature about the quantities required, but in each case this literature addresses the broad question of the general behavior of the quantity, but never quite gives the specific formulas required. The work in Sects. 2 and 3 is narrowly focused, but has the merit of actually giving specific answers, with leading constants specified.

The remaining three sections are brief discussions of issues that need to be addressed to extend the theory beyond the case of binary detection. In order for network detection theory to be useful, it needs to handle the temporal aspects of the problem: not only the detection of terrorist networks, but tracking them as well. Section 4 contends that the most basic building block of such a theory, a dynamic network model, is almost entirely undeveloped in the literature, and proceeds to define a natural dynamic analog of the Erdős–Rényi process and give a few basic properties of it. Section 5 demonstrates how to perform exact tracking on a simple time-evolving plan obscured by noise, provided one is able to maintain probability distributions over a large state space. Finally, Sect. 6 addresses the issue of combinatorial explosion of the state space necessary for tracking in the network domain, illustrating the approach with a simple example.

2 Random Intersection Graphs

The Erdős–Rényi random graph, while analytically tractable, has several deficiencies as a model of real-world networks, most notably the absence of correlated local structure. Random intersection graphs, first studied by Singer in [19], address some of the deficiencies of the Erdős–Rényi model while still maintaining some degree of tractability. The parameters to the model are a vertex set N and a set M, of respective cardinalities n and m, and a probability $p \in [0,1]$. To each vertex $v \in N$, we associate a random subset of M by independently choosing each element of M with probability p. We then obtain the intersection graph $\mathscr{G}(N,M,p)$ defined by these subsets; that is, vertices $v,w \in N$ are joined by an edge iff their associated subsets are not disjoint. We can interpret this in the context of social networks by positing that people randomly choose activities or group affiliations (these affiliations are latent variables in the model), and that we observe social or transactional links between people who have an affiliation in common. Although the latent affiliations are chosen independently, the edges of the intersection graph are not independent in general.

An equivalent view that appears in the literature, and which we will adopt in the rest of this discussion, is to take the random bipartite graph $\mathscr{B}(N,M,p)$ obtained by independently instantiating each possible edge between N and M with probability p. Given a bipartite graph $B \sim \mathscr{B}(N,M,p)$, we can construct a new graph G on the vertex set N by joining two vertices iff their 1-neighborhoods in B are not disjoint. (In other words, we set $G = B^2[N]$.) The G constructed in this fashion is distributed according to $\mathscr{G}(N,M,p)$. The original graph B is called a generator for G, and G is sometimes called the unipartite projection of B. We will write $G = \pi(B)$ in this case. A useful technique for reasoning about the random intersection graph $\mathscr{G}(N,M,p)$ is to enumerate its possible generators.

Many properties of random intersection graphs have been studied in the literature. Karónski, Scheinerman, and Singer study subgraph containment and thresholding properties [19, 20]. Fill, Scheinerman, and Singer show convergence to the Erdős–Rényi model in a certain class of asymptotic limits [21]. Stark studies degree distribution [22]. Behrisch and Taraz give an efficient heuristic algorithm to construct generators of G which succeeds a.a.s. in certain parameter regimes [23]. Generalizations proposed by Godehardt and Jaworski [24] and Newman, Strogatz, and Watts [25] alter the degree distribution of the generators. Newman *et al.* also compare to characteristics of real-world networks.

To extend our previous network detection results to random intersection graphs, we require exact expressions for the numerator and denominator in Equation (1). Although it is an easy combinatorial exercise to write down these expressions (see Sect. 2.1), computing them is a more formidable challenge. As a first step toward approximating them, we exhibit in Theorem 1 an exact (to first order) expression for $\mathbb{E}[X_H(\mathscr{G}(N,M,p))]$ in a particular asymptotic limit.

2.1 Induced edge clique covers; exact quantities

Let $B \sim \mathscr{B}(N,M,p)$, and let $V \subseteq N$ be the 1-neighborhood of a vertex $w \in M$. Then if $G = \pi(B)$ is the unipartite projection of B, it is clear that $G[V]$ is a complete graph. Thus unipartite projection naturally induces a multiset of m cliques of G (which are not necessarily maximal, distinct, or even nonempty). These m cliques form an edge clique cover of G (for brevity, we will henceforth use "cover" to mean edge clique cover.)

Now let G be an arbitrary graph with vertex set N. Let $B \sim \mathscr{B}(N,M,p)$. Then $\pi(B) \subseteq G$ iff each of the m cliques induced by the projection is contained in a complete subgraph of G. Thus we obtain by an elementary computation

$$\Pr(\pi(B) \subseteq G) = c^*(G;p)^m, \tag{5}$$

where we define

$$c^*(G;x) = \sum_{i=0}^{n} X_{K_i}(G) x^i (1-x)^{n-i}.$$

Other containment probabilities, such as $\Pr(\pi(B) \supseteq G)$, are readily found using Möbius inversion.

The definition of the likelihood ratio statistic Λ_H for $\mathscr{G}(N,M,p)$ is analogous to that in (2) for the $\mathscr{G}(n,p)$ case, but the exact formula for it is more complex:

$$\Lambda_H(G) = \frac{\sum_{\substack{H' \subseteq G \\ H' \cong H}} \sum_{\substack{L \subseteq G \\ L \supseteq H'}} (-1)^{e(G)-e(L)} c^*(L;p)^m}{X_H(K_n) \sum_{L \subseteq G} (-1)^{e(G)-e(L)} c^*(L;p)^m}, \tag{6}$$

where all summation indices are restricted to graphs on the full vertex set N.

One unfortunate feature of the above expressions is that they are computationally intractable for problems of any reasonable size. For example, computation of $X_{K_i}(G)$ is NP-hard. Thus any practical application of these ideas will require development of approximations that are easier to handle.

2.2 Expected subgraph counts in the constant–μ limit

As a first step toward approximating the quantities in Sect. 2.1, we propose to compute an exact first-order expression for $\mathbb{E}[X_H(\mathscr{G}(N,M,p))]$ under some suitable asymptotic limit. One's first try might be to study the limit where $m,n \to \infty$, the ratio m/n is held constant, and p is adjusted such that the expected probability that any two vertices are connected by an edge is held constant. For the purposes of computing the expectation of $X_H(\mathscr{G}(N,M,p))$ for fixed H, this limit is equivalent to holding $N = V(H)$ fixed and letting $m \to \infty$, with p as described above. However, under this limit, the random intersection graph turns out to converge to the Erdős–Rényi graph

$\mathscr{G}(n,p)$ – specifically, the total variation distance between the two distributions goes to 0. This is a special case of the main result in [21].

An alternative limit that seems worthy of study lets $m,n \longrightarrow \infty$ with m/n constant and $\mu = mp$ constant. A picturesque interpretation of this limit is that we let the number of people and groups go to infinity, while holding constant the expected number of group memberships per person and the expected group size. We call this the constant-μ limit. This limit is particularly tractable and is studied in, for instance, [25]. There is also an extensive discussion of a more general version of this limit in [19] and [20], where thresholding functions are obtained for induced subgraph appearance and disappearance. (Our present notion of "subgraph" does not require that it be an induced subgraph.) However, as Singer observes, the calculations are sometimes laborious, and constant factors are ignored since they are not relevant to the thresholding analysis. The calculations for non-induced subgraphs are additionally burdensome, since they involve calculating thresholds for all possible supergraphs.

Our contribution here is to relieve the computational burden for a specialized subproblem: we show that the asymptotic leading-order term of the expected (non-induced) subgraph count $\mathbb{E}[X_H(\mathscr{G}(N,M,p))]$ is obtainable exactly by a simple calculation, in the case of the constant-μ limit. For the purposes of the present discussion, it suffices to compute $\Pr(\mathscr{G}(N,M,p) \supseteq H)$, where we consider H as a fixed labeled graph on the fixed vertex set N, and we hold μ constant as $m \longrightarrow \infty$. Computation of $\mathbb{E}[X_H(\mathscr{G}(N,M,p))]$ in the general case where n increases is then immediate from the linearity of expectation.

To state our result, we first need to define some terms. Recall that a graph is called biconnected iff it is connected and has no cut vertices. For an arbitrary graph G, there is a unique decomposition of G into maximal biconnected subgraphs, which are called the blocks of G. If two blocks of G are not disjoint, then they meet at exactly one vertex v, which will be a cut vertex of G. We then define the block degree of v to be the number of blocks that contain v.

Theorem 1. *Let G be a fixed graph on the fixed vertex set N, and let n' be the number of non-isolated vertices. Let $\{v_i\}_{i=1}^k$ be an enumeration of the cut vertices of G, and let $\{b_i\}_{i=1}^k$ be the corresponding block degrees. Let $m \longrightarrow \infty$ with $p = \mu/m$, where μ is a constant. Then we have*

$$\Pr(\mathscr{G}(N,M,p) \supseteq G) = \frac{\mu^{n'-k} \prod_{i=1}^k \phi_{b_i}(\mu)}{m^{\operatorname{rank} G}} \left(1 + O(m^{-1})\right), \tag{7}$$

where $\phi_b(x) = \sum_{j=1}^b \left\{ {b \atop j} \right\} x^j$.

In contrast to the intractable expressions in 2.1, all quantities in Theorem 1 are readily computable: the rank of G and its block degree sequence can both be found in linear time, and can indeed be readily found by inspection for small G. For example, a toy application of the theorem is to compute the transitivity of the random intersection graph in the constant-μ limit. We obtain

$$T = \frac{\Pr(\mathscr{G}(N,M,p) \supseteq K_3)}{\Pr(\mathscr{G}(N,M,p) \supseteq P_2)} = \frac{\mu^3/m^2}{\mu^2(\mu+\mu^2)/m^2}\left(1+O(m^{-1})\right) = \frac{1}{1+\mu}\left(1+O(m^{-1})\right), \tag{8}$$

which agrees with Equation (88) of [25].

Although the complete proof of Theorem 1 is too long to present here, we will sketch the argument to give some justification for the appearance of rank G and the Stirling numbers of the second kind. For simplicity, let us assume that G has no nonisolated vertices (these do not materially change the analysis, but are a minor nuisance to handle).

We tackle the proof by enumerating all generators of G. Although this is difficult to do exactly, it is easier if we perform an approximate enumeration that is correct up to first order. To this end, it turns out to be convenient to classify generators by the covers of G that they induce. In order for the enumeration to make sense, we restrict to irreducible covers, by which we mean a cover of G with no proper subcover. Now, if $\mathscr{C}$ is some irreducible cover of G, we would like to enumerate generators B_i of G, such that the covers they induce have $\mathscr{C}$ as a subcover. To first order, the probability that $B \sim \mathscr{B}(N,M,p)$ will have $\mathscr{C}$ as an irreducible subcover is $\mu^{s(\mathscr{C})}/m^{\mathrm{wt}(\mathscr{C})}$, where we define $s(\mathscr{C}) = \sum_{C\in\mathscr{C}} v(C)$ and the "weight" of the cover as $\mathrm{wt}(\mathscr{C}) = \sum_{C\in\mathscr{C}}(v(C)-1)$. It is therefore clear that we need only consider covers $\mathscr{C}$ having the least possible weight. By summing over all least-weight irreducible covers of G, we obtain the asymptotic probability $\Pr(\mathscr{G}(N,M,p) = G)$. The development to this point is the same as that given in [19] and [20], which can be consulted for further details.

For a graph G, define its weight $\mathrm{wt}(G)$ to be the minimal weight of an irreducible cover of G. We then show that $\mathrm{wt}(G) \geq \mathrm{rank}\, G$, and that equality is achieved iff every block of G is a complete graph. To compute the quantity of interest, $\Pr(\mathscr{G}(N,M,p) \supseteq G)$, it therefore suffices to enumerate all supergraphs $G' \supseteq G$, such that $V(G') = V(G)$ and $\mathrm{wt}(G') = \mathrm{rank}\, G' = \mathrm{rank}\, G$ (this is the best possible, since adding edges to G cannot decrease the rank). This enumeration can be achieved by the following procedure:

1. List all the cut vertices $\{v_i\}_{i=1}^k$ of G, in any order.
2. Consider the set S of blocks of G that contain the cut vertex v_1. Choose some partition of S. For each class $\{B_1, B_2, \ldots, B_j\} \subseteq S$, form the complete graph on $\bigcup_{i=1}^j V(B_i)$. Take the union of all the resulting complete graphs with G to get a supergraph $G_1 \supseteq G$. This operation preserves the block degrees of all other cut vertices of G, which are now also cut vertices of G_1.
3. Repeat step 2 using the cut vertex v_2 and the graph G_1 (instead of v_1 and G, respectively) to get a new supergraph G_2. Continue in similar fashion to get a sequence of supergraphs $G_1, G_2, G_3, \ldots, G_k$.

All supergraphs $G' \supseteq G$ with $\mathrm{wt}(G') = \mathrm{rank}\, G' = \mathrm{rank}\, G$ can be obtained as G_k from this procedure, using a suitable choice of partitions. Furthermore, the set of final supergraphs G_k is in bijective correspondence with the set of all possible choices of the k partitions used in step 2.

To complete the proof, it remains to compute the probability $\Pr(\mathscr{G}(N,M,p) = G_k)$; our desired result will then follow by summing over all possible G_k. Since each G_k has a unique irreducible cover $\mathscr{C}$ (given by the set of maximal cliques of G_k), and since we know $\mathrm{wt}(\mathscr{C}) = \mathrm{rank}\, G$ by construction, it suffices to determine $s(\mathscr{C})$. We find that $s(\mathscr{C}) = n - k + \sum_{i=1}^{k} j_i$, where j_i is the number of classes in the ith partition chosen in step 2. This, then, is the reason for the appearance of the Stirling numbers $\left\{ {b \atop j} \right\}$ in the theorem. With this observation and some routine bookkeeping, we are finished.

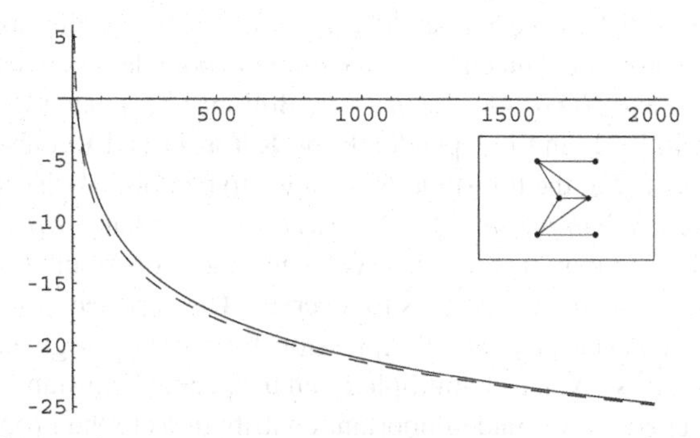

Fig. 1 Plot of $\ln\left[\Pr(\mathscr{G}(V(G),M,\mu/m) \supseteq G)\right]$ versus m, where the graph G is as displayed in the inset box, and where $\mu = 5$. The solid line is the exact probability, whereas the dashed line is the asymptotic approximation.

3 Subgraph Count Variance

The formula (2) for the likelihood ratio $\Lambda_H(G)$ gives no indication of how this statistic behaves. For example, if we assume *a priori* that evidence graphs are equally likely to be generated by $\mathscr{G}(n,p)$ and $\mathscr{G}_H(n,p)$, what is the (posterior) probability that a given graph G was generated by $\mathscr{G}_H(n,p)$ given $\Lambda_H(G) = 2$? To answer questions like this one, we need to know the distribution of Λ_H under both processes, $\mathscr{G}(n,p)$ and $\mathscr{G}_H(n,p)$. From (2) we have, trivially, that $\mathbb{E}[\Lambda_H(G) \mid G \sim \mathscr{G}(n,p)] = 1$, but it is unclear what the variance of Λ_H is under $\mathscr{G}(n,p)$, let alone what the entire distribution is. For the process $\mathscr{G}_H(n,p)$, it is not even immediately obvious what the *mean* of $\Lambda_H(G)$ is.

It is easy to show [26] that $\mathbb{E}[\Lambda_H(G) \mid G \sim \mathscr{G}_H(n,p)] = 1 + \mathrm{var}[X_H]/\mathbb{E}[X_H]^2$ (all statistics being for $\mathscr{G}(n,p)$ for the remainder of the section unless otherwise indicated). Thus, the computation of $\mathrm{var}[X_H]$ provides a partial solution to the full problem, yielding the first moment of $\Lambda_H(G)$ for $G \sim \mathscr{G}_H(n,p)$, as well as the second moment of $\Lambda_H(G)$ for $G \sim \mathscr{G}(n,p)$. Unfortunately, even computing $\mathrm{var}[X_H]$ is a difficult problem. There are various theorems that address this [4, 5], which are analogous to those of Singer *et al.* from Sect. 2 in the following sense: they give the broad outlines of the behavior in all the various asymptotic limits, and for the special cases which are particularly nicely behaved. These formulas typically specify an asymptotic growth rate for statistics of X_H as $n \to \infty$, but with an unspecified leading constant. The results here, as in Sect. 2, specify this constant, revealing interesting structural details of the problem as a by-product of the analysis.

A key quantity in studying the statistics of X_H (in the Erdős–Rényi case) is the *maximum density* $m(H)$ of a graph H, defined by $m(H) = \max\{e(F)/v(F) \mid F \subseteq H\}$. A graph H is *balanced* if $m(H) = e(H)/v(H)$, and is, furthermore, *strictly balanced* if the only subset $F \subseteq H$ for which $e(F)/v(F) = m(H)$ is $F = H$. Erdős and Rényi [1] established the threshold $p(n) = n^{-1/m(H)}$ for the appearance of a copy of H, in the case of balanced H, delineating the boundary between the regimes where $\mathbb{E}[X_H] \to 0$ and $\mathbb{E}[X_H] \to \infty$ as $n \to \infty$. Bollobás [27] extended this result to all graphs. Also in [27], and independently by Karoński and Ruciński in [28], the number of copies of H at the threshold was shown to be Poisson distributed when H is strictly balanced. When $p(n)n^{1/m(H)} \to \infty$ as $n \to \infty$ (so that $\mathbb{E}[X_H] \to \infty$), one can use Stein's method to demonstrate that X_H is normally distributed [29] (provided $p(n)$ is not so large that $n^2(1-p)$ fails to diverge). The variance of this distribution is not specified however. A general expression for $\mathrm{var}[X_H]$ is given in Janson *et al.* [4], but, again, it is given as a multiple of an unspecified constant.

All the work cited above made important contributions to the problem of understanding the distribution of X_H under a wide range of conditions, though none of it results in an approximation to $\mathrm{var}[X_H]$ except for very special forms of H. We will now outline a formula for $\mathrm{var}[X_H]$ which includes an explicit expression for the constant in the leading order term as $n \to \infty$, and applies to any connected graph H. The trade-off for encompassing this generality in H is that the result applies only for a fairly sparse regime of asymptotic behaviors of $p(n)$, only slightly less sparse than $p = O(1/n)$ (see [26] for details).

Restricting the range of applicability to $p = O(1/n)$ for the presentation here allows us to simplify the formula for $\mathrm{var}[X_H]$ given in [26] to

$$\mathrm{var}[X_H] = \mathbb{E}[X_H]\hat{M}(H,np)(1+o(1)). \tag{9}$$

To derive an expression for the generating function $\hat{M}(H,x)$ we consider a fixed embedding of H on n vertices, and all possible embeddings of a graph H' isomorphic to H on the same vertex set. For $p = O(1/n)$, and for all connected graphs H, the leading contribution to $\mathrm{var}[X_H]$ comes from those graphs $H' \cong H$ which have at least one vertex in common with H, and for which $H' \backslash H$ is a forest (i. e., a union of disjoint trees). This condition decomposes the problem into two cases: (1) H is not a tree; and (2) H is a tree.

In the first case, in order for $H' \backslash H$ to be a forest, it is necessary that $H' \cap H$ contain the 2-core of H, or, equivalently, $\mathrm{cr}(H) = \mathrm{cr}(H')$ (where the k-core of a graph is its maximal k-connected subgraph). For this reason, we consider a decomposition of H into its 2-core and the rooted trees at each vertex of $\mathrm{cr}(H)$ (where the root of each tree is the vertex contained in $\mathrm{cr}(H)$). Two rooted trees are isomorphic if there is a root-preserving graph isomorphism between them. We color each vertex of $\mathrm{cr}(H)$ by the isomorphism class of the (possibly trivial) rooted tree there. Now consider the automorphism group $\mathrm{Aut}(H)$ of H, and the subgroup of $\mathrm{Aut}(H)$ comprising the color-preserving automorphisms. We let S denote an arbitrary set of representatives of the left cosets of $\mathrm{Aut}(H)$ modulo its color-preserving subgroup. Then we find the following formula for $\hat{M}(H,x)$:

$$\hat{M}(H,x) = \sum_{\pi \in S} \prod_{i \in V(\mathrm{cr}(H))} B(T_i, T_{\pi(i)}; x), \tag{10}$$

where T_i is the tree rooted at i, and $B(T,T';x)$ is the rooted tree overlay polynomial defined by

$$B(T,T';x) = \sum_{T_1 \subseteq T, T_1' \subseteq T'} \frac{|\mathrm{Iso}(T_1,T_1')|}{|\mathrm{Aut}(T')|} x^{v(T')-v(T_1')}. \tag{11}$$

Here, the sum is over all rooted subtrees T_1 of T sharing the same root as T (and similarly for T_1'), and $\mathrm{Iso}(T_1,T_1')$ is the number of (rooted-tree) isomorphisms from T_1 to T_1'.

In the second case, where H is a tree, $\mathrm{cr}(H)$ is empty, so there is no particular subgraph of H which must be shared with H'. In this case, we may essentially pick any vertices i and j in $V(H)$ as roots for computing the rooted tree overlay polynomial $B(T_i,T_j;x)$. This results in over-counting, but in a orderly, removable manner. The correct form of $\hat{M}(H,x)$ in the tree case is

$$\hat{M}(H,x) = \left(x^k \to \frac{x^k}{v(H)-k} \right) \sum_{[i],[j]} |[i]| B(H_i,H_j;x), \tag{12}$$

where: $[i]$ is the equivalence class of vertices in $V(H)$ defined by $i_1 \cong i_2$ if there is an automorphism of H sending i_1 to i_2; H_i is the tree H rooted at vertex i; and $x^k \to x^k/(v(H)-k)$ is a polynomial transformation which divides the coefficient of x^k by $v(H)-k$ for each k.

An efficient recursive algorithm for computing $B(T,T';x)$ is given in [26]. With this algorithm, the formulas (10) and (12) for $\hat{M}(H,x)$ provide an efficient method for computing the correct asymptotic approximation to $\mathrm{var}[X_H]$ for any connected graph H, in a certain sparse limit of $p(n)$. Although this result and those of Sect. 2 are narrow in scope, they indicate the nature of the mathematical research necessary to answer the questions raised by the inserted subgraph binary detection problem. We now turn to broader questions of detection and tracking on networks.

4 Dynamic Random Graphs

Generalizing the network detection problems in the previous sections to network tracking problems requires a theory of dynamic random graphs. Unfortunately, there are few results in this area. There is work on dynamic behavior on random graphs (e. g., [30]): non-linear systems of differential equations coupled along edges, for example. There is work on geometric random graphs when the underlying vertices obey some motion model. This work is typically motivated by the study of networks of mobile sensors (e. g., [31]). There is also a body of literature on what appear to be dynamic generalizations of the Barabási–Albert preferential attachment model [9], notably [12], but here the dynamics are simply a means of constructing a static, asymptotically large graph with interesting properties; the interest is in the end result of the process, not the process itself.

To embark on a general theory of dynamic random graphs, we first ask, "What is the dynamic analog of the Erdős–Rényi random graph model $\mathcal{G}(n,p)$?" No one appears to have defined such a thing, but there is a natural way to do so. We require that each potential edge location turn on and off independently and according to the same process. The only question, then, is what process should govern the dynamics of a single edge. We believe that a continuous-time edge process is more interesting and useful than a discrete one. Given this, there is a single natural candidate for the edge process: the only (0,1)-valued continuous-time Markov process, which is known as the *telegraph process* [32].

4.1 The telegraph process

Let $\mathsf{T}(\lambda,\mu)$ denote the telegraph process with rate parameters λ and μ. It is defined by stipulating that the wait times to transition from state 1 to 0, and from 0 to 1, are exponentially distributed with means $1/\lambda$ and $1/\mu$, respectively. The steady-state probability of being on is $p=\mu/(\lambda+\mu)$, and knowledge of initial conditions is lost on a time scale $\tau=1/(\lambda+\mu)$. I.e., if $P(t)$ denotes the probability of being on at time t, then $P(t)$ is governed by

$$\frac{d}{dt}P(t) = -\lambda P(t)+\mu\big(1-P(t)\big) = \frac{p-P(t)}{\tau}. \tag{13}$$

We also let $\tilde{\mathsf{T}}(p,\tau)$ denote an alternative parameterization of the telegraph process. To denote the distribution of a process at a given time we use notation such as $\{X_t \mid t\in\mathbb{R}\} = \tilde{\mathsf{T}}(p,\tau)$. In this case, each $X_t \sim \mathscr{B}(p)$, where here $\mathscr{B}(p)$ denotes the Bernoulli distribution.

4.2 The dynamic Erdős–Rényi process

We can use the telegraph process to define $\mathsf{G}(n,p,\tau_E)$, a simple, stochastic analog of the Erdős–Rényi random graph model on a fixed set of n vertices: simply let each possible edge on n vertices be on or off according to independent $\tilde{\mathsf{T}}(p,\tau_E)$ processes. Letting $\{G_t \mid t \in \mathbb{R}\} = \mathsf{G}(n,p,\tau_E)$, each $G_t \sim \mathscr{G}(n,p)$.

To obtain the natural analog of $\mathscr{G}(n,p)$ on a dynamic vertex set, we let V^+ be a large, fixed set of n^+ vertices, each of which obeys a telegraph process $\tilde{\mathsf{T}}(\bar{n}/n^+,\tau_V)$ to determine whether to include it in the smaller, time-varying vertex set V. We then define the stochastic process $\mathsf{G}_{n^+}(\bar{n},p,\tau_E,\tau_V)$ by taking induced subgraphs of $\mathsf{G}(n^+,p,\tau_E)$ on V. The limiting case of $\mathsf{G}_{n^+}(\bar{n},p,\tau_E,\tau_V)$ as $n^+ \to \infty$ is well defined, and we denote it $\mathsf{G}(\bar{n},p,\tau_E,\tau_V)$. Letting $\{G_t \mid t \in \mathbb{R}\} = \mathsf{G}(\bar{n},p,\tau_E,\tau_V)$, each $G_t \sim \mathscr{G}(\mathscr{P}(\bar{n}),p)$, where $\mathscr{P}(\bar{n})$ is a Poisson distribution with mean $\bar{n}$.

There are a variety of questions one might ask about the processes $\mathsf{G}(n,p,\tau_E)$ and $\mathsf{G}(\bar{n},p,\tau_E,\tau_V)$. One class of questions involves structures that persist for a duration T. To address such questions, we define $\mathscr{G}_T^\cap(n,p,\tau_E)$ to be the intersection of an instance of $\mathsf{G}(n,p,\tau_E)$ over a time interval of duration T, and $\mathscr{G}_T^\cup(n,p,\tau_E)$ to be the union. We define $\mathscr{G}_T^\cap(\bar{n},p,\tau_E,\tau_V)$ and $\mathscr{G}_T^\cup(\bar{n},p,\tau_E,\tau_V)$ similarly. The following identities are then easy to verify:

$$\mathscr{G}_T^\cap(n,p,\tau_E) = \mathscr{G}(n,p_\cap), \qquad p_\cap = pe^{-(1-p)T/\tau_E}, \tag{14}$$

$$\mathscr{G}_T^\cup(n,p,\tau_E) = \mathscr{G}(n,p_\cup), \qquad p_\cup = 1-(1-p)e^{-pT/\tau_E}, \tag{15}$$

$$\mathscr{G}_T^\cap(\bar{n},p,\tau_E,\tau_V) = \mathscr{G}(\mathscr{P}(\bar{n}_\cap),p_\cap), \quad \bar{n}_\cap = e^{-T/\tau_V}\bar{n}. \tag{16}$$

However, $\mathscr{G}_T^\cup(\bar{n},p,\tau_E,\tau_V)$ is not an Erdős–Rényi graph because the existence of an edge makes neighboring edges more likely.

A classic result of Erdős is that $p^* = (\log n)/n$ is the (sharp) threshold for an instance of $\mathscr{G}(n,p)$ to be connected. We may combine this with (16), for example, to show that the threshold for $\mathsf{G}(\bar{n},p,\tau_E,\tau_V)$ containing a spanning tree which persists over a given interval of duration T is

$$p_T^* = \frac{\log \bar{n}}{\bar{n}} e^{T(\tau_E^{-1}+\tau_V^{-1})}. \tag{17}$$

One application for continuous-time dynamic random graph models is to provide a more realistic, time-correlated network noise structure for studies like the one in Sect. 5, which currently employs a discrete-time model in which noise is temporally uncorrelated.

5 Tracking on Networks

Communication networks can contain evidence of threatening behavior hidden by an enormous number of benign activities. Thus these, and similar, networks are an example of a classic low signal-to-noise situation. Likelihood Ratio Detection and Tracking (LRDT) [33] is a method to track physical objects such as submarines and swimmers in very low signal-to-noise environments. Building on the results in [15] we extend this traditional LRDT method to detect and track sub-networks within a network.

Consider the following simple but illustrative example: The leader of a terrorist cell contacts two henchmen, ordering them to carry out a task. The completion of the task requires the two henchmen to communicate with each other. When the task is complete the two henchmen both report back to the leader. Between each of these communication phases are periods when no communications are made. Representing the members of the terrorist cell as vertices and the communications between them as edges, this scenario is depicted in Fig. 2 as a sequence of subgraphs in a communication network.

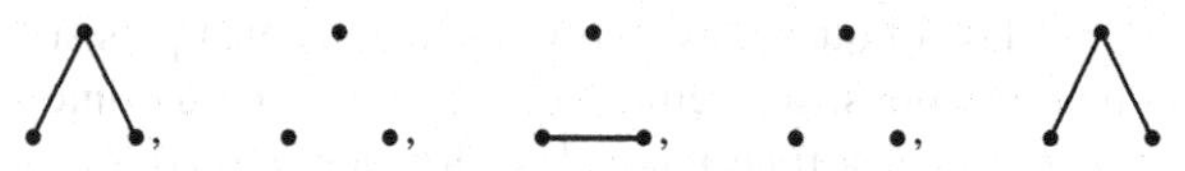

Fig. 2 A simple example of communication within a terrorist cell.

When obscured by other activity, identifying the stages of this scenario even in a small communication network is extremely difficult. However, the LRDT framework allows one to exploit the weak signal of persistent communication to not only identify the members of the terrorist cell but to also identify which stage of their plans they are currently executing.

5.1 The LRDT Framework for Static Networks

The LRDT framework consists of four parts: the state space, the motion model, the measurement model, and the update equation. Since our goal is to identify both the cell members and to track the evolution of their activities through time we define a state space which maintains both the possible cell members and the possible stages of their activities. We begin with a diffuse prior probability distribution on the state space. The motion model captures the natural evolution of the activities due to the progression of time. The measurement model quantifies the likelihood of the evidence given the states in the state space. Finally, the update equation combines these aspects of our model to update the probability distribution on the states.

5.1.1 The State Space

The states in the state space must track both the members of the terrorist cell and the phase of execution of their activities. Thus we define the state space to be $\mathbf{X} =$

$\{(\tilde{H},\tau)\}_{\tilde{H},\tau} \cup \{\varnothing\}$ where $\tilde{H}$ is a choice of cell members and their leader (that is, $\tilde{H}$ is a choice of embedding of the sequence $\{H_k\}$ of stages); τ is the evolution in time of the cell's activities; and $\varnothing$ is the null state, corresponding to the absence of a terrorist cell in the network. The time is $1 \leq \tau \leq T$ for some fixed T when the activities are ongoing; we set $\tau = \alpha$ when the cell has not yet started its activities; and we set $\tau = \omega$ when the cell has completed its activities. In total, there are $T+2$ time states. Denote the number of possible terrorist cells (including the choice of leader) by $X_{\{H_k\}}(K_n)$. Thus, the number of states in $\mathbf{X}$ is $(T+2)X_{\{H_k\}}(K_n)+1$. We assume a diffuse prior probability distribution on $\mathbf{X}$. That is, we define the prior probability distribution on $\mathbf{X}$ to be

$$P^0(x) = \begin{cases} p^0_\varnothing & x = \varnothing, \\ \frac{p^0_\alpha}{X_{\{H_k\}}(K_n)} & x = (\tilde{H},\alpha), \\ \frac{p^0_T}{X_{\{H_k\}}(K_n)\cdot T} & x = (\tilde{H},\tau), 1 \leq \tau \leq T, \\ 0 & x = (\tilde{H},\omega). \end{cases}$$

The constants $p^0_\varnothing$, p^0_α and p^0_T are the prior probabilities of the null state, the probability that the terrorists' activities have not started, and the probability that the terrorists are active, respectively. We assume that there is no probability that the terrorists have already completed their tasks as this state is the same as the null state as far as the data are concerned. These probabilities must be chosen so that $p^0_\varnothing + p^0_\alpha + p^0_T = 1$.

5.1.2 The Motion Model

Absent of any interference, the scenario will evolve over time. We model this evolution as follows. Let $P(x,t)$ denote the probability of state $x \in \mathbf{X}$ at time t. We denote by $P^-(x,t)$ the updated probability of the state x due to the passage of time from $t-1$ to t. We assume that a non-null state cannot evolve into the null state and vice versa. Thus, $P^-(\varnothing,t) = P^-(\varnothing,t-1)$. The case of a non-null state is more complicated. We define the motion updated probability of non-null states to be

$$P^-((\tilde{H},\tau),t) = \begin{cases} (1-S(t))P((\tilde{H},\alpha),t-1) & \tau = \alpha, \\ S(t)P((\tilde{H},\alpha),t-1) & \tau = 1, \\ P((\tilde{H},\tau-1),t-1) & \tau = 2,\ldots,T, \\ P((\tilde{H},T),t-1) + P((\tilde{H},\omega),t-1) & \tau = \omega. \end{cases}$$

The $S(t)$ in the first two expressions of (27) is the probability that the terrorists begin their activities at the next time step. Two reasonable choices for the distribution of $S(t)$ are a uniform distribution and an exponential distribution. The third expression in (27) indicates that the terrorists' plan progresses linearly with time. So P^- for $(\tilde{H},\tau)$ depends on the prior probability for $(\tilde{H},\tau-1)$. To understand the fourth expression in (27), observe that the terrorists could have finished previously,

$\tau = \omega$, or they could be in the final time step of their plan. Thus, P^- for $(\tilde{H}, \omega)$ depends on both the prior probability for $(\tilde{H}, T)$ and the prior probability for $(\tilde{H}, \omega)$.

For ease of understanding we have assumed a deterministic evolution of the terrorist cell's activities; i. e. once the cell begins its activities it proceeds in a deterministic fashion to completion. The motion model described by (27) can be generalized to the situation where the actions of the terrorist cell evolve in a non-deterministic fashion.

5.1.3 Measurement Model

The measurement space $\mathbf{Z}$ is the set of all graphs J on n vertices. We assume that the noise in the underlying network is drawn from the Erdős–Rényi model independently at each time step. Additionally, we assume that, due to measurement error or efforts to evade detection, the individual communications between members of the terrorist cell are not always visible in the evidence graphs, even when the cell is active. Let p_{H_k} denote the probability that an edge (communication) of a stage H_k is observed. The probability that a communication is observed is independent of observations on other edges of H_k.

Assuming $\mathscr{G}(n, p_{ER})$ for the noise in the underlying network, the measurement likelihood function for measurements arising from the null state is $L(J \mid \varnothing) = p_{ER}^{e(J)} q_{ER}^{N-e(J)}$. Here p_{ER} is the probability that an individual edge is present and $q_{ER} = 1 - p_{ER}$ is the probability that an individual edge is not present. The number of edges in J is $e(J)$ and the number of possible edges on the n vertices is $N = n(n-1)/2$.

The likelihood function for the non-null states depends on the cell members $\tilde{H}$ and the stage of execution τ. Let the scenario be in stage H_k at time τ and let $\tilde{H}_k$ be the embedding $\tilde{H}$ restricted to the subgraph H_k. For an edge of $\tilde{H}_k$ to fail to appear, both the noise process and the signal process (i. e., the plan) must fail to produce the edge. Thus, the probability q_* of an edge not appearing is $q_* = q_{ER} q_{H_k}$, the product of the probability q_{H_k} that the plan fails to produce the edge and the probability q_{ER} that the noise process fails to produce the edge. The measurement process may therefore be viewed as two random graph processes: one operating on the edges of $J \cap \tilde{H}_k$ with edge instantiation probability $p_* = 1 - q_*$ and one operating on the edges in $J \backslash \tilde{H}_k$ with edge instantiation probability p_{ER}. With this notation, the likelihood function for measurements arising from the state $(\tilde{H}, \tau)$ is $L(J \mid (\tilde{H}, \tau)) = p_{ER}^{e(J \backslash \tilde{H}_k)} q_{ER}^{N-e(J \cup \tilde{H}_k)} p_*^{e(J \cap \tilde{H}_k)} q_*^{e(\tilde{H}_k \backslash J)}$. This expression may be rewritten as $L(J \mid (\tilde{H}, \tau)) = p_{ER}^{e(J)} q_{ER}^{N-e(J)} p_{H_k}^{e(\tilde{H}_k)} r_*^{e(J \cap \tilde{H}_k)}$ where $r_* = 1 + \frac{p_{H_k}}{p_{ER} q_{H_k}}$.

5.1.4 The Update Equation

Given a motion updated probability distribution P^- on the state space $\mathbf{X}$ and an evidence graph J_t observed at time t, we update the probability distribution by taking the product of the motion updated probability $P^-(x,t)$ of the state x and the measurement likelihood $L(J_t\,|\,x)$ of x. This product is then normalized to obtain the probability $P(x,t)$ of state x at time t: $P(x,t) = \frac{1}{C}L(J_t\,|\,x)P^-(x,t)$ where C is the normalization factor $C = \sum_{x' \in \mathbf{X}} L(J_t\,|\,x')P^-(x',t)$. Note that x can be $\varnothing$ or $(\tilde{H},\tau)$.

The significance of this approach can be appreciated when the results it yields on simulated data are compared to *ad hoc* attempts to solve the same problem. Naturally, the *ad hoc* methods perform worse: they are being compared to the optimal method. What is striking, however, is how much worse they perform – it is difficult to produce an *ad hoc* method which yields even marginally acceptable performance on problems that the optimal method solves easily. One reason for this is that the optimal method has the luxury of full representation of the underlying state space. But as the problem size grows, the size of the state space grows rapidly. Section 6 investigates a method for coping with this problem.

6 Hierarchical Hypothesis Management

In the previous section, we described an application of the LRDT framework in a network setting. A large class of problems can be formulated in similar terms. In such a formulation, a random graph process $\mathscr{R}$ is used to model background noise, and a particular target subgraph or sequence of subgraphs is inserted according to some signal model. Depending on the precise formulation of the problem, the state space roughly consists of all possible configurations of the target. For example, in the previous section, the state space consisted of all possible terrorist cells, with a designated leader, in a network on n vertices, together with all possible time states of an inserted communication plan, and the null hypothesis, i. e., no terrorist cell. While this is a simple scenario, it serves to show the applicability of LRDT to the transactional domain.

One of the notable features of any such problem is the combinatorial nature of the state space, and in particular the finite but (from a computational point of view) potentially vast number of possible target configurations. In the communication plan example, recall that there were $|\mathbf{X}| = (T+2)X_{\{H_k\}}(K_n)+1$ possible target configurations, where T is the total number of time steps in the plan and $X_{\{H_k\}}(K_n)$ is the number of possible terrorist cells with a designated leader. For example, with $n = 30$ and $T = 40$ we obtain $|\mathbf{X}| = 511560$, and with $n = 500$, $T = 40$, we get $|\mathbf{X}| \approx 2.6 \times 10^9$. This is already a difficult number of hypotheses to handle, and the growth in this case is only cubic becase there are only three members of the terrorist cell. In general, maintaining a complete probability distribution over all states is manageable only when the number of vertices is relatively small, and quickly becomes problematic as n and/or $|H_k|$ increases. We refer to this phenomenon as the "combinatorial explosion" of the state space. The combinatorial explosion calls into question the practicality of using LRDT in the transactional domain. Is the approach only applicable for small n, or can it be extended to massive transactional data, where, for example, each vertex represents one of thousands of individuals?

The overwhelming size of the LRDT state space is not new. In the application of LRDT to the tracking of physical targets, the state space is of course infinite, at least in theory, since the possible states are all possible physical configurations of the target. In practice, some discrete approximation of the state space must be used to implement the algorithm, but the number of possible states is still vast. The physical application of LRDT, however, can make use of the topology or metric structure of the state space, for example, through the use of a particle filter. The combinatorial state space of the transactional domain does not easily lend itself to the same techniques.

In this section we will discuss some preliminary attempts to manage the computational complexity of LRDT in the transactional domain. We describe what we mean by "Hierarchical Hypothesis Management (HHM)," which is a loosely-defined paradigm for reducing the complexity of the state space by approximating it at various levels of fidelity. We also give an example of a particular scenario or "challenge problem" which was designed to test the feasibility of HHM, and our implementation of HHM in this example.

6.1 The Hypothesis Lattice

Consider a combinatorial state space $\mathbf{X}$. The HHM paradigm consists of two parts: A lattice of hypotheses which represent the state space at various levels of precision, and an algorithm which exploits this structure. For concreteness, we will keep in mind the simple example of the possible insertions of a subgraph H onto n labeled vertices. In this case $\mathbf{X}$ is the set of all subgraphs of K_n which are isomorphic to H, so $|\mathbf{X}| = \binom{n}{v(H)} \frac{v(H)!}{|\mathrm{Aut}(H)|}$. Our goal is to approximate the state space by considering various subsets of the state space as hypotheses in their own right, and then winnowing our way down to more specific hypotheses as the likelihood of particular subsets increases. In our example, one such generalized hypothesis is the set $\mathbf{X}_e$ of all insertions of H which contain a particular edge e. More generally, we could consider the set $\mathbf{X}_S$ of all insertions of H which contain all edges in a set S. In this way a lattice $\mathscr{H} = \{\mathbf{X}_S \mid |S| \leq e(H)\}$ of hypotheses emerges. We order $\mathscr{H}$ by reverse inclusion so that $\mathbf{X}_S \leq \mathbf{X}_T$ iff $\mathbf{X}_T \subseteq \mathbf{X}_S$, which places coarse, less specific hypotheses lower in the hierarchy. In our example, note that $\mathbf{X}_S \leq \mathbf{X}_T$ iff $S \subseteq T$.

6.2 The HHM Algorithm

The philosophy of HHM is that lower-level hypotheses should be computationally more manageable, and that an approximation algorithm should exploit this structure. At this point it is necessary to flesh out the problem in more detail. In our example, we will assume an Erdős–Rényi random graph noise model $\mathscr{R} = \mathscr{G}(n, p)$, which is represented by a sequence of random subgraphs of K_n. Our signal model is the total insertion of the target graph H into some fixed, but unknown location in K_n on top of each random instance independently with probability p_I. (Thus, in the LRDT framework, the motion model in this example is trivial.) This process results in a sequence of evidence graphs $J_1, \ldots, J_t \subseteq K_n$. While computing the exact likelihood ratio for a hypothesis $\mathbf{X}_S$ versus the null hypothesis can be quite difficult and depends on the particular isomorphism type of the target graph H, it is clear that $P(S \subseteq J_i \mid \mathbf{X}_S) > P(S \subseteq J_i \mid \neg\mathbf{X}_S)$. A simple but potentially effective algorithm in this case goes as follows: First, for each edge e in $G_0 = K_n$, count the number of appearances $c(e)$ of e in the sequence $J_1, \ldots, J_t$. Let $G_1 = \{e \mid c(e) \geq T_1\}$ for some threshold T_1: that is, discard all edges which do not appear "frequently enough" in the sequence of evidence graphs. Next, determine the count $c(S)$ for all $|S| = 2$ such that $S \subseteq G_1$. Discard all such edge-sets S from G_1 such that $c(S) < T_2$ for some threshold T_2 to form a new graph G_2. Continue this process until no more than $e(H)$ edges remain. Obviously this is an approximate algorithm, and it is far from clear how to choose the correct thresholds T_i, or how effective it is likely to be given a particular target graph H. Note that there are $\binom{n}{2}$ "edge" hypotheses $\mathbf{X}_e$, $\binom{n}{3}$ hypotheses of the form $\mathbf{X}_S$ where $|S| = 3$, etc., and so such a scheme could provide significant computational savings over explicit maintenance of all individual hypotheses.

6.3 An Example

We considered a particular case of the above problem and implemented a version of the above algorithm to find the target subgraph. In our case, we took $n = 100$ and $p = 0.5$. For H, we chose an embedding of the Petersen graph, which has 10 vertices and 15 edges (as well as a certain social network significance, being an optimal covert configuration under certain security and information-flow conditions [34]). The probability of insertion was $p_I = 0.2$, and 125 evidence graphs were generated. Note that, even in this simple example, the total number of hypotheses is $|\mathbf{X}| \approx 5.2 \times 10^{17}$, far too many for an exhaustive approach. Note that 1.6×10^{13} copies of H are expected to arise in each random graph instance just from noise, but that the probability of any given copy of H arising just from noise is only about 3×10^{-5}. In this example, we only considered hypotheses of the form $\mathbf{X}_S$ where S is connected; this further reduced the number of hypotheses which needed to be considered. One can also show that connected edge sets generally discriminate better between $\mathbf{X}_S$ and $\neg\mathbf{X}_S$ than arbitrary edge sets. Thresholds were determined experimentally. The method worked well in this scenario – we were able to find the target subgraph using this version of the algorithm using a lattice of depth 6.

7 Conclusion

Detection theory provides a principled approach for leveraging the results of network theory to produce a methodology for finding and tracking terrorist activities. The approach is difficult however. Even the simple binary detection problem discussed in Sects. 2 and 3 leads quickly to some rather involved formulas. Sections 4–6 broaden the scope of the problem, indicating the kinds of mathematics necessary for further progress along various lines.

The results in this chapter constitute only a brief, initial exploration of what appears to be a deep, rich subject. A great deal of development is required before the subject will become a useful tool in counterterrorism. In particular, the networks discussed in this chapter have no *attributes*: all the vertices and edges are the same type. In the real world, attribute information is often crucial. The modeling of attributes is generally a messy, highly domain-specific procedure, but we believe that they could be incorporated into network detection theory in a systematic way once solid theoretical underpinnings have been established for the case of no attributes or a few representative ones. Hopefully, this chapter has convinced the reader that the mathematics involved in constructing such an edifice is not only important, but intrinsically interesting as well.

Acknowledgments

This work was supported by AFRL contract FA8750-06-C-0217 and ONR contract N0014-05-C-0236.

References

1. Erdős, P., Rényi, A.: On the evolution of random graphs. Magyar Tud. Akad. Mat. Kutató Int. Közl **5** (1960) 17–61
2. Szemerédi, E.: On sets of integers containing no k elements in arithmetic progression. Acta Arithmetica **27** (1975) 199–245
3. Green, B., Tao, T.: The primes contain arbitrarily long arithmetic progressions. Annals of Mathematics **167** (2008) 481–547
4. Janson, S., Łuczak, T., Ruciński, A.: Random Graphs. John Wiley & Sons, New York (2000)
5. Bollobás, B.: Random Graphs. Cambridge University Press, New York (2001)
6. Milgram, S.: The small world problem. Psychology Today **2** (1967) 60–67
7. Wasserman, S., Faust, K.: Social Network Analysis. Cambridge University Press (1994)
8. Newman, M.E.J.: The structure and function of complex networks. SIAM Review **45** (2003) 167–256
9. Barabási, A.L., Albert, R.: Emergence of scaling in random networks. Science **286** (October 1999) 509–512
10. Barabási, A.L.: Linked: How Everything Is Connected to Everything Else and What It Means. Plume (2002)
11. Watts, D.J.: Six Degrees, The Science of a Connected Age. W. W. Norton & Company (2003)

12. Bollobás, B., Janson, S., Riordan, O.: The phase transition in inhomogeneous random graphs. Random Structures and Algorithms **31**(1) (2007) 3–122
13. Krebs, V.E.: Uncloacking terrorist networks. First Monday (2001)
14. Mifflin, T.L., Boner, C.M., Godfrey, G.A., Skokan, J.: A random graph model for terrorist transactions. In: 2004 IEEE Aerospace Conference Proceedings. (March 2004)
15. Mifflin, T., Boner, C., Godfrey, G., Greenblatt, M.: Detecting terrorist activities in the twenty-first century: A theory of detection for transactional networks. In Popp, R.L., Yen, J., eds.: Emergent Information Technologies and Enabling Policies for Counter-Terrorism. Wiley-IEEE Press (2006) 349–365
16. Kay, S.M.: Fundamentals of Statistical Signal Processing, Vol. II, Detection Theory. Prentice Hall, NJ (1998)
17. DeGroot, M.H.: Optimal Statistical Decisions. McGraw–Hill, New York (1970)
18. Ferry, J., Lo, D.: Fusing transactional data to detect threat patterns. In: Proceedings of the 9th International Conference on Information Fusion. (July 2006)
19. Singer, K.: Random Intersection Graphs. PhD thesis, Johns Hopkins University, Baltimore, Maryland (1995)
20. Karónski, M., Scheinerman, E.R., Singer-Cohen, K.B.: On random intersection graphs: The subgraph problem. Combinatorics, Probability, and Computing **8** (1999) 131–159
21. Fill, J.A., Scheinerman, E.R., Singer-Cohen, K.B.: Random intersection graphs when $m = \omega(n)$: an equivalence theorem relating the evolution of the $G(n,m,p)$ and $G(n,p)$ models. Random Structures and Algorithms **16**(2) (March 2000) 156–176
22. Stark, D.: The vertex degree distribution of random intersection graphs. Random Structures and Algorithms **24**(3) (May 2004) 249–258
23. Behrisch, M., Taraz, A.: Efficiently covering complex networks with cliques of similar vertices. Theoretical Computer Science **355** (2006) 37–47
24. Godehardt, E., Jaworski, J.: Two models of random intersection graphs and their applications. Electronic Notes in Discrete Mathematics **10** (2001) 129–132
25. Newman, M.E.J., Strogatz, S.H., Watts, D.J.: Random graphs with arbitrary degree distribution and their applications. Physical Review **64** (2001) 026118
26. Ellis, R.B., Ferry, J.P.: Estimating variance of the subgraph count in sparse Erdős-Rényi random graphs. submitted to Discrete Applied Mathematics (2008)
27. Bollobás, B.: Threshold functions for small subgraphs. Math. Proc. Cambridge Philos. Soc. **90**(2) (1981) 197–206
28. Karoński, M., Ruciński, A.: On the number of strictly balanced subgraphs of a random graph. In: Graph theory. Volume 1018 of Lecture Notes in Mathematics. Springer, Berlin (1983) 79–83
29. Ruciński, A.: When are small subgraphs of a random graph normally distributed? Prob. Theory Related Fields **78** (1988) 1–10
30. Durrett, R.: Random Graph Dynamics. Cambridge University Press, Cambridge (2007)
31. Díaz, J., Mitsche, D., Pérez-Giménez, X.: On the connectivity of dynamic random geometric graphs. In: SODA '08: Proceedings of the nineteenth annual ACM-SIAM symposium on Discrete algorithms, Philadelphia, PA, USA, Society for Industrial and Applied Mathematics (2008) 601–610
32. Parzen, E.: Stochastic processes. Holden-Day (1962)
33. Stone, L.D., Corwin, T.L., Barlow, C.A.: Bayesian Multiple Target Tracking. Artech House, Inc., Norwood, MA, USA (1999)
34. Lindelauf, R., Borm, P.E.M., Hamers, H.J.M.: The influence of secrecy on the communication structure of covert networks. CentER Discussion Paper 2008-23, Tilburg University.

Part III
Communication/Interpretation

Security of Underground Resistance Movements

Bert Hartnell and Georg Gunther

Abstract In an attempt to better understand the possible lines of communication in a terrorist group, we consider a graph theoretic model and the competing demands of the desire for ease of communication versus the danger of betrayal of subverted or captured members. We first examine what the design should be when the attack on the network is optimal as well as when it is random but in both cases with no restriction on the surviving network. Then the question of what the design should be to minimize damage but to ensure that the average size of the surviving components is as large as possible is considered. Finally the restriction that all the survivors are capable of communicating is examined.

1 Introduction

A terrorist organization can be modeled very naturally by a graph $G(V,E)$ whose vertex set V consists of the individuals in the organization. Two vertices A, B will be joined by an edge if the individuals know each other so that they are each in a position to betray the other.[1] In such an organization, messages are passed along the edges, and so we can refer to this graph as a communications network. For such a network to be effective, it is essential that any two of its members should be able to communicate; this means that the graph G needs to be connected. Since speed of communication is clearly an asset when planning communal action, ideally any two

Bert Hartnell
Saint Mary's University, Halifax, Nova Scotia, Canada, e-mail: bert.hartnell@smu.ca

Georg Gunther
Sir Wilfred Grenfell College, Memorial University of Newfoundland, CornerBrook, Newfoundland, Canada, e-mail: ggunther@swgc.mun.ca

[1] This model can be refined by looking at digraphs, where a directed edge from A to B would mean that A could betray B, but B could not betray A.

N. Memon et al. (eds.), *Mathematical Methods in Counterterrorism*,
DOI 10.1007/978-3-211-09442-6_11,

agents in the network should be able to communicate. In this case the underlying graph G would be a complete graph.

However, ease of communication is not the sole factor to be considered. When agents in the network are subverted or betrayed or arrested, there is the danger that they will quickly betray everyone they know. So from the perspective of network security, agents should know as few others as possible.

In designing an effective terrorist organization, these two considerations work against each other. The objective is to design a network that contains enough connectivity to facilitate quick communications without making it overly vulnerable to the effects of subversions or betrayals.

In this chapter, we address several of the problems that arise out of these conflicting demands of connectivity versus security. There exists a substantial literature relating to questions of this type; specific references can be found in the bibliography at the end. Specifically, we will look at the following questions.

1. Suppose a graph G representing a terrorist organization has p vertices, and suppose that b of these vertices are subverted, so that they betray everyone they know in the network. How much damage do these b subversions do to the graph G?
2. Suppose a network G is attacked by subverting a single agent, and suppose that this subversion disconnects the network. Can we make any predictions about the sizes of the remaining disconnected pieces?
3. Suppose that a terrorist organization G is attacked by the subversion and deletion of a single vertex A and its closed neighborhood. How should G be designed so that the survivor graph G' remains connected?

2 Best defense against optimal subversive strategies

Let G be a graph on p vertices, and suppose that we are going to attack G by subverting b of the vertices, with the understanding that these subverted vertices will subsequently betray everyone they know in the graph. In other words, when a vertex A is subverted, it will betray its closed neighborhood $N[A]$. We will define a subversive strategy on G to be any subset C of b vertices in G. For such a strategy C, we define the symbol

$$K_C(G,b) = \text{total number of betrayals resulting from strategy } C$$

Since each vertex in C can be considered to be 'self-betrayed', we see that $K_c(G,b) \geq b$ for any strategy C. For a given graph G, it is obvious that some subversive strategies will inflict more damage than others. In view of this, we define the KGB-number of the graph G by the symbol $K(G,b)$:

$$K(G,b) = \max\{K_C(G,b) | C \text{ is a subversive strategy on } b \text{ vertices}\}$$

It is also equally obvious that this number $K(G,b)$ depends heavily on the particular graph G. For example, if $G = K_p$, is the complete graph on p vertices, then $K(G,1) = p$; on the other hand, if G is the completely disconnected graph on p vertices, then $K(G,b) = b$. In view of this, suppose that we are given parameters p and b. We now define the symbol

$$K(p,b) = \min\{K(G,b)|G \text{ is a graph on } p \text{ vertices}\}$$

Our intention is to determine this number $K(p,b)$, and to describe graphs on which these numbers are realized. Let us make this clear. Suppose you are planning to recruit p individuals to be members of your terrorist organization, represented by the graph G. You anticipate that b of your agents will be subverted. How should you construct your lines of communication so that, in the event the adversary manages to inflict the maximum damage on your network, this maximum is kept as small as possible?

Suppose now that G is a graph on p vertices and T is a spanning tree of G. Clearly, for any subversive strategy C, we will have $K_C(T,b) \leq K_C(G,b)$. In view of this, we may restrict ourselves to an examination of trees.

In the analysis of this problem, three families of trees play a special role. The first are the paths on p vertices, denoted by the symbol P_p. The second are the superstars, denoted by the symbol S_n. Such a superstar is formed by taking n paths with 3 vertices and joining them at a common end-vertex Z. The third special family of trees consists of the fifth-column graphs F_n. To form such a fifth-column graph, we take n paths with 5 vertices; let us denote these 5-paths by $\{A_1,B_1,Z_1,C_1,D_1\},\{A_2,B_2,Z_2,C_2,D_2\},\ldots,\{A_n,B_n,Z_n,C_n,D_n\}$. We then insert the edges from the center of each 5-path to the center of an adjacent 5-path. These three types of graphs are shown in Fig. 1.

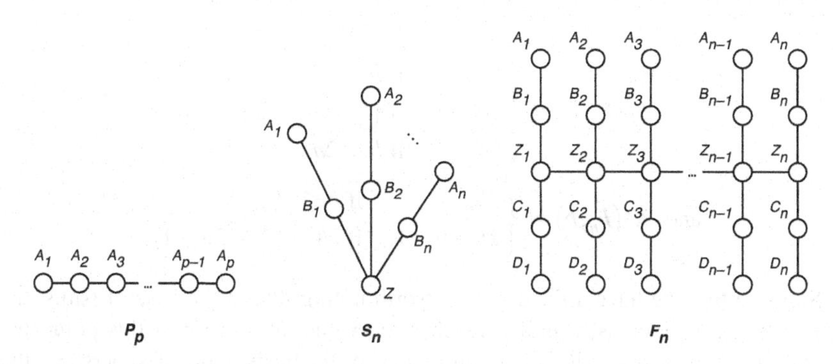

Fig. 1 P_p, S_n and F_n

Suppose we have one of these trees, and we now recruit a new agent B into our network. In the case of the path, we simply adjoin the new vertex to the end of the path. In the case of a superstar, we adjoin the new vertex B to the center Z of the superstar to form the tree we will denote by the symbol S'_n. In this situation, if we now recruit another vertex A, we will adjoin this to B to form S_{n+1}. If instead, we wish to add vertices, one at a time, to a fifth-column graph F_n, we do this by adding the vertices $Z_{n+1}, B_{n+1}, C_{n+1}, A_{n+1}, D_{n+1}$, in that order, to obtain the graphs denoted $F_n^1, F_n^2, F_n^3, F_n^4$ and finally F_{n+1}.

Suppose now that we attack one of these graphs by subverting b vertices. For the path P_p, it is clear that an optimal subversive strategy would include every third vertex of the path. Such a strategy picks up 3 vertices per subversion. For a superstar S_n, we have a choice of two strategies $C_1 = \{Z_1, A_1, \ldots, A_{b-1}\}$ and $C_2 = \{B_1, \ldots, B_b\}$. When attacking a fifth-column graph F_n, it is clear that any optimal strategy should subvert at most 2 vertices in each 5-path (since subverting B_i and C_i will betray the entire 5-path). If b is small compared to n, it may be necessary to subvert some center vertices Z_i in order to betray some of the vertices in adjacent 5-paths. A detailed analysis of these strategies yields the following results.

For the path P_p: Attacking P_p with b subversions gives us

$$K(P_p, b) = \begin{cases} p, & \text{if } \lfloor \frac{p}{3} \rfloor < b; \\ 3b, & \text{if } \lfloor \frac{p}{3} \rfloor \geq b. \end{cases}$$

For the superstar S_n (or S_n'): Attacking S_n (or S_n') with b subversions gives us

$$K\ (S_n, b) = \begin{cases} n+b, & \text{if } b < n \text{ (use } C_1) \\ 2n+1, & \text{if } b \geq n \text{ (use } C_2) \end{cases}$$

$$\text{and } K\ (S_n', b) = \begin{cases} n+b+1, & \text{if } b \leq n+1 \text{ (use } C_1) \\ 2n+2, & \text{if } b > n+1 \text{ (use } C_2) \end{cases}$$

For the fifth-column graphs F_n (or F_n^i): Attacking F_n (or F_n^i) with b subversions gives us

$$K\ (F_n, b) = \begin{cases} 5b, & \text{if } b \leq \lfloor \frac{n}{3} \rfloor \\ 2b+n, & \text{if } \lfloor \frac{n}{3} \rfloor < b \leq 2n \\ 5n, & \text{if } b > 2n \end{cases}$$

$$\text{and } K\ (F_n^i, b) = \begin{cases} 5b, & \text{if } b \leq \lfloor \frac{n}{3} \rfloor) \\ 2b+n+1, & \text{if } \lfloor \frac{n}{3} \rfloor < b \leq 2n-1) \end{cases}$$

Suppose that we have a fixed b. It is worth noting that in the case of paths, the number $K(P_p, b)$ increases until p reaches the value $3b$; once past this point, the number $K(P_p, b)$ stays at its maximum value of $3b$. In the case of superstars, the number $K(S_n, b)$ increases without bound as n increases; for fifth-column graphs, on the other hand, the number of betrayals increases by 1 for every 5-path we attach until there are at least three times as many 5-paths as subversions; at this point, the number of betrayals remains constant at the value $5b$.

These comments suggest that, for some fixed value b of subversions, as the number p of agents increases, there will come two transitional values. First, a point will be reached at which the fifth-column structure yields a smaller number of total betrayals than does the superstar structure; subsequent to that, a second point will be reached at which paths do better than fifth-column graphs. Specifically, we can see this in the following example.

Suppose that $b = 10$ and $p = 31$. If these 31 agents are arranged in the path P_{31}, then we know that $K(P_{31}, 10) = 30$. If instead, we use a superstar structure, then we will obtain the graph S_{15}, and using strategy C_1 to attack this graph will give us $K(S_{15}, 10) = 25$ subversions in total. Finally, if we decide to arrange these agents in a fifth-column configuration, we would obtain the graph F_6^1; attacking this graph with 10 subversions would yield $K(F_6^1, 10) = 27$ subversions. Hence for these values of p and b, the superstar structure provides the best security.

Now suppose we recruit 9 additional agents, so that $p = 39$, and suppose that b is still 10. For the path (which for 10 betrayals reached its maximum at $p = 30$), we still have $K(P_{30}, 10) = 30$. We now compare the graphs S_{19} and F_7^4. When these two graphs are attacked by 10 subversions, we obtain $K(S_{19}, 10) = 29$, whereas $K(F_7^4, 10) = 28$. Thus, for these values of p and b, we find that now the fifth-column structure is better at minimizing the total number of betrayals.

If we now recruit an additional 7 agents, so that $p = 46$, and attack the resulting fifth-column graph with $b = 10$, we see that $K(F_9^1, 10) = 30$. From this point on, as the number of agents continues to increase, the path minimizes the total number of betrayals.

This example illustrates a general result which can be stated informally as follows: if b is large compared to p, superstars provide better defense against the effects of subversions; if on the other hand, b is small compared to p then the fifth-column graphs become more secure. In approximate terms, this change takes place when the b becomes smaller than $\frac{3}{10}p$.

A careful analysis shows that for many choices of p and b, the value $K(p,b)$ is realized on either a superstar or a fifth-column graph (see [1] for details of this). This means that these two families of graphs are optimal solutions to this problem in the sense that no other tree structure will provide greater security against the effects of b subversions followed by the resultant betrayals. The main theorem in [1] is proved by a double induction on the values of p and b; the proof splits into three cases, depending on the value of $b(mod3)$. This result is summarized here in the following way.

Theorem 1. *For a fixed p and b the following hold:*

i. *$K(p,b) = p$, if $p \leq 2b+1$*
ii. *For $b \leq 3$, $K(p,b) = 3b$ if $p \geq 4b$; for $b \geq 4$, $K(p,b) = 3b$ if $p \geq 5b-4$*
iii. *Case 1: $b = 3t+1$*

   a. *For $p \leq 6t+3$, $K(p,b) = p$ (this is case (i) above).*
   b. *For $6t+3 < p \leq 10t+5$, the optimal trees are the superstars S_k and S'_k, where $b = 3t+1 \leq k \leq 5t+2$. As p increases by 2's from $6t+3$ to $10t+5$, the number $K(p,b) = K(p,3t+1)$ increases at half the rate of p, from $6t+3$ to $8t+3$.*

c. *For* $10t \leq p \leq 15t$, *the optimal trees are fifth-columns* F_k *and their intermediate forms* F_k^i, *where* $2t+1 \leq k \leq 3t$. *For every increase by* 5 *of* p, *the value of* $K(p,b)$ *increases by* 1, *going from* $8t+4$ *to* $9t+2$.
d. *For* $p \geq 15t+1 = 5b-4$, $K(p,b) = 3b$ *(this is (ii) above).*

iv. *Case 2:* $b = 3t+2$

a. *For* $p \leq 6t+5$, $K(p,b) = p$.
b. *For* $6t+5 < p \leq 10t+10$, *the optimal trees are the superstars* S_k *and* S_k', *where* $b = 3t+2 \leq k \leq 5t+4$. *As* p *increases by* $2'$*s from* $6t+5$ *to* $10t+10$, *the number* $K(p,b)$ *increases at half the rate of* p, *from* $6t+5$ *to* $8t+6$.
c. *For* $10t+11 \leq p \leq 15t+5$, *the optimal trees are fifth-columns* F_k *and their intermediate forms* F_k^i, *where* $2t+2 \leq k \leq 3t+1$. *For every increase by* 5 *of* p, *the value of* $K(p,b)$ *increases by* 1, *going from* $8t+7$ *to* $9t+5$.
d. *For* $p \geq 15t+6 = 5b-4$, $K(p,b) = 3b$.

v. *Case 3:* $b = 3t+3$

a. *For* $p \leq 6t+7$, $K(p,b) = p$.
b. *For* $6t+8 < p \leq 10t+15$, *the optimal trees are the superstars* S_k *and* S_k', *where* $3t+3 \leq k \leq 5t+7$. *As* p *increases by* $2'$*s from* $6t+8$ *to* $10t+15$, *the number* $K(p,b)$ *increases at half the rate of* p, *from* $6t+7$ *to* $8t+9$.
c. *For* $10t+16 \leq p \leq 15t+10$, *the optimal trees are fifth-columns* F_k *and their intermediate forms* F_k^i, *where* $2t+3 \leq k \leq 3t+2$. *For every increase by* 5 *of* p, *the value of* $K(p,b)$ *increases by* 1, *going from* $8t+10$ *to* $9t+8$.
d. *For* $p \geq 15t+11 = 5b-4$, $K(p,b)$=$3b$.

These results can be summarized in Fig. 2, which shows the p-axis:

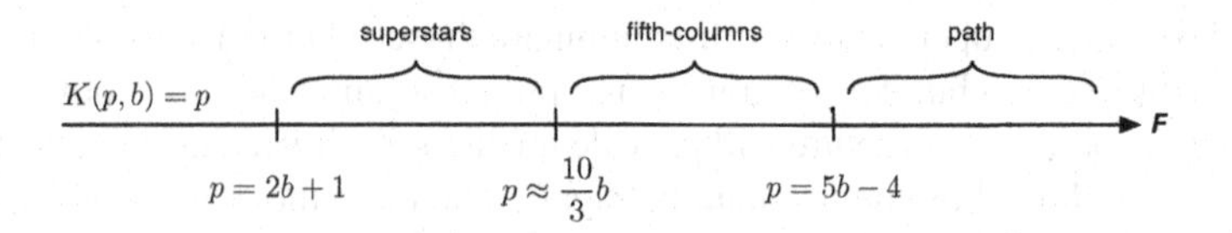

Fig. 2

What this tells us is that if you wish to minimize the damage done by b subversions, then you should use the following tree structure: for p between b and $\frac{10}{3}b$ (approximately), use a superstar, for p between $\frac{10}{3}b$ and $5b-4$, use a fifth column graph, and for p greater than $5b-4$, use a path.

3 Best defense against random subversive strategies

The preceding analysis is based on the assumption that the terrorist organization is being attacked by b subversions, where these subversions are chosen in a way that inflicts the maximum damage to the network. The three families of graphs have the property that they offer the best defense against such optimal subversive strategies.

In cases where the graph G models a real terrorist organization, it is unrealistic to assume that the lines of communication would be set up in a way that assumes such a 'worst-case' scenario. After all, if the adversary knows enough about the terrorists to be able to pick an optimal subversive strategy, then presumably the structure of the entire network is already known. In view of this, it is reasonable to be concerned about a more realistic threat to such a terrorist organization. Instead of making the paranoid assumption that the adversary will attack with the most damaging subversive strategy, it is far more realistic to assume that the attack will be of a more random nature. More specifically, if G is the modeling graph on p vertices, then there exist $\binom{p}{b}$ different subversive strategies on b vertices. If we make the assumption that all of these strategies are equally likely, we can ask about the expected value of the total number of betrayals, following such a random attack on G. We illustrate this idea by looking at the specific example in which a tree T on 6 vertices is attacked by the random subversion of two of these vertices. Fig. 3 shows the six possible trees on 6 vertices, which have been labeled $T_1, T_2, \ldots, T_6$.

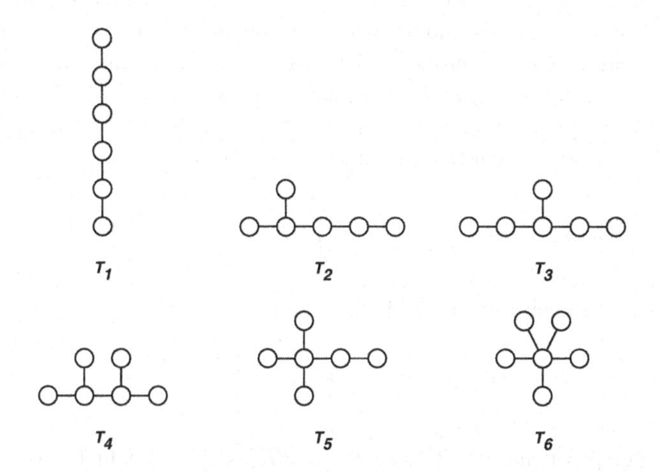

Fig. 3

There exist 15 possible subversive strategies in this case, and it is easy to compute the average number of betrayals in each case. In the case of T_1, this number is 4.4, in the case of T_2 and T_3 it is $4.\overline{3}$, in the case of T_4 it is $4.2\overline{6}$, in the case of T_5 it is 4.2, and finally, in the case of T_6, it is 4. So we see that for these numbers, the star (which is T_6) is best since it yields the smallest average number of betrayals, while the path (which is T_1) fares the worst, giving the largest number.

Motivated by these considerations, we now define the random KGB-number of a graph G, denoted by the symbol $RK(G,b)$, as this average number of betrayals. Specifically, we write

$$RK(G,b) = \frac{1}{\binom{p}{b}} \sum K_C(G,b)$$

where this sum is taken over all possible strategies C of size b.

We are interested in characterizing those graphs G that minimize the number $RK(G,b)$. Once again, if G is any connected graph, and S is a spanning tree of G, then it is clear that $RK(G,b) \geq RK(S,b)$; hence if we are looking for graphs that minimize the random KGB-number, we may restrict our attention to trees.

Suppose now that T is a tree on p vertices, and we attack T by subverting a single vertex (so $b = 1$). Then we see that for an edge AB that strategy A betrays B, and strategy B betrays A (in addition to the 'self-betrayals' that occur). Hence we see that each edge in the tree contributes two betrayals to the sum $\sum K_C(T,1)$. Since there are $p-1$ edges in the graph, we can therefore conclude that $\sum K_C(T,1) = 2(p-1)+p$, where the additional p betrayals come from the 'self-betrayals' referred to above. This analysis tells us that for any tree T, we have $RK(T,1) = 3 - \frac{2}{p}$.

When $b > 1$, it becomes difficult to give a direct calculation of the random KGB-number for arbitrary trees. However, such a calculation is straightforward in the case where the given tree is the star $\overline{S_{p-1}}$ on p vertices. This star consists of a central vertex Z which is adjacent to $p-1$ leaves $A_1, \ldots, A_{p-1}$. The possible strategies on this star fall into two sets S_1 and S_2, where S_1 consists of those strategies C which consist of b leaves, and S_2 consists of the strategies which include the center Z and $b-1$ leaves. Clearly, we have $\binom{p-1}{b}$ strategies in S_1, and $\binom{p-1}{b-1}$ strategies in S_2. Every S_1-strategy yields a total of $b+1$ betrayals, while every S_2-strategy yields p betrayals. Hence we see that for the star $\overline{S_{p-1}}$, we obtain

$$RK(\overline{S_{p-1}},b) = \frac{1}{\binom{p}{b}} \left[\binom{p-1}{b} \cdot (b+1) + \binom{p-1}{b-1} \cdot p \right].$$

This can easily be simplified to yield the formula

$$RK(\overline{S_{p-1}},b) = (2b+1) - \frac{b(b+1)}{p}$$

Note that when $p = 6$ and $b = 2$, this gives us $RK(\overline{S_5},2) = 4$, which is of course the value we computed earlier for the star T_6.

For all the possible trees T on six vertices, we found that $RK(\overline{S_5},2) \leq RK(T,2) \leq RK(P_6,2)$. So in this case, the star provides the best solution, while the path is the worst. In [2], it is shown that this result is not a consequence of the particular choice of parameters $p = 6$ and $b = 2$, but is in fact general. Specifically, the following theorem holds.

Theorem 2. *Let T be any graph on p vertices. Then we have $RK(\overline{S_{p-1}},b) \leq RK(T,b) \leq RK(P_p,b)$.*

This theorem tells us that the optimal defense against b randomly chosen subversions is the star $\overline{S_{p-1}}$. In spite of this result, a terrorist organization would be more than foolish if their lines of communication were set up along these lines. The problem with the star is that a single lucky subversion at the center of the star eliminates the entire network. This then suggests that it would be more realistic to set up communications in such a way that two criteria are met. First, the network should keep to a minimum the expected number of betrayals resulting from an attack by a random subversive strategy on b vertices, and second, that the b subversions followed by the resultant betrayals should not totally destroy the network. Graphs with this property will be called b-secure.

The question of characterizing b-secure graphs is addressed in [4]. A careful analysis shows that among all b-secure graphs of a given order p, the ones for which the random KGB-number is minimal are a blend of star and superstar. Specifically, using the notation of [4], these are the graphs denoted by the symbol $S(p,b)$, illustrated in Fig. 4.

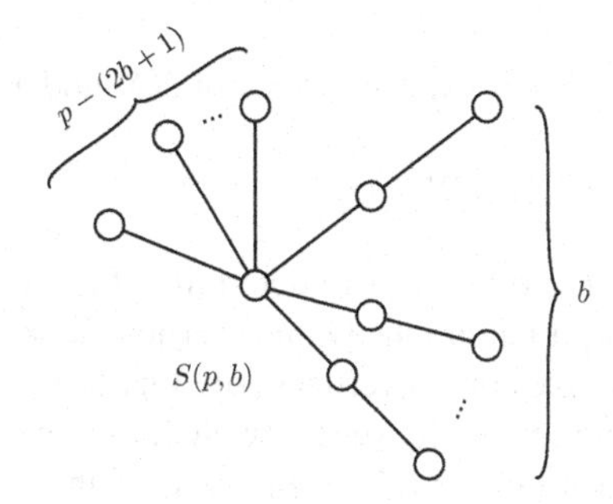

Fig. 4

4 Maximizing the size of surviving components

In the preceding discussions, we considered the effectiveness of attacks on a terrorist organization by determining the number of betrayals resulting from b subversions. There is another way to assess the effectiveness of an attack on such an organization. Instead of looking at the number of betrayals, we can ask questions of the following type: suppose the terrorist organization G is attacked by some subversive strategy C. After deleting all the betrayed vertices from the network, what can we say about the network of surviving agents?

Suppose the terrorist organization modeled by graph G is attacked by a subversive strategy C. Ideally (from the perspective of the terrorists), the surviving network should still be connected; however, it is quite conceivable that such an attack will disconnect the network and result in a number of disjoint components. These considerations prompt the following two questions.

Question 1: Suppose the surviving network is disconnected. What can we say about the average size of these surviving components?

Question 2: How should the lines of communication be set up in the initial graph G so that the survivor graph is still connected?

The analysis of both questions becomes extremely complicated when the number of betrayals increases. For this reason, we consider what happens in the simplest case where we subvert a single agent. We begin with a brief discussion of question 1. Full details can be found in [3].

Suppose then that graph G on p vertices is attacked by the subversive strategy $C = \{X\}$, where X is some vertex in G. This attack leaves the survivor graph $G - N[X]$, containing $n(X) = p - deg(X) - 1$ surviving vertices. Letting $d(X)$ be the number of surviving components, we now define the following parameters:

$$\overline{c}(X) = \frac{n(X)}{d(X)} = \text{average order of the surviving components;}$$

$$\overline{c}(G) = \frac{1}{p} \sum_{X \in V(G)} \overline{c}(X) = \text{expected value of } \overline{c}(X) \text{under random attacks;}$$

$$\overline{c}(p) = \max\{\overline{c}(G) \mid |V(G)| = p\}.$$

The question we are interested in is to determine, for a given p, what the value of $\overline{c}(p)$ is and what the optimal graphs G on p vertices look like for which $\overline{c}(G) = \overline{c}(p)$. The following examples on 10 vertices (see Fig. 5) will illustrate these ideas. For the path P_{10}, depending upon the vertex deleted, the surviving components will have sizes ranging from 8 down to 3; in the case of the tree TS_3, the surviving components will range in size from 8 down to 1; in the case of the cycle C_{10}, all surviving components will have size 7. For these examples, it is easy to compute that $\overline{c}(P_{10}) = 5\frac{1}{10}$, $\overline{c}(TS_3) = 5\frac{4}{10}$ while $\overline{c}(C_{10}) = 7$. So for these three examples, the path is worst, while the cycle is optimal. The optimality of the cycle should not come as a surprise. If G contains vertices X for which $deg(X) > 2$, then attacking G by strategy X will result in smaller survivor graphs, and will hence lower the value of the parameter $\overline{c}(G)$. This insight is expressed in the following theorem.

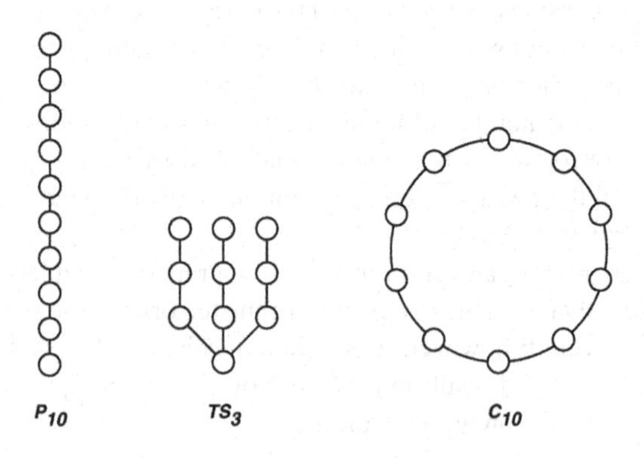

Fig. 5

Theorem 3. *If $p > 5$, then $\overline{c}(p) = p - 3$, and this optimal value is achieved only on the cycle C_p.*

This problem becomes more complicated and interesting in the case where the initial graph G is a tree on p vertices. To illustrate the problems that arise, it is constructive to look once more at the tree TS_3 (see Fig. 5). In this tree, the three leaves are far enough away from vertices of high degree so that a subversive strategy on a leaf will only delete two vertices, leaving a surviving component of high order. Such leaves can only be present if somewhere, deeper within the tree, there are vertices of higher degree whose deletion disconnects the tree into a larger number of disconnected components of small order. This means that the search for optimal trees is governed by two off-setting criteria. First, we would like to have many leaves whose adjacent stem is of degree 2; second, we would like to have a tight central core of high degree vertices.

Note that the tree consists of a central 'core' vertex to which we have attached three 3-arms. In [3], it is shown that optimal trees T (trees for which the number $\overline{c}(T)$ is as large as possible) must have precisely this sort of structure: they must consist of a core of higher degree vertices where each of the core vertices has a number of 3-arms attached to it. Knowing this result, it is then a question of determining what sort of 'core' tree gives us the optimal solution to the question of maximizing the average size of surviving components. The following theorem provides a partial answer to this.

Theorem 4. *Let T be the tree which consists of a core K on $k+1$ vertices with the property that to each vertex of the core, we have attached m 3-arms for some fixed value m. Then, provided that m is sufficiently large, the optimal core is the $k+1$-star $\overline{S_k}$.*

We shall use the symbol $T(k,m)$ to denote the tree whose core is the star $\overline{S_k}$, and which has m 3-arms attached to each core vertex. Fig. 6 shows the tree $T(3,2)$ (the core-vertices have been colored black). An explicit formula can be worked out for the number $\overline{c}(T(m,k))$ [see [3] for details]. It is of interest to note that $\lim_{m\to\infty} \overline{c}(T(k,m)) = \frac{2p}{3} + \frac{k-4}{3}$, where p, the number of vertices in the tree $T(k,m)$, is given by $p = (k+1)(3m+1)$. This asymptotic limit suggests that if we compare trees $T(k,m)$ and $T(k',m')$, then, provided that they have the same order and provided that the m-values are large enough, the one with the greater k-value should be better.

The following specific examples will help to illustrate the ideas presented here. Note that the trees $T(23,m)$ and $T(5,4m+1)$ have the same order $p = 24(3m+1)$.

When $m = 1$, we obtain $\overline{c}(T(23,1)) = 58.9$ while $\overline{c}(T(5,5)) = 63.3$.

When $m = 2$, we obtain $\overline{c}(T(23,2)) = 110.6$ while $\overline{c}(T(5,9)) = 111.7$.

When $m = 3$, we obtain $\overline{c}(T(23,3)) = 160.5$ while $\overline{c}(T(5,13)) = 159.8$.

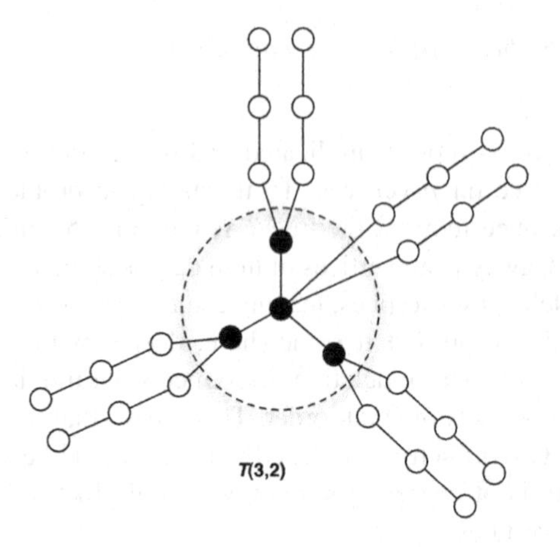

Fig. 6

5 Ensuring that the survivor graph remains connected

In the preceding discussion, the optimal solutions turned out to be tree structures. For real terrorist organizations, it is unlikely that the lines of communication would give rise to a tree. The classical solution to the problem of setting up communications within such an illegal organization is to set up small cells or cliques in which all the individuals know each other, and then set up some system of communication between these cells.

In the wake of the terrorist activity over the past 10 years, much effort has gone into devising ways of detecting communication activities among potential terrorist cells. A summary of these efforts can be found in [5]. These findings bear out the fact that currently operating terrorist organizations appear to be structured along these classical lines, where small cliques have loosely structured lines of communications amongst each other.

It therefore appears that in any graph G modeling a terrorist organization, the edges in the graph are either edges that connect members of a clique, or are edges that facilitate communication between cliques. In the latter case, an edge could either directly connect individuals belonging to two different cliques, or alternatively, an edge could connect vertex a in one clique to some 'intermediary' vertex B, where B does not belong to a clique, but acts as a 'courier' between cliques. This idea can be illustrated by the simple example of the cycle C_6. We present the edge-structure of this cycle in two different ways, as illustrated in the figure. In Fig. 7 (a), we interpret the 6-cycle as consisting of three 2-cliques (indicated by solid edges), with intra-clique lines of communications shown by the dotted edges. Note that in this interpretation of the 6-cycle, every vertex belongs to a 2-clique. In Fig. 7 (b), on the other hand, we have two 2-cliques (solid edges) connected via dotted edges to two couriers whose task in the organization is to pass messages between cliques. Of course, in the case of the 6-cycle, these distinctions are somewhat artificial, since both interpretations yield the same graph. However, the example is instructive since it allows us to think about two ways of setting up communications.

We can take these ideas to devise two non-isomorphic graphs on 12 vertices, as shown in the next figure. Fig. 8 (a) consists of four 3-cliques (solid edges), where each member of a clique has one external line of communication (dotted edges) to another clique. In Fig. 8 (b), cliques are no longer able to communicate directly with each other; instead, communications are channeled through three couriers, where each courier is able to contact exactly one individual in each of the three cliques.

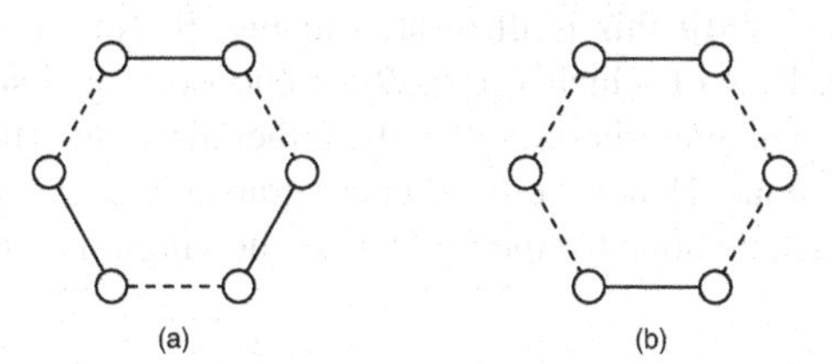

Fig. 7

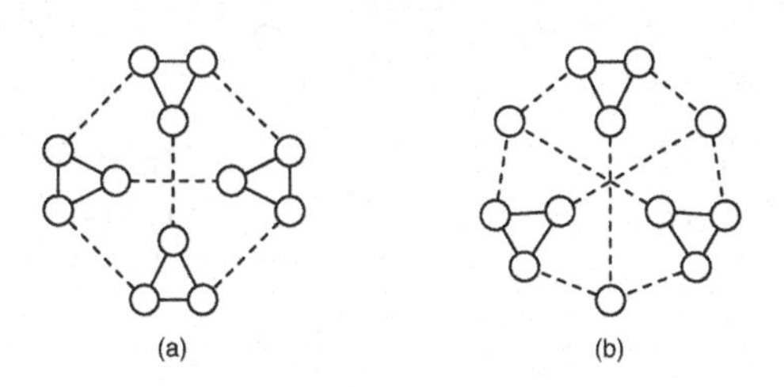

Fig. 8

Motivated by these background considerations, we are now in a position to formulate the following problem. We wish to design a graph G (modeling a terrorist organization) which has the property that it remains connected after it has been attacked by a subversive strategy C in which k agents are subverted, with the resultant betrayals of everyone they know in the network.

Specifically, we make the following definition.

Definition 1. Given any natural number $k \geq 2$, a graph G is k-neighbor-connected (or k-n-connected) if for any subversive strategy C on fewer than k vertices, the survival subgraph left after the deletion of the subverted vertices and their neighborhoods is still connected and is not a clique.[2]

[2] This last criterion is included for both technical and practical reasons. Technically, insisting that the survivor graph is not a clique removes one degree of variability from the analysis, and makes the problem more tractable; practically, cliques are highly vulnerable to attack since the subversion of any vertex will eliminate the entire clique.

In order to obtain a feel for what this means, consider again the examples discussed above. If G is the 6-cycle C_6, then the deletion of any single closed neighborhood leaves the path P_3 as the survivor graph. If A is one of the leaves of this surviving 3-path, then the deletion in P_3 of the closed neighbourhood $N[A]$ would leave a single surviving vertex, which can be thought of as a 1-clique. What this shows is that the 6-cycle is 2-neighbour-connected (but not 3-neighbour-connected).

Suppose now we begin with the graph Cl_3 (see Fig. 9). We attack this graph by the deletion of the neighborhood $N[A]$ (since this graph is 1-transitive on the vertices the choice of vertex A is without loss of generality). When we do this, we obtain the survivor graph $Cl_3 - N[A]$; this is illustrated in Fig. 9. Now it is straightforward to verify that, regardless of which vertex B we choose in this survivor graph, the deletion of the closed neighborhood $N[B]$ will neither disconnect this survivor graph, nor leave behind a clique. Hence we see that the graph Cl_3 is 3-neighbor-connected. A similar analysis can be done for the graph Co_3 showing it is 3-n-connected.

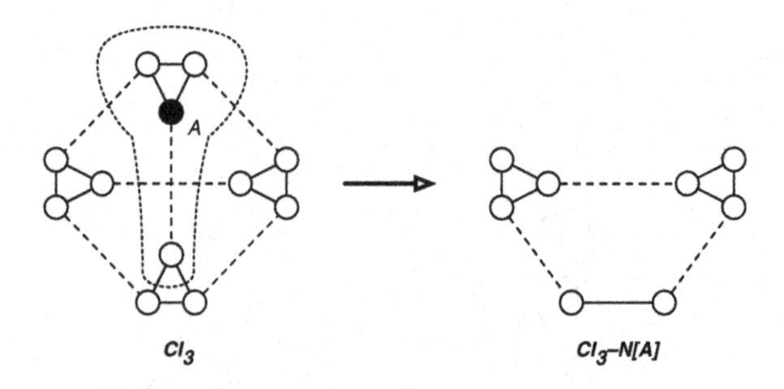

Fig. 9

These examples generalize in a very natural way to give us 'clique' and 'courier' graphs Cl_k and Co_k, for all $k \geq 2$. To construct the clique graph Cl_k, we start with $k+1$ k-cliques $C_1, \ldots, C_{k+1}$. We let the vertices in clique C_i be denoted by $A^i_1, A^i_2, \ldots, A^i_{i-1}, A^i_{i+1}, \ldots, A^i_{k+1}$. We then join separate cliques by joining vertex A^i_j in clique C_i to vertex A^j_i in clique C_j. To construct the courier graph Co_k, we start with k k-cliques $C_1, \ldots, C_k$, and with k couriers $Z_1, \ldots, Z_k$. We now let the vertices in clique C_i be denoted by $B^i_1, \ldots, B^i_k$, and we join each courier Z_j to the clique-vertices $B^1_j, \ldots, B^k_j$.

It is not difficult to show that both the clique graph Cl_k and the courier graph Co_k are k-neighbor-connected. It is also worth noting that these graphs are k-regular, and of order $k^2 + k$.

Much effort has gone into the attempt to learn more about k-regular, k-n-connected graphs. Here, insisting on k-regularity is a reasonable minimality restriction; it arises out of the fact that in a k-n-connected graph, every vertex is of degree $\geq k$. One immediate consequence of this useful structural fact is that graphs G may not contain either a 4-cycle $ABCD$ or a 4-cycle $ABCD$ with a diagonal BD. For in either case, subverting neighbourhood $N[C]$ would yield a survivor subgraph G' in which vertex A would be of degree $k-2$, and so this subgraph could not be $(k-1)$-n-connected.

The two families of graphs Cl_k and Co_k are of interest because of the following result [6].

Theorem 5. *Let G be a k-regular k-n-connected graph which contains a k-clique and which is of order k^2+k. Then G is either Cl_k or Co_k.*

In the context of this chapter, it is well worth remembering that terrorist organizations do not remain static. Instead, they grow by the recruitment of new individuals into the organization. Suppose then that G is a graph modeling such an organization, and suppose further that G has been set up to be k-n-connected. How can new vertices be added to G so that the resultant graph G' remains k-n-connected? Several ways exist for building new k-n-connected graphs from known examples; see [7] and [8] for details. Specifically, two methods are described here.

Method 1 (adding one vertex at a time): Let G be a k-n-connected graph, and choose any vertex V in G. Now construct the new graph G' by joining a new vertex V' to every vertex in $N[V]$. Then this new graph G' is also k-n-connected.

Method 2 (adding an m-clique): Let G be a k-regular k-n-connected graph which contains some k-clique C. Now construct the new graph G' by joining every vertex of a new m-clique D to every vertex in C. Then the new graph is also k-n-connected.

In constructing both families of graphs Cl_k and Co_k, we proceeded by stringing together a number of k-cliques in a very precise way. The result was that to produce examples of large order, we needed to begin with cliques of large size. This of course begs the question to what extent these graphs serve as realistic models for terrorist organizations, where internal network security is necessarily a predominant criterion. Viewed from this perspective, it would be desirable to construct graphs that are k-n-connected but in which the maximal clique size is kept small. This question is addressed in [9], [10] and [11].

It is not our intention here to give a detailed summary of results; we will however highlight some of the relevant ideas. In the following we will work with the assumptions that G is a k-regular, k-n-connected graph, and that G does not contain any chordless 5-cycles (i.e., no 5-cycles without diagonals). Suppose further that $C = Z_1, \ldots, Z_m$ is a maximum m-clique in G. Then (as is shown in [9]), G must have the macrostructure shown in Fig. 10. Here, $B_i = \{X \in V(G) | X \in N(Z_i)/C\}$and E_i consists of those vertices not in the central clique C that are connected to some vertex in B_i. A careful count now allows us to establish bounds on the number of vertices in G. This is made explicit in the following result.

Assume that G is a k-regular, k-n-connected graph containing no chordless 5-cycles, and containing maximum m-cliques. Then

$$-\alpha m^3 + \beta m^2 + \gamma m \leq |V(G)| \leq -km^2 + (k^2+k+1)m$$

where $\alpha = \lambda^2 + 3\lambda + 1$, $\beta = 2\lambda(k+\lambda+3) + (k+1)$, $\gamma = 1 - \lambda(\lambda + 2k + 3)$ and $\lambda = \lfloor \frac{k}{m-1} \rfloor - 1$.

In particular, when $k=3$ and $m=2$, this tells us that $14 \leq |V(G)| \leq 14$ and when $k=4$ and $m=3$, then $21 \leq |V(G)| \leq 27$.

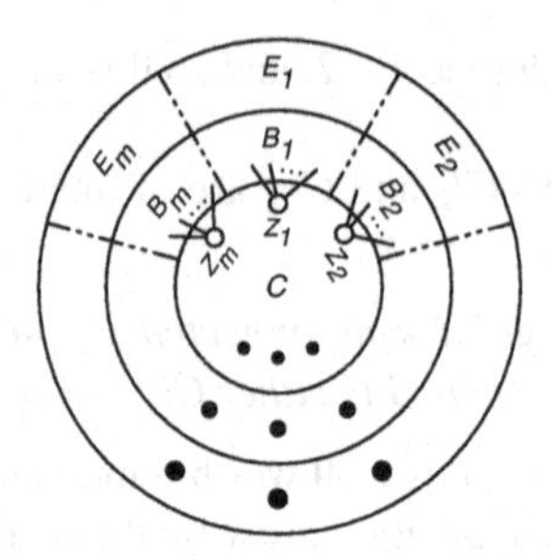

Fig. 10

Much of the literature has focused on attempts to determine specific information about the structure of k-regular, k-n-connected graphs that are of minimal order. There exists an intimate relation between this question and balanced incomplete block designs (b,v,r,k,l) consisting of v objects and b blocks such that each block consists of k objects, each object occurs in r blocks and each pair of objects occurs in exactly one block. Such a design can be associated in a natural fashion with a bipartite graph $G(b,v,r,k,l)$, where the blocks and objects form respectively the two color classes of vertices, and where the adjacency matrix of G is derived from the incidence matrix of the design. As is shown in [11], the following result holds.

Theorem 6. *If a (b,v,r,k,l) design exists and $k \geq 3$, then the graph $G(b,v,r,k,l)$ is k-n-connected.*

In particular, if the design is a finite projective plane of order $k-1$ (so that we have a $\left(k^2-k+1,k^2-k+1,k,k,1\right)$ design), then the associated graph is also k-regular, of girth 6, and is of order $|V(G)| = 2(k^2-k+1)$. Indeed, it can be shown, for $k \geq 3$, that every k-regular graph of order $|V(G)| = 2(k^2-k+1)$ must be the graph $G\left(k^2-k+1,k^2-k+1,k,k,1\right)$.

We show the smallest graph obtained by this construction. This is the graph associated with the minimal projective plane $PG(2,2)$ of order 2, (which is the design $(7,7,2,2,1)$). Fig. 11 shows both the minimal projective plane $PG(2,2)$ and the associated 3-regular, 3-n-connected graph $G(PG(2,2))$, which we recognize as the familiar Heawood graph on 14 vertices.

There is another way to obtain k-regular, k-n-connected graphs from finite geometries. Here we can now start with either a finite projective or affine plane geometry Π. A flag in such a finite geometry is any pair (X,x) where X is a point, x is a line and X lies on x. We now construct the associated flag-graph $Fl(\Pi)$ whose vertices are the flags in the geometry and where vertices (X,x), (Y,y) are adjacent if either $X=Y$ or $x=y$. It is shown in [9] and [10] that the flag-graphs constructed in this way have the neighbor-connectivity properties summarized in the following theorem.

Theorem 7. *Let Π be finite projective or affine, planes of order n, and let $Fl(\Pi)$ be the associated flag-graph constructed in the manner described above. Then the following holds:*

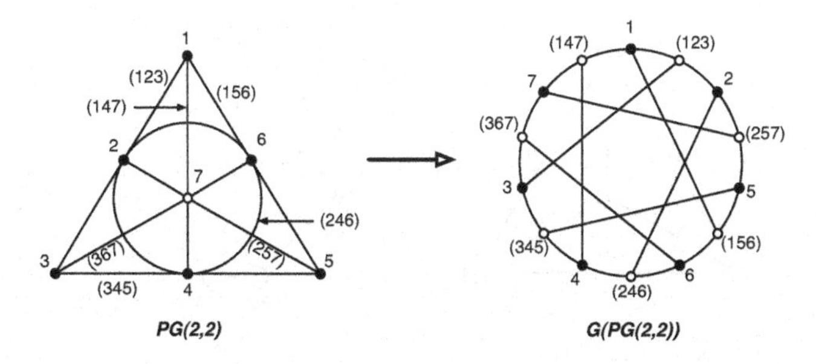

Fig. 11

1. *When Π is a projective plane: $Fl(\Pi)$ is a $2n$-regular, $2n$-neighbour connected graph on $(n+1)(n^2+n+1)$ vertices, containing maximum $(n+1)$-cliques. Every vertex belongs to exactly two $(n+1)$-cliques.*
2. *When Π is an affine plane: $Fl(\Pi)$ is a $(2n-1)$-regular, $(2n-1)$-neighbor connected graph on $n^2(n+1)$ vertices, containing maximum $(n+1)$-cliques. Every vertex belongs to exactly one $(n+1)$-clique and one n-clique.*

For example, if Π is the minimal projective plane $PG(2,2)$ illustrated above, then we see, for example, that the flag $(1,(123))$ is adjacent to the flags $(1,(147))$, $(1,(156))$, $(2,(123))$ and $(3,(123))$. Since the geometry is 1-transitive on the points, we see that the associated graph $Fl(PG(2,2))$ is 4-regular. As well, we see that the vertices $(1,(123))$, $(2,(123))$ and $(3,(123))$ form a 3-clique in the graph. Since the projective plane $PG(2,2)$ contains 7 points, with 3 points on each line we conclude that the associated flag graph $Fl(PG(2,2))$ is a 4-regular graph on 21 vertices which contains maximum 3-cliques, and which is 4-n-connected. This graph is shown in Fig. 12.

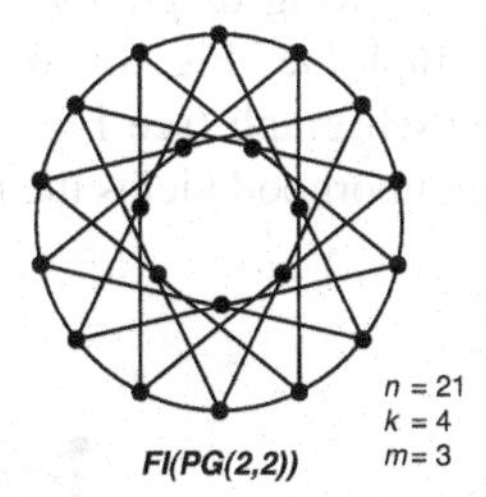

Fig. 12

We can now form the affine plane $AG(2,2)$ from $PG(2,2)$ in the usual way by deleting one line of $PG(2,2)$. For instance, if we delete the line labeled (246), and build up the graph $Fl(AG(2,2))$ we obtain the graph shown in Fig. 13. Note that in this case, we recognize the graph $Fl(AG(2,2))$ as the familiar clique graph Cl_3; when we go to finite geometries of higher orders, this construction yields a new family of graphs.

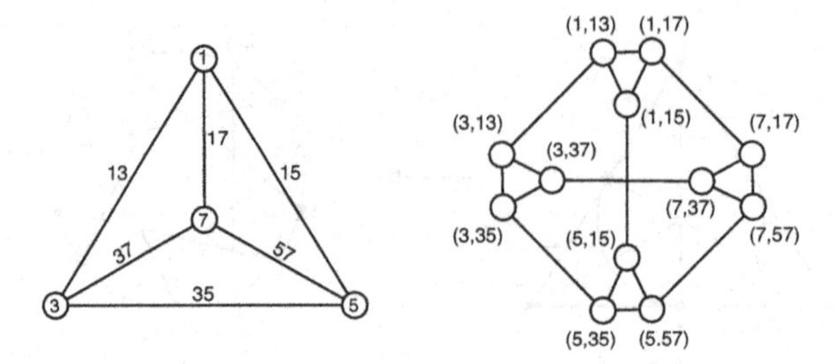

Fig. 13

The definition of neighbor connectivity implies that when a graph G is attacked by some subversive strategy C, then all of the subverted neighborhoods are to be removed simultaneously. It is arguable whether such an assumption is realistic in the context of this chapter. If G is meant to model some terrorist organization, it might be more realistic to assume that this organization will be attacked by successive deletions of closed neighborhoods. For example, a single agent A might be arrested with the subsequent betrayal of the agents he knows. Questioning of these agents will then give the counter-terrorist agency the information needed to make a second arrest of someone who was not in the closed neighborhood of A.

This reasoning suggests a useful variant of the concept of neighbor-connectivity. We now insist that the vertices in the subversive strategy C should be independent in the graph (in other words, none of the vertices lies within the neighborhood of another). We can then look for graphs G that are k-independent-neighbor-connected (or k-INC) , which means that for any set C of $k-1$ independent vertices in G, the deletion of the closed neighborhoods of the vertices in C the survivor graph is connected but not complete. An initial analysis of this problem (see [7] for details) shows that this results in different graphs. Specifically, we include the smallest 3-INC and 4-INC graphs in the following diagrams. Note that the smallest 3-INC graph is the familiar Peterson graph (see Fig. 14) of order 10, while the smallest 4-INC graph known as the Clebsch graph (see Fig. 14) has the property that the deletion of any single closed neighborhood yields the Peterson graph as its survivor graph.

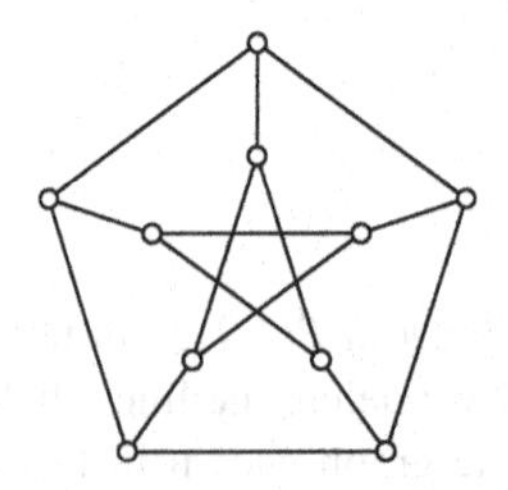
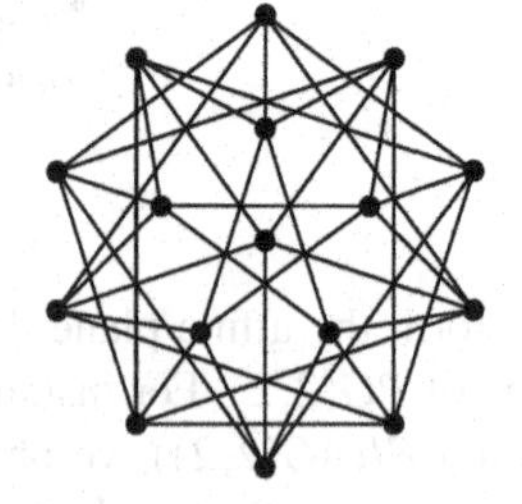

Fig. 14 Peterson Graph and Clebsch Graph

In concluding this discussion, it is worth making the following point. If the graph G serves as the model of a terrorist organization, then all of the questions have been of the type that optimize some operational aspect of such an organization. In a real scenario, it is unlikely that the organization would have been set up along such optimal lines. However, it is of value to know what such optimal networks might look like; this kind of knowledge provides valuable insight into how effective an attack by a counter-terrorist agency might be when it is carried out against a real terrorist network.

References

1. Gunther G., Hartnell, B. L.: On minimizing the effects of betrayals in resistance movements, Proceedings of the Eighth Manitoba conference on Numerical Mathematics and Computing, 285–306 (1978)
2. Hartnell, B. L.: The optimum defense against random subversions in a network, Proc. 10th S-E Conf. Combinatorics, Graph Theory and Computing, 493–499 (1979)
3. Gunther, G., Hartnell, B. L., Rall, D.: On maximizing the average order of surviving components, Congressus Numerantium 92, 97–104 (1993)
4. Gunther, G., Hartnell, B. L.: Optimal k-secure graphs, Discrete Applied Mathematics 2, 225–231 (1980)
5. Hayes, B.: Connecting the dots, American Scientist 94, 400–404 (2006)
6. Gunther, G.: Neighbor-connectivity in regular graphs, Discrete Applied Mathematics 11, 233–243 (1985)
7. Gunther, G., Hartnell, B. L.: On m-connected and k-neighbour-connected graphs, Proceedings of the Sixth Quadrennial International Conference on the Theory and Applications of Graphs Western Michigan University, Kalamazoo, MI, 585–596 (1991)
8. Gunther, G.: On the existence of neighbour-connected graphs, Congressus Numerantium 54, 105–110 (1986)
9. Gunther, G., Hartnell, B. L.: Flags and neighbour connectivity, Ars Combinatoria 24, 31–38 (1987)
10. Gunther, G.: Minimum k-neighbour connected graphs and affine flags, Ars Combinatoria 26, 83–90 (1988)
11. Gunther, G., Hartnell, B. L., Nowakowski, R.: Neighbor-connected graphs and projective planes, Networks 17, 221–247 (1987)
12. Hartnell, B. L., Kocay, W.: On minimal neighbourhood-connected graphs, Discrete Mathematics 92, 95–105 (1991)
13. Boesch, F., Tindell, R.: Circulants and their connectivities, J. Graph Theory 8, 487–499 (1984)
14. Doty, L.L., Goldstone, R.J., Suffel, C.L.: Cayley graphs with neighbor connectivity one, SIAM J. Discrete Math. 9, 625–642 (1996)
15. Goldstone, R. J.: The structure of neighbor disconnected vertex transitive graphs, Discrete Mathematics 202, 73–100 (1999)
16. Doty, L. L.: A new bound for neighbor-connectivity of abelian Cayley graphs, Discrete Mathematics 306, 1301–1316 (2006)
17. Cozzens, M.: Extreme values of the edge-neighbor connectivity, Ars Combinatoria 39, 199–210 (1995)
18. Gambrell, M. J.: Vertex-neighbor-integrity of magnifiers, expanders, and hypercubes, Discrete Math. 216, 257–266 (2000)
19. Cozzens, M., Wu,S. Y.: Relationships between vertex-neighbor-integrity and other parameters, Ars Combin. 55, 271–282 (2000)

20. Cozzens, M., Wu, S. Y.: Vertex-neighbor-integrity of powers of cycles, Ars Combin. 48, 257–270 (1998)
21. Cozzens, M., Wu, S. Y.: Bounds of edge-neighbor-integrity of graphs, Australas. J. Combin. 15, 71–80 (1997)
22. Cozzens, M., Wu, S. Y.: Vertex-neighbor-integrity of trees, Ars Combin. 43, 169–180 (1996)
23. Cozzens, M., Wu, S. Y.: Edge-neighbor-integrity of trees, Australas. J. Combin. 10, 163–174 (1994)

Intelligence Constraints on Terrorist Network Plots

Gordon Woo

Abstract Since 9/11, the western intelligence and law enforcement services have managed to interdict the great majority of planned attacks against their home countries. Network analysis shows that there are important intelligence constraints on the number and complexity of terrorist plots. If two many terrorists are involved in plots at a given time, a tipping point is reached whereby it becomes progressively easier for the dots to be joined and for the conspirators to be arrested, and for the aggregate evidence to secure convictions. Implications of this analysis are presented for the campaign to win hearts and minds.

1 Introduction

Societies where there are strong professional law enforcement and intelligence forces are very different in their susceptibility to terrorist attack from societies where the police and security services are weak, corrupt or compromised. In the former case, terrorism is subject to a significant degree of state control resulting from an effective counter-terrorism environment that is hostile to terrorist plotting. In the latter case, terrorists may have considerable opportunity to plot and attack at will, with the style and frequency of attacks being less determined by external counter-terrorism constraints than by terrorist campaign strategy and military logistics. In keeping with the terrorist aspiration for success in the resort to violence, national variations in security tend to encourage international terrorist threat shifting from one country to another.

Intelligence is the first and most important defence against terrorist plots. Even in the absence of adequate protection of potential targets, losses can be avoided if plots are interdicted, or if the risk of interdiction curtails attack preparation and rehearsal,

Gordon Woo
Risk Management Solutions Ltd., 30 Monument Street, London EC3R 8NB, UK, e-mail: gordon.woo@rms.com

N. Memon et al. (eds.), *Mathematical Methods in Counterterrorism*,
DOI 10.1007/978-3-211-09442-6_12,

thus inducing plot failure. Since 9/11, there have been more than fifty planned Jihadi attacks against the principal coalition home countries: USA, Canada, Western Europe, Australia and Singapore. It is a tribute to the diligent professionalism of the security services of these countries, and the solidarity of international security cooperation, that the great majority of these plots have failed: some notable successes have been the Madrid rail bombings of March 11, 2004, and the London transport bombings of July 7, 2005.

Irrespective of adaptive inventive behavior of terrorists in target and weapon selection, novel types of terrorist attack can be thwarted through meticulous intelligence information gathering. The level of intelligence gathering that is acceptable in a democratic society varies according to the perceived threat level. Surveillance and other counter-terrorism measures that might infringe upon civil liberties must be kept commensurate with the threat. Whilst it is the responsibility of civic authorities to establish the appropriate balance, it is the duty of libertarians to challenge whether counter-measures are more than proportionate.

Champions of civil rights warn that excessively authoritarian actions targeted against a specific population group can cause alienation and hence be counterproductive [1]. Instead of ever repressive surveillance, emphasis should be placed on winning over the hearts and minds of dissidents providing support for militants. The presence of a significant body of community support for the cause espoused by militants sustains prolonged campaigns of violence by non-state groups. In protracted separatist, racial and sectarian conflicts, populist support is continuously drawn from the underlying nationalist, ethnic and religious communities.

Even if only a hard core minority of a community might be motivated to participate directly in acts of violence, a much more substantial proportion might be prepared to turn a blind eye to evidence of terrorist plotting, if not provide active support to facilitate plots, such as with financing, safe housing, target surveillance, weapon materiel procurement and other operational logistics, etc. Problematic for counter-terrorism forces is that, within this sizeable group of active supporters of a militant cause, there may be no effective profiling system capable of selecting and identifying those more likely to be involved in facilitating a plot. As is clear from the profiles of known Islamist militants, terrorists may have any socio-economic, educational and religious background; they may be male or female, and span a wide range of age, nationality and ethnicity. The randomness factor in the emergence of the population of active support considerably broadens the social network of terrorist links.

2 Tipping Point in Conspiracy Size

As far as the counter-terrorism security services are concerned, the group of active supporters constitutes a very large essentially random social network, interlocking sub-networks of which may be involved in terrorist plots at any given time. Key to the disruption of plots is the discernment of links between nodes of a terrorist sub-network. The higher the likelihood of identifying a link between terrorist two nodes, the clearer a pattern of connections will become, and the easier it is to 'join the dots' to disrupt a terrorist plot, and gain the necessary corroborative evidence for successful criminal proceedings.

The Renaissance political theorist, Nicolò Machiavelli [2], noted that conspiracies have to be kept small if they are to elude the attention of the authorities. The larger a conspiracy is, the greater the chance that it will fail haphazardly due to a tip-off from an inside informant or an observant member of the public, through indiscretion or betrayal by one of the conspirators, or the investigative diligence and detective skill of law enforcement authorities. After a hotel bomb explosion in October 1984 narrowly failed to kill the British Prime Minister, Margaret Thatcher, the IRA famously warned the British authorities that *"we only have to be lucky once, you will have to be lucky always"*. The individual responsible for the bombing, Pat Magee, was arrested shortly afterwards, but the comparatively small size of the IRA attack cell precluded the prior leakage of information to the British intelligence services. The larger the size of a conspiracy is, the likelier that the authorities will be "lucky" in foiling it.

As the most notorious example, the 9/11 Al Qaeda operation came close to failure when Zacharias Moussaoui was taken into FBI custody on August 16, 2001. According to George Tenet [3], CIA learned a week later via the French security service, DST, that Moussaoui had Chechen terrorist links. Despite this information, and the FBI Minneapolis field office's suspicion over Moussaoui's flight training intent, lawyers at FBI headquarters believed a search warrant could not be authorized. A week after 9/11, CIA belatedly found evidence that Moussaoui had connections with a Malaysian company managed by Yazid Sufaat, whose condo in Kuala Lumpur was used for a 9/11 planning meeting, attended by two of the hijackers, Khalid Al Midhar and Nawaf Al Hazmi. Al Qaeda's good fortune in keeping the 9/11 plot secret continued with the release, following freeway detention, of Flight 93 pilot Ziad Jarrah on his penultimate journey.

Audacious though 9/11 was, there was an earlier Al Qaeda version of the plot, which was even more ambitious, but the leadership prudently scaled it down, realizing they would be pushing their luck. A devastating follow-up wave attack planned by Al Qaeda on embassies and other iconic targets in Singapore failed through astute intelligence gathering. Al Qaeda's luck with grand scale operations had run out: a member of the public tipped off the Singapore authorities about a Singaporean of Pakistani ancestry who boasted that he knew Osama bin Laden. The monitoring of close associates led to the audacious and menacing six truck bomb plot being gradually unraveled [4].

The capability of intelligence services to detect links between possible terrorist nodes is restricted by the cost of enhanced surveillance, as measured in terms of the depletion of both financial and human spying resources, as well as civil liberties. The practical need for the UK security service MI5 to prioritize surveillance resources was highlighted in the aftermath of the Operation Crevice trial in London, which concluded on April 30th, 2007. The conviction of five terrorists for plotting bomb attacks in England in 2004 was an undoubted success for MI5, but one muted by the public disclosure that the ringleaders of this plot were known by MI5 to have met with Siddique Khan and Shehzad Tanweer, the ringleaders of the July 7th, 2005 London transport bombings.

Out of 55 associates of the Operation Crevice plotters, 15 were classified as essential targets to be put under surveillance, the remaining 40, including Khan and Tanweer, being classified only as desirable targets. A link surveillance ratio of 15 out of 55 associates is 0.27. For this comparatively sizeable surveillance value to be achieved, Operation Crevice involved every police and MI5 surveillance team in southern Britain, and targeted homes, phones, cars and people. With the full allocation of such resources, the probability of detecting terrorist links was high, although not high enough to thwart the 2005 London bombings. But had Khan and Tanweer been actively preparing their plot in 2004, MI5 might have assessed them to be more than petty criminals on the fringes of terrorist activity, and kept them under surveillance.

As a conspiracy expands in size, or as a series of conspiracies are interlocked, so the discernible signature of plotting becomes increasingly recognizable. Given the demanding burden of proof required to bring criminal charges and then secure terrorism convictions in the courts of justice of the western democracies, evidence gathered from a cluster of suspects is of compounded value in making arrests to interdict plots. Not only are there multiple witnesses, but cell phone and computer records gathered for a cluster of suspects may reveal mutually incriminating plot evidence, which collectively is far stronger than if only one or two isolated arrests were made. The intuition that too many terrorists spoil the plot, can be expressed in a quantitative way utilizing modern developments by network cluster theorists working on complex systems in the physical, biological and social sciences.

Suppose that, within the large disparate population of terrorist supporters, there is a group who are actively involved in operational planning at a particular time. It is in the general terrorist interest to tend to randomize network connections, to keep security services guessing about plot involvement, and defeat efforts at profiling. However, the disconcerting news for terrorists is that, even with randomization, there is a tipping point in the size of the group, beyond which the presence of conspiratorial plotting should become increasingly manifest to the security services. Even the most careful plotting may be dashed by excessive numbers of operatives.

Population clusters forming within random networks may be analyzed graphically in terms of a basic clique of three people, who all are interlinked. Erdõs-Rényi graph theory percolation analysis [5] reveals a dramatic change in the topological features of a graph, i. e., a percolation or tipping point transition, when the link detection probability attains the value: $p = (2N)^{-1/2}$, where N is the size of the group. Rearranging this formula, the tipping point arises where the size of group exceeds one half of the inverse-square of the link detection probability. There is an intuitive argument underlying this percolation transition criterion. In extending a chain of cliques, the expectation value of the number of adjacent cliques, differing in only one vertex, is equal to unity at the tipping point. A smaller expectation value would result in the size of clusters decaying; a larger expectation value would lead to a series of bifurcations, indicative of a giant conspiratorial cluster.

The nonlinearity embedded in the expression for the link detection probability embodies in a concise and convenient manner the rapidly escalating dependence of counter-terrorism performance on surveillance capability. The greater the link detection probability is, the smaller the size of conspiracies that can be disrupted. Conversely, the smaller the link detection probability, the more tenuous is the prospect of identifying plots.

To illustrate this rule, if the link detection probability is as high as 1/2, then the tipping point is at 2; if it is 1/10, then the tipping point is at 50; and if the link detection probability is as small as 1/50, then the tipping point is 1250. In circumstances where the link detection probability is very slim, e. g., 1/100, then the tipping point is too far removed at 5000 to be of practical interest.

3 Tipping Point Examples

The London courtroom disclosure of information on surveillance ratios relating to Operation Crevice has helped bring out into the public domain the important role of link detection in plot interdiction. Insight into the significance of link detection probability can be gained from considering some relevant examples. Commenting after the London bombings of 2005, the director-general of MI5, Eliza Manningham-Buller, remarked that *"MI5 would have to be the size of the Stasi to have the chance of stopping every possible attack, and even then, it would be unlikely that it would succeed."* This is a stark commentary on the superior capabilities and powers that police states have in combating dissidents. A deeper understanding of the meaning of this statement can be gained by examining it in simple arithmetic terms.

In the German Democratic Republic of the repressive Honecker regime, the Stasi kept records on about 5 million of 17 million East German citizens. Even though a quarter of the total population were under Stasi scrutiny, the proportion that might have been actively supportive of a regime change, and prepared to risk arrest, was smaller. Recognizing that the dissident movement consisted largely of the educated middle-class, up to around 10 % of the population might have been potential activists, in which case the surveillance ratio of potentially radical citizens to 85,000 Stasi agents was about 20:1.

It is known that the Stasi recruited several hundred thousand informers. Assuming that an agent, in conjunction with a team of informers and police officers, could keep reasonable surveillance of approximately five citizens, there would be a quarter chance that a member of a GDR conspiracy would be under surveillance. Since each link has two nodes, the chance of a link being detected between two conspirators would then be about twice this, or 1/2. According to the tipping point rule, for such a high link detection probability, the corresponding tipping point is 2: implying that any conspiracy against the East German state involving more than two people would have had very little chance of escaping detection. The state surveillance apparatus would implicitly have been sized to achieve this extremely high level of state protection against conspiracy.

A contrasting security environment was that of Northern Ireland in 1992, before the initiation of the Ulster peace process. The lack of state control over republican terrorism is apparent from the low link detection probability. There were 600 MI5 officers in Northern Ireland gathering intelligence on the IRA [6], which had the active support of about 300,000 Irish people. Dating as far back as the 1916 Easter Rising, the IRA had the advantage of working amidst a population the greater proportion of which was friendly [7]. With each MI5 agent being aided by informers and the local constabulary, as well as tip-offs from the general public, so as to keep reasonable surveillance of approximately five citizens, a poor surveillance ratio of approximately 500 to one equates to a pretty slim link detection probability of 1/50. This corresponds to a large tipping point value of 1250. With such a high tipping point, terrorists had considerable operational latitude to plan and perpetrate numerous attacks, even though a high proportion might yet have been interdicted. Indeed, the terrorism catalogue for 1992 is grim: apart from hundreds of terrorist shootings, there were 371 explosions in Northern Ireland in that year.

An intermediate regime is that of UK in 2008. According to UK opinion polls, about 6 % of adult British Muslims thought that the 7/7 London bombings were justified. Assuming this ratio approximately holds for the half million Muslim visitors, most of whom are from Pakistan, an estimated 100,000 Muslims within Britain might actively support the Jihad. The roots of Islamist militancy in UK are to be found in the liberal asylum policy of the past decades, which afforded refuge to many radical Islamists [8], at a time when the terrorism focus of MI5 was on the IRA. Countering the rising UK threat of Islamist militancy since 9/11 through the doubling of its pre-9/11 staff, MI5 have around a thousand personnel for Islamist terrorism surveillance operations. Following the same argument as above, a surveillance ratio of 1/100 equates to a link detection probability of about 1/10, and a moderate tipping point of 50.

This would considerably constrain the annual frequency of planned macro-terror plots. Detailed information on terrorist plots emerging from the plethora of UK terrorism court cases shows the interlocking pattern of Jihadi cells, arising from various common social network factors such as camp training in Pakistan and mosque attendance. This has been directly confirmed by a spokesman for the radical group Al Muhajiroun, Hassan Butt, who has since renounced violence, and made public new information about Islamist militants in UK. As a British-Pakistani Jihadi, he was trained in Pakistan, where he made network connections with many others, including Mohammed Saddique Khan, the 7/7 London bomb leader. If the aggregate number of terrorists involved in concurrent planning were to exceed 50, plotting would be jeopardized through the detection of occasional links. However, it would still be possible for a moderate number of major terrorist attacks to be planned, including spectaculars using CBRN weapons.

By contrast with their British Muslim brethren, American Muslims are much better integrated into society, and are widely accepted as citizens rather than as perpetual foreigners. Reciprocally, they are more likely to show greater loyalty to their fellow US citizens than to the Muslim community around the world. Few American Muslims would have thought that 9/11 was justified, but according to a detailed Pew

Research Center survey [9], 5 % of adult American Muslims have a favorable attitude towards Al Qaeda. This equates to about 75,000 people. Of these, some might be prepared to be involved in some active way with a Jihadi conspiracy. This may be a fairly moderate number amounting to a few tens of thousands. But as tragically demonstrated by 9/11 itself and by revelations of later plots, apart from American Muslims, a major security concern is the community of Muslim visitors and immigrants to the USA, including over a hundred thousand on special registration under the NSEER (National Security Entry-Exit Registration System).

Altogether, it has been estimated by Muslim rights advocates that the size of the 'virtual internment camp' under watch is about 200,000. Comparing this with some 2000 FBI field agents in the Joint Terrorism Task Force, the ratio of potential surveillants for each such agent is about 100 to 1. Following the same argument as above, a surveillance ratio of 1/100 equates to a link detection probability of about 1/10, and a moderate tipping point of 50, similar to that in UK. As before, this is a notable constraint on the annual frequency of planned macro-terror plots: American cities will not be experiencing the daily bombings which have been witnessed in Iraq. But still possible would be the annual planning of a moderate number of major terrorist attacks, including spectaculars using CBRN weapons. The notion that there is some practical limit of this kind for terrorist operations against the US homeland would have been intuitive to a prudent leader, such as Osama bin Laden. Excessive complexity and size can be counter-productive. In May 2001, in the notorious Al Farooq training camp, Osama bin Laden had advised his followers that, *"There are 50 men willing to bear their souls in their hands for the Jihad on a mission to attack America."*

Compared with both UK and the USA, the inquisitorial legal system in France allows for more state control of potential terrorist activities, not least the power to detain suspects for several years, pending the completion of a magistrate's investigation. The need for a strong central French state has evolved over centuries of dealing with sedition and terror. Whereas across the English Channel, the compulsory carrying of an identity card is seen as an infringement of civil liberties, in France, an identity card is the seal of a strong, centralized government.

The flexible French state apparatus for monitoring and investigating the terrorist threat in France includes the Direction de la Surveillance du Territoire (DST), the Direction Centrale des Renseignements Généraux (DCRG), and the Division Nationale Anti-terroriste (DNAT). Across the various security agencies, the number of surveillance officers tracking the Jihadi terrorist threat to France is two to three times larger than in UK. But then the Muslim population resident or visiting France is also two to three times as large as that in UK. This population is mainly of North African origin, where radical Islam has a strong foothold as in Pakistan, from where most British Muslims originate. The Pew Research Center survey indicates as high a level of French support for suicide bombing as in UK. Accordingly, the ratio of potential surveillants for each French intelligence agent is about 100 to 1, indicating an approximately similar link detection probability of 1/10.

In France, as with UK and the USA, state resources are adequate to ensure that intelligence operations have a significant effect in controlling terrorism. Most plans are disrupted, and active cells are infiltrated or otherwise neutralized, in effective ways which have been charted by a past informer [10]. Given that the Jihadi threat may shift from one of these leading western nations to another, in response to differential counter-terrorism pressure, it is to be expected that the respective threat levels would be broadly similar. However, the occurrence of a successful attack in any of these countries is still a question of when it will happen, not if it might ever happen: sooner or later, an attempted attack will succeed. On June 29 and 30, 2007, gas bomb attacks in London and Glasgow evaded MI5 surveillance, but failed for technical reasons. One of the reasons why this plot evaded the surveillance net was the small size of the conspiracy. In particular, the multiplicity of the attack was comparatively low: two car bombs in London, and one in Glasgow.

4 Stopping Rule for Terrorist Attack Multiplicity

The multiplicity of a coordinated attack is another aspect of terrorism attack planning which is constrained by the threat of detection by the intelligence and law enforcement services. Multiple points of attack are a defining hallmark of Al Qaeda. The larger the swarm of synchronous attacks is, the greater the scale of potential loss, and the more spectacular the impact. However, there is an optimal stopping point in multiple attack planning, similar to the predicament that crafty burglars face in hoarding stolen goods before disposing of them for gain [11]. Beyond some multiplicity, the risk of the plot being discovered, with the wastage of all the planning resources expended to date, outweighs the marginal benefit of enlarging the plot still further. With increasing resources allocated to counter-terrorism services, the optimal stopping point may be getting earlier, with the consequence that the multiplicity of spectacular attacks is constrained.

Suppose that a terrorist organization implements the "burglar's" optimal stopping rule as defined by Haggstrom. This formulation by Haggstrom provides a convenient model for studying some extensions of optimal stopping problems which were first considered by Gilbert and Mosteller [12]. In the city in which multiple synchronous attacks are planned, the potential yields from strikes are assumed to be samples from an exponential distribution with mean loss L. This is a reasonable assumption, based on knowledge of terrorist modus operandi and the spectrum of plausible attack modes.

Let the probability of being caught while preparing for another strike be p, and let us write q as the complement $1-p$. Then the optimal stopping rule is that the planned attack multiplicity should be ratcheted up until the sum of the sampled losses from strikes exceeds the threshold: $Z = qL/p = (1/p-1)L$. We are interested in the probability distribution for the multiplicity of successful attacks, which are not interdicted in the preparation process. Let N be the number of strikes causing losses building up towards, but not exceeding, the threshold Z. Given that strike losses have an exponential distribution, which represents arrivals for a Poisson distribution, the number of strikes causing aggregate losses not exceeding Z has a Poisson distribution with mean $(1/p-1)$. Adopting the burglar's rule, the attack multiplicity distribution is then obtained simply by shifting upwards the number of strikes by one; this corresponding to that ultimate strike which takes the aggregate loss over the threshold Z. The probability that the attack multiplicity is M is thus just the Poisson distribution probability: $P(N=M-1)$ where $M=1,2,3,\ldots$. For specific attack modes, the Poisson distribution can be parameterized from data on planned and actual attacks.

5 Preventing Spectacular Attacks

The cornerstone of UK counter-terrorism strategy is the set of four P's: *protecting, preparing, pursuing and preventing*. But whatever the defensive resources allocated for protecting against, and mitigating the consequences of, a chemical, biological, radiological or nuclear (CBRN) attack, prevention is still the best policy. Accordingly, since 9/11, western security services have been granted ever increasing resources for interdicting CBRN plots. Are these resources adequate? A minimum number of terrorists required for a CBRN spectacular operation is about a dozen. For plots of this size to have a high chance of interdiction, the link detection probability would need to be about 0.2. At current Jihadi threat levels, it would take a doubling of surveillance operations in the US to double the link detection likelihood from 0.1 to 0.2. Whilst some further rise in surveillance may be acceptable in a free society, draconian increases in surveillance would not be tolerable, and may so alienate the Muslim community as to be counter-productive. Ultimately, a residual level of public exposure to a spectacular terrorist attack is the price paid for maintaining civil rights in a democratic society – a free society can never be risk-free.

For the terrorist threat to be lowered in democracies of the western alliance, heightened surveillance is not the answer; there has to be a reduction in the Muslim (Umma) sub-population who consider terrorism against these countries to be justified. Psychology has been an art of war for millennia, but the technological means of strategic persuasion have reached their zenith in the 21st century. Al Qaeda's chief strategist, Ayman Al Zawahiri, wrote as follows to Musab Al Zarqawi in July 2005: *"I say to you: that we are in a battle, and that more than half of this battle is taking place in the battlefield of the media. And that we are in a media battle in a race for the hearts and minds of our Umma."* Overseeing propaganda is the head of Al Qaeda's Media Committee, Abu Abdel Rahman Al Maghrebi, the son-in-law of chief strategist Dr Ayman Al Zawahiri. Under dynamic leadership, Al Qaeda's media arm, Al Sahab, has substantially increased its annual output of audio and video messages.

For security in UK, with its traditional colonial links with Pakistan and large diaspora community, there is a need to win hearts and minds of young Pakistanis. This was recognized in an address on July 27, 2007, to the Pakistan Youth Parliament, by the British Foreign Secretary. David Miliband delivered the following qualitative message on international relations, which is supported by the quantitative terrorism risk analysis presented here: *"Diplomacy needs to change to be about winning hearts and minds, instead of bureaucrats holding meetings behind closed doors."* As an indication of the potential of this approach, after the planned terrorist attacks on Glasgow airport on 30 June 2007, one prominent Anglo-Pakistani extremist, Hassan Butt, reversed his terrorist perspective: *"Muslim scholars must go back to the books and come forward with a refashioned set of rules and a revised understanding of the rights and responsibilities of Muslims whose homes and souls are firmly planted in what I'd like to term the land of co-existence."* The more that this moderate message of co-existence comes across clearly, and drains general Muslim support for militancy, the tighter will be the intelligence constraints on terrorist plots.

References

1. Chakrabarti S. The first victim of war – compromising civil liberties, in Britain and Security, (Ed., P. Cornish). The Smith Institute (2007)
2. Machiavelli N., Discorsi sopra la prima deca di Tito Livio (1517)
3. Tenet G., At the Center of the Storm. HarperCollins (2007)
4. Bell S. The Martyr's Oath, John Wiley & Sons (2005)
5. Derenyi I., Palla G., Vicsek T. Clique percolation in random networks. Phys. Rev. Lett., Vol.94 (2005)
6. Coogan T.P. The I.R.A. Harpers Collins (2000)
7. Foy M.T. Michael Collins's Intelligence war. Sutton Publishing (2006)
8. Wiktorowicz Q. Radical Islam Rising. Rowman & Littlefield (2005)
9. Pew Research Center, American Muslims. May (2007)
10. Nasiri O. Inside the Jihad. Perseus books group. November (2006)
11. Haggstrom G.W. Optimal sequential procedures when more than one stop is required. Annals of Mathematical Statistics, Vol.38, No.6, pp.1618-1626 (1967)
12. Gilbert J.,Mosteller F. Recognizing the maximum of a sequence. J. Amer. Stat. Assoc., 61, 35–73 (1966)

On Heterogeneous Covert Networks

Roy Lindelauf, Peter Borm, and Herbert Hamers

Abstract Covert organizations are constantly faced with a tradeoff between secrecy and operational efficiency. Lindelauf, Borm and Hamers [13] developed a theoretical framework to determine optimal homogeneous networks taking the above mentioned considerations explicitly into account. In this paper this framework is put to the test by applying it to the 2002 Jemaah Islamiyah Bali bombing. It is found that most aspects of this covert network can be explained by the theoretical framework. Some interactions however provide a higher risk to the network than others. The theoretical framework on covert networks is extended to accommodate for such heterogeneous interactions. Given a network structure the optimal location of one risky interaction is established. It is shown that the pair of individuals in the organization that should conduct the interaction that presents the highest risk to the organization, is the pair that is the least connected to the remainder of the network. Furthermore, optimal networks given a single risky interaction are approximated and compared. When choosing among a path, star and ring graph it is found that for low order graphs the path graph is best. When increasing the order of graphs under consideration a transition occurs such that the star graph becomes best. It is found that the higher the risk a single interaction presents to the covert network the later this transition from path to star graph occurs.

Roy Lindelauf
Military Operational Art & Science, Netherlands Defence Academy, P.O. Box 90002, 4800 PA Breda, The Netherlands, e-mail: rha.lindelauf.01@nlda.nl

Peter Borm
CentER and Department of Econometrics and OR, Tilburg University, P.O. Box 90153, 5000 LE Tilburg, The Netherlands, e-mail: p.e.m.borm@uvt.nl

Herbert Hamers
CentER and Department of Econometrics and OR, Tilburg University, P.O. Box 90153, 5000 LE Tilburg, The Netherlands, e-mail: h.j.m.hamers@uvt.nl

N. Memon et al. (eds.), *Mathematical Methods in Counterterrorism*,
DOI 10.1007/978-3-211-09442-6_13,

1 Introduction

Recently an increasing interest in the application of methods from social network analysis to the study of terrorism can be observed. For instance in counterinsurgency social network analysis is recognized to be one of the most important tools to describe effects of the operational environment and to evaluate the threat [15]. Moreover, it is realized that methods from several mathematical disciplines are valuable in analyzing covert networks. Sageman [16] discusses the use of applying the network paradigm (clustering, small-world phenomena, etc.) to analyze terror networks. Social network analysis of specific terror events are available, although not abundant, see for instance Koschade [11,12]. There is something to be gained by applying and extending ideas from graph theory and game theory in the analysis of covert networks. Social and affiliation network analysis as well as spatiotemporal point pattern analysis are valuable mathematical methods that certainly warrant further exploration in the analysis of subversive activities, terrorism and guerrilla warfare.

Terrorism is not a topic that is easily researchable on the basis of practical data because of the clandestine nature of terrorist groups [9]. Therefore theoretical frameworks that describe how rational actors should behave in trying to attain certain strategic goals can provide insights into the functioning of terrorist groups. For instance, the strategic interaction between economic actors, within an explicitly given network structure, is modeled in Jackson [8]. In this paper we present and extend theoretical insights into the dilemma of secrecy and operational control in covert networks. In Lindelauf et al. [13] a theoretical framework on the homogeneous communication structure of covert networks is established. A secrecy measure and information measure are defined and the Nash bargaining criterion is adopted to determine optimal covert networks of given order. Several scenarios are analyzed. First, under the assumption of uniform individual exposure probability and high link detection probability it is shown that a star graph is optimal. However, on the assumption of low link detection probability it is shown that the complete graph is optimal. Second, if the exposure probability of individuals depends on their centrality with regard to information exchange it is shown that cellular networks are optimal.

This paper puts the theoretical framework on homogeneous covert networks to the test by applying it to the 2002 Jemaah Islamiyah Bali bombing. The theoretical framework does well in explaining most aspects of the network structure that Jemaah Islamiya adopted to carry out this attack. In addition however it is recognized that the nature of interaction between entities in a covert organization is not necessarily homogeneous. Hence the theoretical framework is extended to incorporate heterogeneity of the network. The most basic heterogeneous network is that in which all but one interaction present similar risks to the organization. The optimal pair of individuals that should conduct the interaction that presents the highest risk to the organization turns out to be the pair that is the least connected to the remainder of the network. In addition, when choosing among a path, star and ring graph with a single high risk interaction pair it is found that for low order graphs the path graph is best. Increasing the order a transition occurs such that the star graph becomes best. It is found that the higher the risk a single interaction presents to the covert network the later this transition from path to star graph occurs. Furthermore, approximate optimal networks given a single risky interaction are determined by simulation. This paper is organized as follows. After presenting some graph theoretical preliminaries in Sect. 2 secrecy and communication in networks and the main theoretical findings of Lindelauf et al. [13] will be reviewed in Sect. 3. The Jemaah Islamiyah 2002 Bali bombing operation will be discussed in Sect. 4 and compared to the theoretical results on optimal covert networks. In Sect. 5 the theoretical framework is extended to incorporate the heterogeneity of interaction between entities in covert networks.

2 Preliminaries

In this section we present graph theoretical preliminaries. A good general overview is given by Bollobas [3].

A graph g is an ordered pair (V,E), where V represents the finite set of vertices and the set of edges E is a subset of the set of all unordered pairs of vertices. An edge $\{i,j\}$ connects the vertices i and j and is also denoted by ij. The order of a graph is the number of vertices $|V|$ and the size equals its number of edges $|E|$. The set of all graphs of order n and size m is denoted with $\mathbb{G}(n,m)$. The set of graphs of order n is denoted by $\mathbb{G}^n$. In this paper we are only interested in connected graphs because we study the organizational form of groups in which the actions of individuals are coordinated. Therefore each graph under consideration is assumed to be connected. The degree of a vertex is the number of vertices to which it is connected. We denote the degree of vertex i in graph g by $d_i(g)$. A graph is called k-regular if all vertices have degree k. A star graph on n vertices is denoted by g^n_{star}. We denote a ring graph of order n by g^n_{ring} and a path graph of order n by g^n_{path}. The complete graph of order n is denoted by g^n_{comp}. See Figure 1 for an illustration of these graphs of order 5. The shortest distance (in number of edges one has to travel) between vertex i and j is called the geodesic distance between i and j. The geodesic distance between vertices i,j in g is denoted by $l_{ij}(g)$. Clearly, $l_{ij}(g) = l_{ji}(g)$. We will write l_{ij} instead of $l_{ij}(g)$ if there can be no confusion about the graph under consideration. The total distance $T(g)$ in the graph $g = (V,E)$ is defined by $\sum_{i,j} l_{ij}(g) = \sum_{i \in V} \sum_{j \in V} l_{ij}(g)$. The diameter $D(g)$ of a graph $g = (V,E)$ is defined to be the maximum over the geodesic distances between all pairs of vertices, i. e., $D(g) = \max_{(i,j) \in V \times V} l_{ij}(g)$. Furthermore, we assume without loss of generality that $n \geq 3$. We denote the set of 'neighbors at distance k' of vertex i by $\Gamma_{i,k}(g)$, i. e., $\Gamma_{i,k}(g) = \{j \in V | l_{ij}(g) = k\}$.

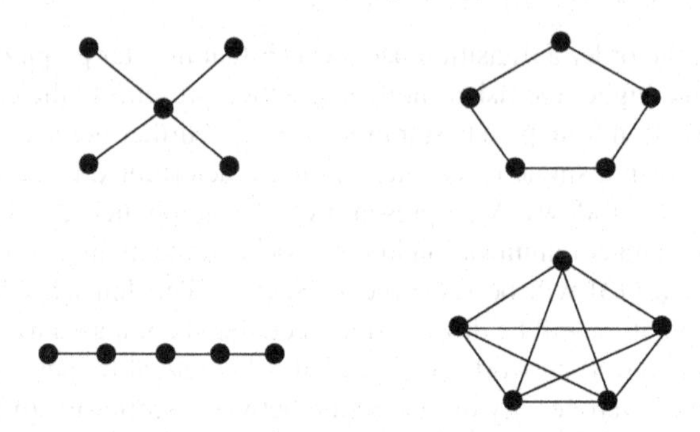

Fig. 1 Star graph of order 5 (top left), ring graph of order 5 (top right), path graph of order 5 (down left) and complete graph of order 5 (down right).

3 Secrecy and Communication in Homogeneous Covert Networks

Covert networks are constantly challenged with the tradeoff between secrecy on one side and operational capability on the other side. For instance, consider the failed Israeli covert operation in Egypt known as Operation Susannah [9]. Israel's Military Intelligence decided to set up a network of sleeper agents in Egypt which was activated in 1954 to prevent the British withdrawal from Egypt. However, after the arrest of a suspect names of accomplices of the operation were found in his apartment and the network was subsequently dismantled. This example shows that every member of a covert network presents a risk to the secrecy of the network in the sense that upon his exposure potentially others are exposed.

Another example of a covert network that was exposed is the following. During the 1950's a group of lawyers and legal experts that turned against the communist regime in East Berlin were selected by the CIA to be converted to an underground armed resistance group consisting of cells of 3 individuals each [18]. However, the network topology of this organization equalled that of a complete graph (the worst possible in the sense of secrecy) because all individuals were acquainted with each other. Upon the exposure of one network member the Soviets discovered and arrested all other members. The operation was a failure.

Another more recent example is that of Al Qaeda. It is currently widely known that Al Qaeda morphed from a bureaucratic, hierarchical organization into an ideological umbrella for loosely coupled jihadi networks [14]. We argue that the changing environment pressured the Al Qaeda leadership into adopting network topologies that maintain secrecy while simultaneously providing some possibility to coordinate and control. Videotapes of lecture series from the summer of 2000 show Abu Musab al Suri (Mustafa Setmariam Nasar) explicitly discussing (and providing critique of) hierarchical network structures [2,5]. Ideally the network should consist of small autonomous cells with limited strategic guidance. However, it is known that in reality there still exist weak bonds between local groups and experienced jihadists or Al Qaeda operatives, such was the case for instance in the Madrid and London attacks [17]. What is important is the fact that current covert organizations definitely take the secrecy versus operational efficiency dilemma explicitly into account.

The examples above show that it is important for a covert organization to take the network structure explicitly into account. Operation Susannah and CIA's East Berlin operation illustrated that failing to do this may result in failure of the operation. It appears that current terrorist organizations take their network structure explicitly into account as the example of Al Qaeda shows. In absence of further information we are interested in what structure these organizations actually adopt. In Lindelauf et al. [13] a theoretical framework for the analysis of the communication structure of covert networks was given. The optimal network structure was derived considering one of several scenarios. Below we recapitulate the theoretical framework and present the main results.

Imagine two agents, one responsible for network secrecy and the other one for information efficiency, bargaining over the set $\mathbb{G}^n$ of connected graphs of given order n. The information measure $I(g)$ of $g \in \mathbb{G}^n$ is given by

$$I(g) = \frac{n(n-1)}{T(g)}. \tag{1}$$

The secrecy measure $S(g)$ of a graph $g \in \mathbb{G}^n$ is defined as the *expected* fraction of the network that remains unexposed under the assumption of exposure probability of individual $i \in V$ being equal to α_i. The fraction of the network that individual i exposes (including himself) is defined to be $1 - u_i$. Then,

$$S(g) = \sum_{i \in V} \alpha_i u_i. \tag{2}$$

The balanced tradeoff between secrecy and information is modeled as a game theoretic bargaining problem. Hence, the optimal graph in the sense of the Nash bargaining solution is the graph $g \in \mathbb{G}^n$ that maximizes

$$\mu(g) = S(g)I(g). \tag{3}$$

Two scenarios are considered in Lindelauf et al. [13]. In the first scenario it is assumed that the probability of exposure of an individual in the organization is uniform over all network members, i. e., $\alpha_i = \frac{1}{n}$. Additionally it is assumed that the fraction of the network that individual i exposes is equal to the expected number of neighbors that will be detected if communication on links is detected independently and identically with probability p, i. e., we set $1 - u_i = \frac{pd_i+1}{n}$. The main result is that for a low value of p the complete graph is optimal and for a high value of p the star graph is optimal:

Theorem 1.

1. *If* $p \in [0, \frac{1}{2}]$, *then* $\mu(g^n_{comp}) \geq \mu(g)$ *for all* $g \in \mathbb{G}^n$,
2. *If* $p \in [\frac{1}{2}, 1]$, *then* $\mu(g^n_{star}) \geq \mu(g)$ *for all* $g \in \mathbb{G}^n$.

As an illustration consider the network structure of the former Dutch National Clandestine Service's so-called 'stay behind organization'. After the Second World War it was decided that precautionary measures should be taken such that in the event of a sudden invasion of the Netherlands a covert organization would be present to assist in subversive and covert activities to support the overthrow of the occupying forces [6]. This covert organization was divided into two groups: group A and B. Support group 'A' consisted of single agents all equipped with radio systems to connect to the Allied Clandestine Base (ACB). These single agents were not aware of each other because the chosen network structure equalled that of a star graph. Due to the extreme covert nature of this network (which was finally disbanded after the end of the Cold War in 1992) the initial exposure probability of network members may be assumed to be uniform. Communicating with the ACB presented a high link detection probability (high value for p) hence it can be argued that the star network design was an optimal choice. However, after operating for an extended period of time the exposure probability of the single agents would not be uniform anymore but would start to depend on their 'activity' in exchange of information. This alternative scenario is also analyzed in Lindelauf et al. [13]. In the second scenario it is assumed that the probability of exposure of an individual in the network $g \in \mathbb{G}(n,m)$ depends on his centrality with regard to the exchanging of information in the network. It is argued that setting $\alpha_i = \frac{d_i+1}{2m+n}$ for all $i \in V$ is an adequate choice. The optimal networks for low order graphs are presented in Fig. 2.

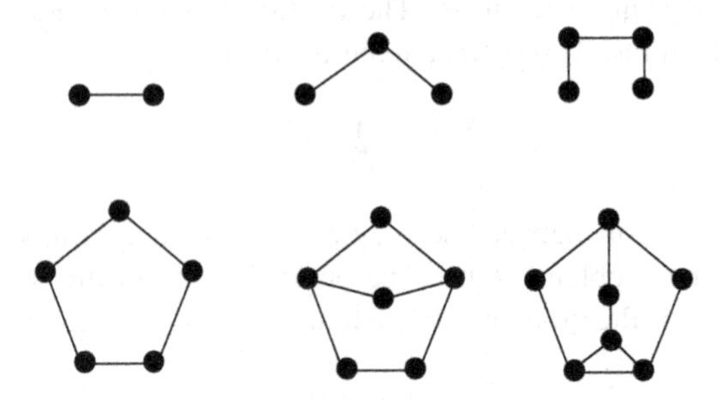

Fig. 2 Optimal graphs for $n \in \{2,\ldots,7\}$.

Optimal graphs for larger order were approximated by computer simulation and are presented in Fig. 3. Generally speaking it can be seen that cellular structures emerge: each individual is connected to a limited member of network members.

4 Jemaah Islamiya Bali bombing

In this section we analyze an organization that faced the tradeoff between secrecy and operational efficiency. In doing this we put the theoretical framework of Linde-

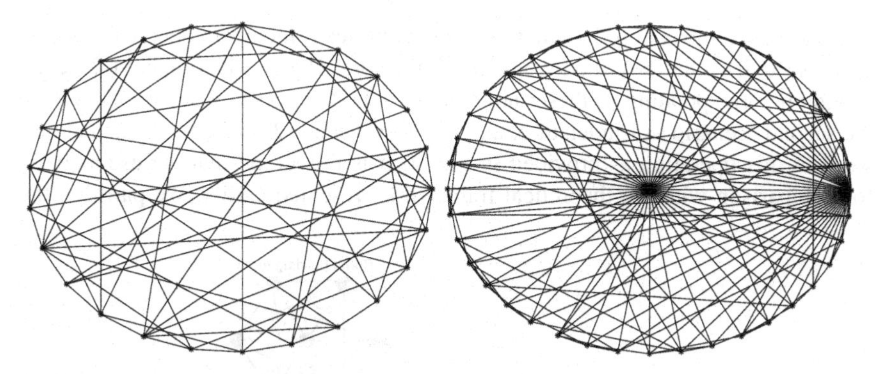

Fig. 3 Approximate optimal graphs for $n = 25$ (left) and $n = 40$ (right).

lauf et al. [13] to the test. We analyze how this organization dealt with this tradeoff by studying and comparing the network structure that they adopted to the theoretical framework. Jemaah Islamiya started as an Indonesian Islamist group and is a loosely structured organization characterized by four territorial divisions (mantiqis) corresponding to peninsular Malaysia and Singapore; Java; Mindanao, Sabah, and Sulawesi; and Australia and Papua [12]. Abdullah Sungkar, motivated by the need for a new organisation that could work to achieve an Islamic State in Indonesia, started JI in Malaysia around 1995. Al Qaeda infiltrated JI during the 1990's and JI subsequently developed into a pan-Asian network extending from Malaysia and Japan in the north to Australia in the south [7]. By doing this Al Qaeda set out to link these groups into a truly transnational network [1].

The tactical operation of the Bali attack that was conducted by Jemaah Islamiyah's Indonesian cell is described in Koschade [12]. The attack was carried out on October 12, 2002, by having a first operative explode a vest of explosives in Paddy's bar. This caused people to flood to the streets, which triggered the second attack by a vehicle based improvised explosive (VBIED) of about 1000 kilograms of TNT and ammonium nitrate. Consequently 202 people were killed. The operational setting consisted of a team of bomb builders located in a safe-house, a separate support team split over two safe-houses and a command team. The individuals in the safe-houses were thoroughly aware of the need for secrecy. This is indicated by the fact that each member used their Balinese alias and that communication occurred in code words. The individuals in the safe-houses rarely left these houses and used methods to reduce the probability of link detection: they only communicated by SMS and they changed their sim cards frequently. Hence, due to the similarity of these individuals from the viewpoint of secrecy the probability of exposure of those individuals may be assumed to be uniform. In terms of the theoretical framework by Lindelauf et al. [13] described in the earlier section the setting in which these individuals operated reflects the first scenario. Hence, the actual subgraph corresponding to these individuals is best compared to the results obtained for this scenario. To coordinate the operation a command team consisting of five individuals was set up. The operational commanders were highly active with regard to exchange of information. Hence the setting in which the command team members operated fits best to the second scenario of the theoretical framework. Hence we compare the actual subgraph corresponding to these individuals to the theoretical results obtained for the second scenario in Lindelauf et al. [13]. Koschade [12] presents the actual network of this operation as provided in Fig. 4. It is this graph that we use as basis for comparison with the theoretical framework presented earlier. We partition the

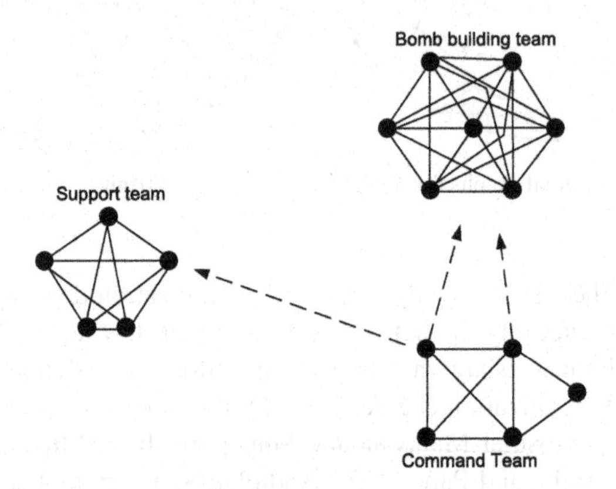

Fig. 4 Social Network of Jemaah Islamiyah cell that conducted the Bali Operation on October 6, 2002.

network into three subnetworks corresponding to the groups of individuals with intrinsically different goals. The Bali Bombing cell can be split into the bomb making team (cell 18), the support team (team Lima) and the command team. It can be seen that cell 18 as well as team Lima adopted the structure of a complete graph. That is, by choosing a location with tight security, never leaving the house and having someone on guard they tried to lower the exposure probability and link detection probability as much as possible. Both cells obtained the optimal graph according to the theoretical framework. The command team visited both cells and coordinated the operation.

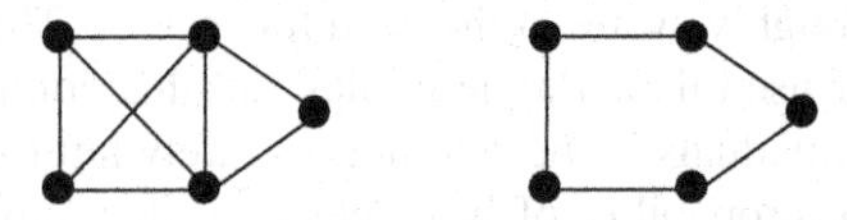

Fig. 5 JI Command Team (left) and the theoretically optimal command team (right)

The theoretical framework of Lindelauf et al. [13] only considered a homogeneous communication structure, not the nature of interaction that this communication represents. In his analysis Koschade [12] considered a weighting function on the edges by scaling the frequency and duration of interaction between 1 and 5. This already indicates that the nature of interaction among individuals in the network is not homogeneous. The frequency and duration of interaction differed most among the members of the subgraph corresponding to the command team. This non-homogeneity of interactions will be incorporated into the theoretical framework in the next section.

5 A First Approach to Heterogeneity in Covert Networks

An organization conducting a covert operation not only has to consider the communication structure of its network but also has to take into account that the nature of interaction between individuals is not homogeneous. For instance, the act of delivering a pre-manufactured bomb to the triggerman is potentially more dangerous than the internal communication (possibly through codewords) discussing the planning of an attack. Therefore in this section we will extend the theoretical framework on covert networks by differentiating between the nature of interaction among network individuals. Two questions come to mind. First, given a network structure which pair of individuals should conduct the interaction that presents the highest risk to the organization? Second, given the fact that there is a pair of individuals conducting an interaction that presents a high risk to the organization which network structure is optimal?

5.1 The Optimal High Risk Interaction Pair

We consider the situation that the interaction between individuals in the network is not completely homogeneous. This among others because the frequency, duration and nature of interaction differs between individuals. Hence, certain interactions present a higher risk to the organization than others. We model this by assigning 'weights' to the links, representing the risk of that interaction. For graph $g = (V, E)$ we define the weighting function $w : E \mapsto [1, \infty)$ such that $w_{ij} > w_{kl}$, $ij, kl \in E$, is interpreted as interaction between individual i and j presenting a higher risk to the organization than interaction between k and l. Denote the set of all such weighting functions by $\mathbb{W}$. Explicitly we denote a graph g with weight $w \in \mathbb{W}$ assigned to its edges by $g(w)$. The interpretation of this weighting function forces us to adjust the definition of secrecy. The information measure needs not to be adapted: one either interacts with an individual or not. However, risky interactions provide an enhanced security threat to the organization.

We adjust the secrecy measure corresponding to the second scenario in Lindelauf et al. [13]. For $g \in \mathbb{G}(n, m)$ we again set $u_i = 1 - \frac{d_i+1}{n}$ but adjust the probability of detection of an individual. This probability of detection not only depends on *the number* of individuals this individual is connected to but also on the nature of that interaction. Let $w_i = \sum_{j \in \Gamma_i(g)} w_{ij}$ where $\Gamma_i(g) = \{j \in V | ij \in E\}$ and define,

$$W = \sum_{i \in V} w_i = 2 \sum_{ij \in E} w_{ij}. \tag{4}$$

Motivated by the fact that a risky interaction increases the relative probability of exposure of an individual we set $\alpha_i = \frac{w_i+1}{W+n}$. In case $w_{ij} = 1$ for all $ij \in E$, α_i reduces to the one in Lindelauf et al. [13], i. e., $\alpha_i = \frac{d_i+1}{2m+n}$. Secrecy is again defined by

$$S(g) = \sum_{i \in V} \alpha_i u_i.$$

It can be seen that the secrecy measure of a graph g is the expected fraction of the network that survives upon exposure of an individual in the network according to probability distribution $(\alpha_i)_{i \in V}$. It is easily derived that

$$S(g) = \frac{n^2 - 2m - n + W(n-1) - \sum_{i \in V} d_i w_i}{n(W+n)}. \tag{5}$$

It follows that $S(g^n_{comp}) = 0$. Slightly more general for any k-regular graph $g \in \mathbb{G}^n$ it holds that $S(g) = 1 - \frac{k+1}{n}$. With $I(g) = \frac{n(n-1)}{T(g)}$ we find that,

$$\mu(g) = S(g)I(g) = \frac{(n-1)}{T(g)} \frac{n^2 - n - 2m + W(n-1) - \sum_{i \in V} d_i w_i}{W+n}. \tag{6}$$

The following result is readily obtained,

Lemma 1.

1. $\mu(g^n_{star}) = \frac{n-2}{2n-2} \cdot \frac{n-1+\frac{1}{2}W}{n+W}$ *if the path is given by* $1,2,\ldots,n-1,n$.

2. $\mu(g^n_{path}) = \frac{3}{n+1} \cdot \frac{(n-2)(n-1)+(2n-6)W+w_{12}+w_{n-1,n}}{n(W+n)}$.

3. $\mu(g^n_{ring}) = \begin{cases} \frac{4n-12}{n(n+1)} & \text{if } n \text{ is odd} \\ \frac{4(n-3)(n-1)}{n^3} & \text{if } n \text{ is even} \end{cases}$

Due to the symmetry of g_{ring} and g_{star} the interaction that presents the highest risk can be conducted by any pair of individuals. This can also be seen directly from lemma 5.1. In addition we determine the optimal location of the highest risk interaction for the path graph. The best position (in terms of maximizing μ) in the path graph is between either pair of individuals such that one of these individuals is an endpoint of the path. So, if the path is given by $1,2,\ldots,n-1,n$ w_{12} or w_{n-1n} is maximal. Thus an organization structured as a path graph does best by having either pair of players conducting the risky interaction as far away as possible from the central players. This is in accordance with intuition. In general it is shown that the pair of individuals in the organization that should conduct the interaction that presents the highest risk to the organization is the pair that is the least connected to the remainder of the network.

Theorem 2. *Let* $g = (V,E) \in \mathbb{G}(n,m)$ *and* $\{kl\} = argmin_{ij \in E}\ (d_i + d_j)$. *Set* $\hat{w}_{kl} = W - (m-1)$, $\hat{w}_{ij} = 1$ *for all* $ij \in E \setminus \{kl\}$. *Then* $\mu(g(\hat{w})) > \mu(g(w))$ *for all* $w \in \mathbb{W}$ *with* $\sum_{ij \in E} w_{ij} = W$.

Proof: It can be seen from Equation 5 that, given a graph $g \in \mathbb{G}(n,m)$ and total weight
$W = \sum_{ij \in E} w_{ij}$, maximizing $\mu(g)$ is equal to minimizing $\sum_{i \in V} d_i w_i$. It readily follows that $\sum_{i \in V} d_i w_i = \sum_{ij \in E} w_{ij}(d_i + d_j)$, hence the result follows. □

Given the situation that only a single interaction presents a higher risk to the organization we now compare the optimal path, star and ring graph using these results. We analyze the situation of a slightly riskier interaction ($z = 2$) and the situation of a much more riskier ($z = 100$) interaction. The results are summarized in Fig. 6. The

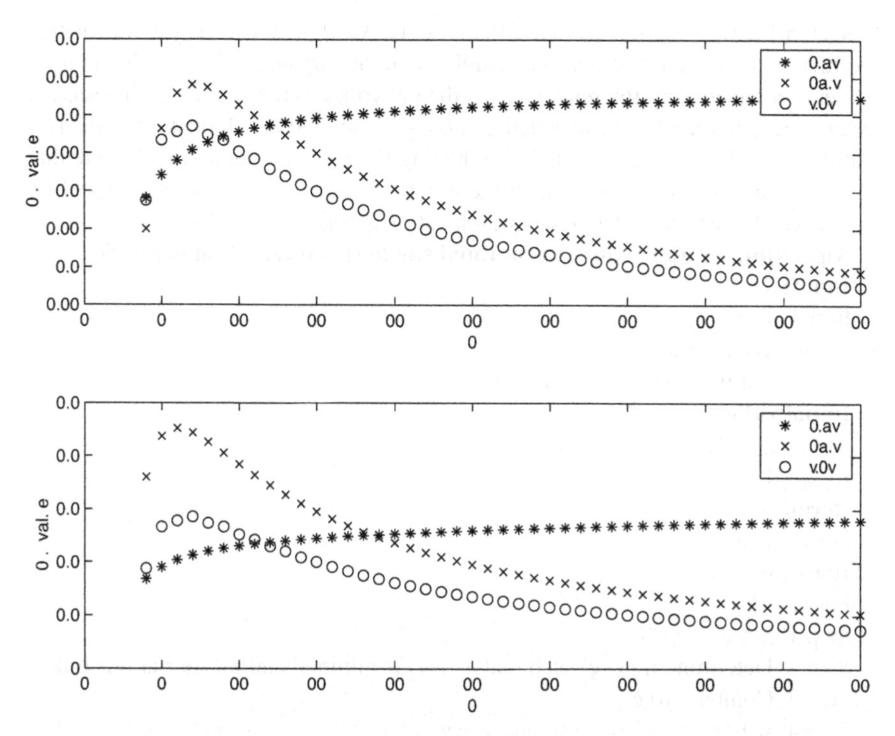

Fig. 6 A comparison between star, path and ring graph for $z = 2$ (top) and $z = 100$ (bottom).

ring graph is always dominated. It can be seen that for low values of n the path has a higher value of μ than the star graph. At a certain value of n a transition occurs such that $\mu(g^n_{path})$ becomes smaller than $\mu(g^n_{star})$. In case of $z = 2$ this transition occurs at $n = 11$. In case $z = 100$ this transition occurs at $n = 18$. Thus it can be seen that the amount of risk an interaction poses to the organization influences this transition point. For instance imagine one has to consider an organizational form that either is very centralized (star graph) or decentralized (path graph). If the number of individuals in the organization, n, is very large the star graph is the better choice. This can be understood intuitively because of the difficulty of information exchange in large path graphs as opposed to star graphs. However, if there is a single interaction that is much more risky relative to the others it still is advantageous to adopt a path graph organizational form. Clearly, this reduces the capability to process information but from the perspective of secrecy has the advantage of reducing the risk to the organization by positioning the risky interaction as far away as possible from the central players.

5.2 Approximating Optimal Heterogeneous Covert Networks

In Sect. 5.1 it was established that if there exists exactly one pair of individuals that conduct an interaction that presents a high risk to the organization they should have the least connection to the remainder of the network (theorem 5.1). In this section we are interested in *which* connected graph $g \in \mathbb{G}^n$ should be adopted given the fact that the pair of individuals $i, j \in V$ conducting the risky interaction is the one that minimizes $d_i + d_j$. We approximate the graphs that are optimal in this respect by simulation. We conducted a greedy optimization algorithm as follows.

Algorithm for approximating optimal single risk interaction network.

Input:

Initial graph $g^n_{initial}$.

Value of risky interaction z.

Number of times edges are added m.

Initialization:

$\bar{g} = g_{initial}$. (Denote $\bar{g} = (V, E)$).

$\mu(g_{help}) = 0$.

Iteration 1:

For i = 1:m

Iteration 2:

For $kl \in E^c$

Step 1. Set $g' = \bar{g} \cup kl$.

Step 2. Determine $i, j \in g'$ such that $d_i + d_j$ is minimal and locate z at this link.

Step 3. Compute $\mu(g')$.

Step 4. If $\mu(g') > \mu(g_{help})$ set $g_{help} = g'$.

End iteration 2.

$\bar{g} = g_{help}$.

End iteration 1.

Output:

$\bar{g}$.

$\mu(\bar{g})$.

The best results of this greedy optimization are presented in Tab. 1 for graphs of order $4 \leq n \leq 10$. The location of the pair of individuals that conduct the interaction that presents a high risk to the organization is presented in bold.

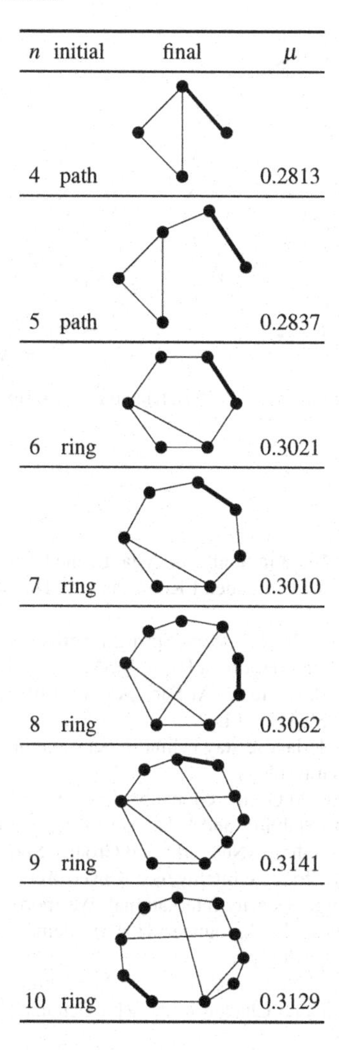

n	initial	final	μ
4	path		0.2813
5	path		0.2837
6	ring		0.3021
7	ring		0.3010
8	ring		0.3062
9	ring		0.3141
10	ring		0.3129

Table 1 Approximate optimal graphs with single high risk interaction, $z = 2$, indicated in **bold**.

As a further illustration optimal graphs of larger order are approximated by using the greedy optimization algorithm, see Fig. 7. It can be seen that cellular structures emerge around a centralized individual. Comparing these to Fig. 3 it can be seen that the networks in Fig. 7 are less dense. In addition it can be seen that the individuals conducting the interaction that presents the highest risk to the organization are members of a cell with limited connectivity to the remainder of the network.

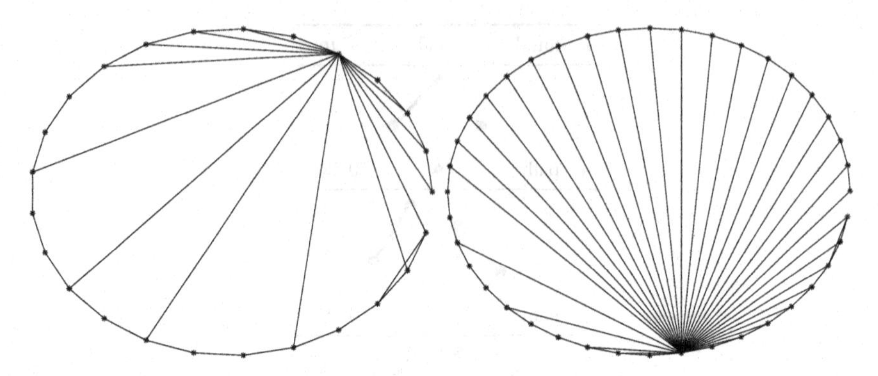

Fig. 7 Approximate optimal graphs for $n = 25$ (left) and $n = 40$ (right).

References

1. Abuza, Z. (2003). Militant Islam in Southeast Asia. Lynne Rienner Publishers: London.
2. Bergen, P. (2006). The Osama Bin Laden I Know: an Oral History of al Qaeda's Leader. Free Press: New York.
3. Bollobas, B. (1998). Modern Graph Theory. Springer-Verlag: New York.
4. Cammaert, A.P.M. (1994). Het Verborgen Front. EISMA B.V.: Leeuwarden.
5. Cruickshank et al. (2007). Abu Musab Al Suri: Architect of the New Al Qaeda. Studies in Conflict and Terrorism Vol 30(1): 1–14.
6. Engelen, D. (2005). De Nederlandse stay behind-organisatie in de koude oorlog, 1945–1992. PIVOT-rapport nr. 166: 's Gravenhage.
7. Gunaratna, R. (2003). Inside Al Qaeda: Global Network of Terror. Berkley Trade.
8. Jackson, M.O. (2001). The Stability and Efficiency of Economic and Social Networks. In: Dutta B. and Jackson M.O., editors. Networks and Groups. Springer-Verlag: Heidelberg.
9. Johnson, L.K. (ed.) (2007). Strategic Intelligence: Covert Action–Beyond The Veils of Secret Foreign Policy, Vol 3. Praeger Security International: Westport.
10. Koschade, S (2002). Indonesia Backgrounder: How the Jemaah Islamiyah Terrorist Network Operates. International Crisis Group.
11. Koschade, S (2005). A Social Network Analysis of Aum Shinrikyo: Understanding Terrorism in Australia. Proceedings Social Change in the 21st Century Conference, QUT Carseldine, Brisbane.
12. Koschade, S. (2006). A Social Network Analysis of Jemaah Islamiyah: The Applications to Counterterrorism and Intelligence. Terrorism and Political Violence 29: 559–575.
13. Lindelauf, R.H.A. , Borm, P. and Hamers, H. (2008). The Influence of Secrecy on the Communication Structure of Covert Networks. CentER Discussion Paper, 2008-23, pp. 1–18.
14. Mishal, S. and Rosenthal, M. (2005). Al Qaeda as a Dune Organization: Toward a Typology of Islamic Terrorist Organizations. Studies in Conflict & Terrorism 28(4): 275–293.
15. Petraeus, D.H. et al. (2007). The U.S. Army Marine Corps Counterinsurgency Field Manual. University of Chicago Press: Chicago.
16. Sageman, M. (2004). Understanding Terror Networks. University of Pennsylvania Press: Philadelphia, Pennsylvania.
17. Vidino, L. (2006). Al Qaeda in Europe: The New Battleground of International Jihad. Prometheus Books: New York.
18. Weiner, T (2007). Legacy of Ashes: The History of the CIA. Doubleday.

Two Models for Semi-Supervised Terrorist Group Detection

Fatih Ozgul, Zeki Erdem and Chris Bowerman

Abstract Since discovery of organization structure of offender groups leads the investigation to terrorist cells or organized crime groups, detecting covert networks from crime data are important to crime investigation. Two models, GDM and OGDM, which are based on another representation model – OGRM are developed and tested on nine terrorist groups. GDM, which is basically depending on police arrest data and "caught together" information and OGDM, which uses a feature matching on year-wise offender components from arrest and demographics data, performed well on terrorist groups, but OGDM produced high precision with low recall values. OGDM uses a terror crime modus operandi ontology which enabled matching of similar crimes.

1 Introduction

Group detection refers to the discovery of underlying organizational structure that relates selected individuals with each other, in broader context; it refers to the discovery of underlying structure relating instances of any type of entity among themselves [29]. Link analysis and group detection is a newly emerging research area which is at the intersection of link analysis, hypertext – web mining, graph mining [10, 11] and social network analysis [32, 36, 48]. Graph mining and social network

Fatih Ozgul
Department of Computing, Engineering & Technology, University of Sunderland, St. Peter's Way, SR6 0DD, Sunderland, UK, e-mail: fatih.ozgul@sunderland.ac.uk

Zeki Erdem
TUBITAK-Marmara Research Centre, Information Technologies Institute, 41470 Gebze, Kocaeli, TURKEY, e-mail: zeki.erdem@bte.mam.gov.tr

Chris Bowerman
Department of Computing, Engineering & Technology, University of Sunderland, St. Peter's Way, SR6 0DD, Sunderland, UK, e-mail: chris.bowerman@sunderland.ac.uk

N. Memon et al. (eds.), *Mathematical Methods in Counterterrorism*,
DOI 10.1007/978-3-211-09442-6_14, © Springer-Verlag/Wien 2009

analysis (SNA) in particular attracted attention from a wide audience in police investigation and intelligence [17, 18]. As a result of this attention, the police and intelligence agencies realized the knowledge about offender networks and detecting covert networks are important to crime investigation [32, 43]. Since discovery of an underlying organizational structure from crime data leads the investigation to terrorist cells or organized crime groups, detecting covert networks are important to crime investigation. Detecting an offender group, a terrorist network or even a part of group (subgroup) is also important and valuable. A subgroup can be extended with other members with the help of domain experts. An experienced investigator usually knows the friends of well-known terrorists, so he can decide which subgroups should be united to constitute the whole network. Specific software like Analyst Notebook [1] and Sentient [44] provide some visual spatio-temporal representations of offender groups in graphs, but they lack automated group detection functionality. In this paper, we make the following contributions for terrorist group detection:

- We introduce two models for terrorist group detection; GDM and OGDM (Section 4 and 5).
- We demonstrate a terrorism modus operandi ontology system for matching similarly skilled terrorists (Section 5.2).
- We show how two models GDM and OGDM performed on detecting terrorist and other offender groups (Section 6.2).
- We discuss how much crime and offender demographics data is important to detect terrorist groups (Section 7).

2 Terrorist Group Detection from Crime and Demographics Data

So far in the literature, there is a difference between detecting terrorist groups and other offender groups, there has been some research in offender group detection. The most remarkable works are COPLINK [2, 3, 4, 5, 6, 7, 23, 24] CrimeNet Explorer, which is developed by Xu et al. [46, 47] and Terrorist Modus Operandi Detection System (TMODS), which is developed by 21st Century Technologies [8, 9, 19, 25, 29, 30, 31, 34, 39].

2.1 COPLINK CrimeNet Explorer

Xu et al. [46, 47] defined CrimeNet Explorer framework for automated network analysis and visualization. Using concept space algorithm, COPLINK connect and COPLINK detect [2, 3, 4, 5, 6, 7, 23, 24] structure to obtain link data from text reports of arrest records and other offender data. CrimeNet Explorer and all COPLINK framework relies on concept space algorithm [47].The concept space algorithm automatically computes the strength of relationships between each possible pair of concept descriptors identified in a document collection. This algorithm is applied to Tucson Police Department's documents to collect documents, doing co-occurrence analysis, and associative retrieval.

Co-occurrence analysis is a basic technique dating back to the 1960s creates a concept space that is a graph of concepts. Co-occurrence analysis uses similarity and clustering functions to weight relationships between all possible pairs of concepts. This net-like concept space holds all possible associations between objects, which mean that all existing links between every pair of concepts is retained and ranked.

Associative retrieval is done when a search term is entered into the concept space algorithm; the system returns a list of co-occurred terms for user analysis. In the COPLINK Concept Space, the associated terms are presented using multiple rank ordered lists in a tabular format. The six tabular columns represent the six information fields used in the co-occurrence analysis. The use of a tabular format creates better summarization and visualization of the retrieved information by allowing officers to target the information field/type that they want.

In order to deal with problems of information integration and ease of access, COPLINK Connect is developed and used at Tucson Police Department [23, 24]. COPLINK Connect employs a consistent and intuitive approach which integrates different data sources (for instance crime records and vehicle database), such that the multiplicity of data sources remains completely transparent to the end user.

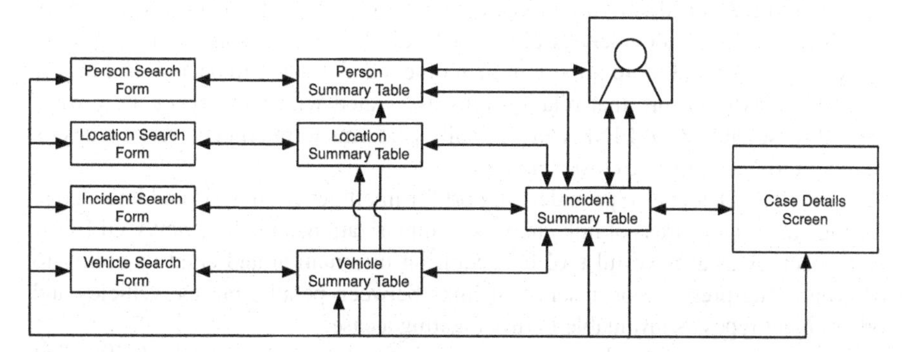

Fig. 1 COPLINK Connect search schema

COPLINK Detect was specifically designed to aid law enforcement investigators and detectives in criminal intelligence analysis, helping to improving efficiency and effectiveness. Concept space or automatic thesaurus was used to identify relationships between objects (terms or concepts) of interest. The technique has been frequently used to develop domain-specific knowledge structures for digital library applications. The concept space was used as the basis for the COPLINK Detect application. In COPLINK Detect, detailed case reports are the underlying space and concepts are meaningful terms occurring in each case. COPLINK Detect provides the ability to easily identify relevant terms and their degree of relationship to the search term.

From a crime investigation standpoint, COPLINK Detect can help investigators link known objects to other related objects that might contain useful information for further investigation, for instance, people and vehicles related to a given suspect. In the COPLINK Detect based upon the concept space algorithm as an alternative investigation tool that captures the relationships between objects in the entire database. Based on careful user requirement analysis, five entity fields from the database were deemed relevant for analysis: person, organization, location, vehicle, and crime incident type. The purpose of this tool is to discover relationships between and among different crime-related entities. It is important not only to know that there is a relationship, but also to know what each relationship is.

There are three main steps in building the domain-specific COPLINK Detect concept space. The first task is to identify collections of documents in a specific subject domain; these are the sources of terms or concepts. The next step is to filter and index the terms. The final step is to perform a co-occurrence analysis to capture the relationships among indexed terms. The resulting output is then inserted into a database for easy manipulation. With the collaboration of personnel from the Tucson Police Department, a set of term types for the COPLINK Detect were identified and created in order to balance performance and comprehensiveness. The index maintains the relationship between a term and the document in which it occurs. Both index and reverse index are required for co-occurrence analysis. The index contains the links from term to document; the reverse index contains the links from document to term. After identifying terms, the term frequency, the document frequency for each term in a document is also computed. In general, some term types are more descriptive and more important than others and deserve to be assigned higher weights so as to ensure that relationships associated with these types are always ranked reasonably. COPLINK Detect finally performs term co-occurrence analysis based on its "asymmetric cluster function".

COPLINK Detect also provides support for intelligence analysis and knowledge management in the areas of link analysis, summarization, and efficiency. COPLINK Detect serves as a powerful tool for acquiring information and cited its ability to determine the presence or absence of links between people, places, vehicles and other object types as invaluable in investigating a case.

The impact of link analysis on investigative tasks is crucial to the building of cases. An officer assigned to investigate a crime has to have enough information to provide a lead before he/she can begin working. Too many cases have to be closed because of lack of information or inability to utilize information existing elsewhere in the records management system. COPLINK Detect manages all the data in the records system in such a way that it can be used as knowledge about the suspect.

CrimeNet Explorer used a Reciprocal Nearest Neighbour (RNN) based clustering algorithm to find out links between offenders, as well as discovery of previously unknown groups. CrimeNet Explorer framework includes four stages: network creation, network partition, structural analysis, and network visualization (Figure 3).

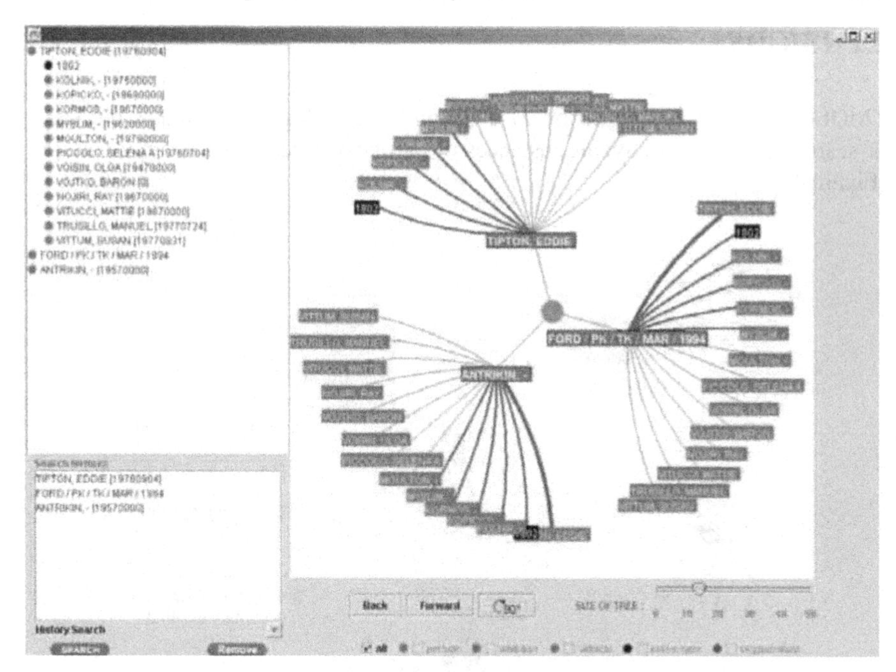

Fig. 2 Hyperbolic tree view of associations in COPLINK Detect search schema

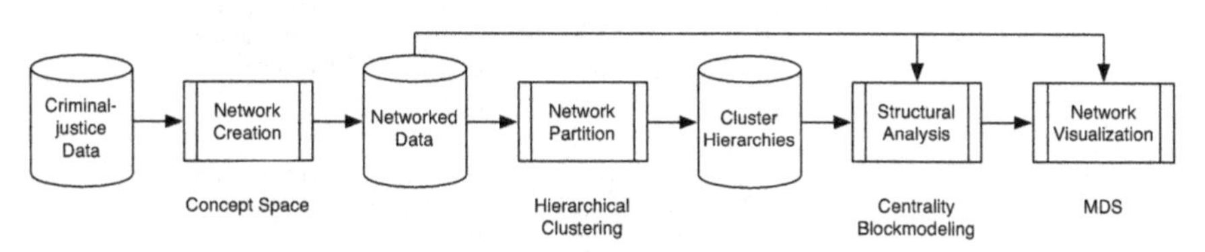

Fig. 3 CrimeNet Explorer framework

CrimeNet Explorer uses concept space, COPLINK Connect and COPLINK Detect for network creation, RNN-based hierarchical clustering algorithm for group detection; social network analysis based structural analysis and Multi Dimensional Scaling for network visualization. CrimeNet Explorer is the first model to solve offender group discovery problem and its success comes from the powerful functionality of overall COPLINK structure. On the other hand, since CrimeNet Explorer was evaluated by university students for its visualization, structural analysis capabilities, and its group detection functionality, the operationally actionable outputs of CrimeNet Explorer on terrorist groups has not been proved.

2.2 TMODS

TMODS, which is developed by 21st Century Technologies, automates the tasks of searching for and analyzing instances of particular threatening activity patterns (Figure 4).

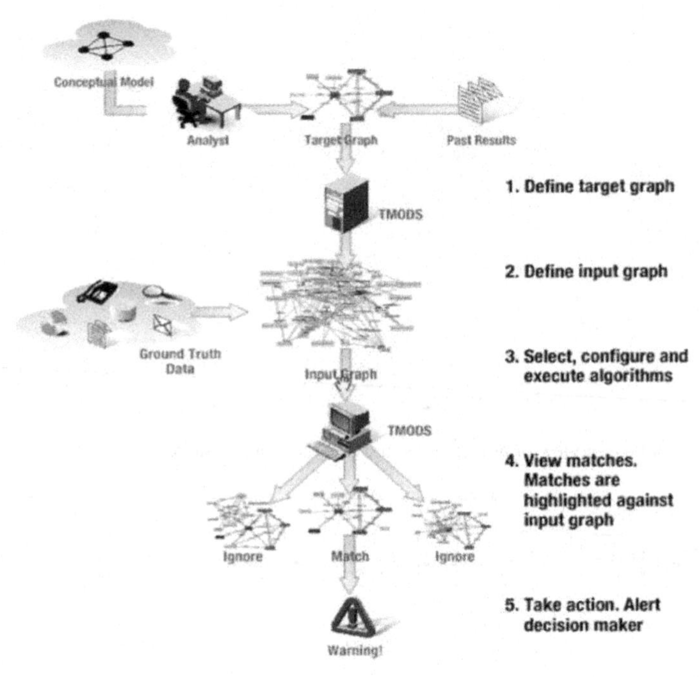

Fig. 4 TMODS framework. A possible search pattern is matched to observed activity by using a pattern in ontology.

With TMODS, the analyst can define an attributed relational graph to represent the pattern of threatening activity he or she is looking for. TMODS then automates the search for that threat pattern through an input graph representing the large volume of observed data. TMODS pinpoints the subset of data that match the threat pattern defined by the analyst thereby transforming a manual search into an efficient automated graph matching tool. User defined threatening activity or pattern graph can be produced with possible terrorist network ontology and this can be matched against observed activity graph. At the end, human analyst views matches that are highlighted against the input graph. TMODS is mature and powerful distributed java software that has been under development since October 2001 [31, 34]. But it needs a pattern graph and an analyst to run the system. Like a supervised learning algorithm, TMODS tries to tailor the results according to pre-defined threatening activity.

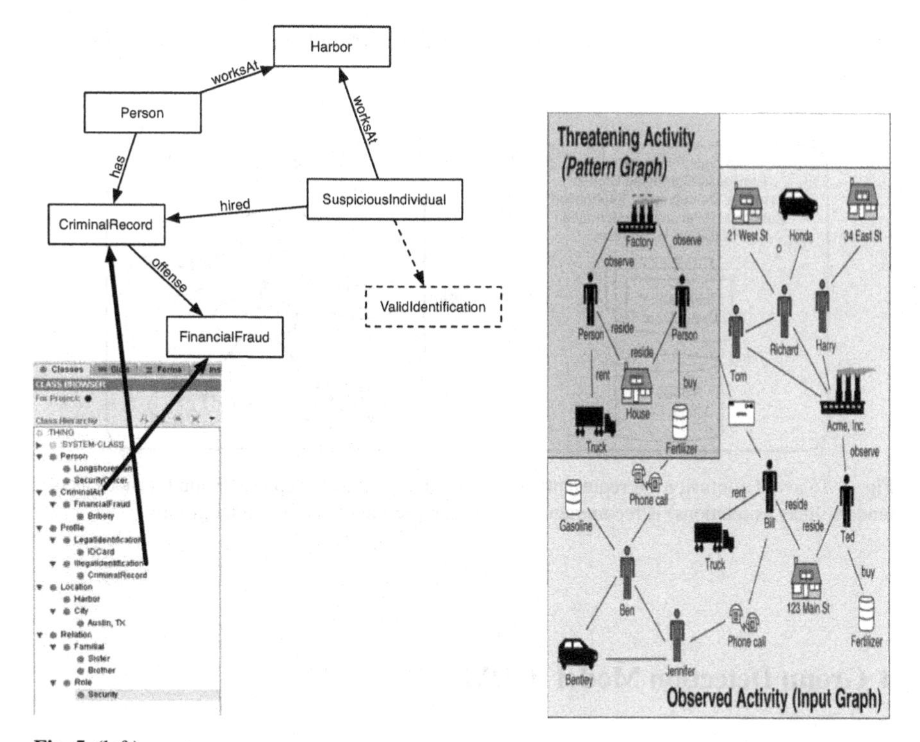

Fig. 5 (left)

Fig. 6 (right) TMODS uses graph matching on multi-mode graphs. But it also uses a case ontology pattern matching technique. Ontology used in TMODS represents a known case not similarity ontology between two or more crimes.

Another possible drawback is graphs used in TMODS are multi-mode and can be disadvantageous for further analysis. Multi-mode graph means that nodes in multi-mode graphs are more than two types of entities. A person, a building, an event, a vehicle are all represented as nodes; when for instance we want to detect key players in multi-mode graph, a building can be detected as key player, not a person. This can be a cause of confusion. To overcome this confusion the definition of a one-mode (friendship) social network should be used rather than representing all entities as nodes.

3 Offender Group Representation Model (OGRM)

It is better to represent actors (offenders) as nodes and rest of the relations as edges in one-mode (friendship) social networks (Figure 7). This can produce many link types such as "co-defendant link", "spatial link", "same weapon link", and "same modus operandi link". Thereby many graph theoretical and SNA solutions can be used on one-mode (friendship) networks effectively such as friendship identification, finding key actors.

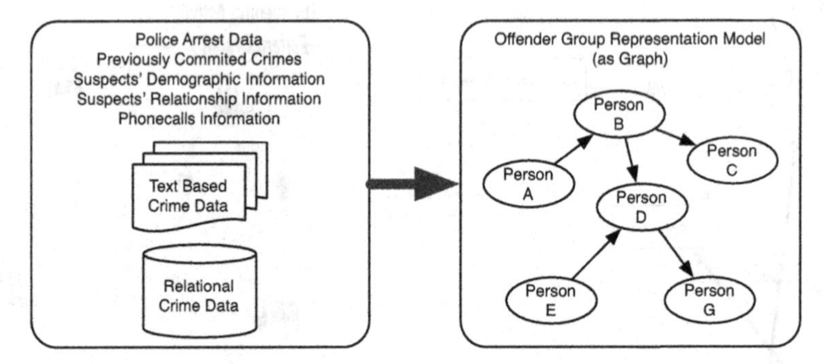

Fig. 7 To avoid confusion on representation, a one-mode network representation (vertices as offenders, links as relations) is recommended for best representation of offender groups.

4 Group Detection Model (GDM)

It is a fact that people who own the same name and surname can mislead the investigator and any data mining model. Minimum requirement to apply GDM is to have a police arrest table or text where it has to include unique crime reference number and unique person number.

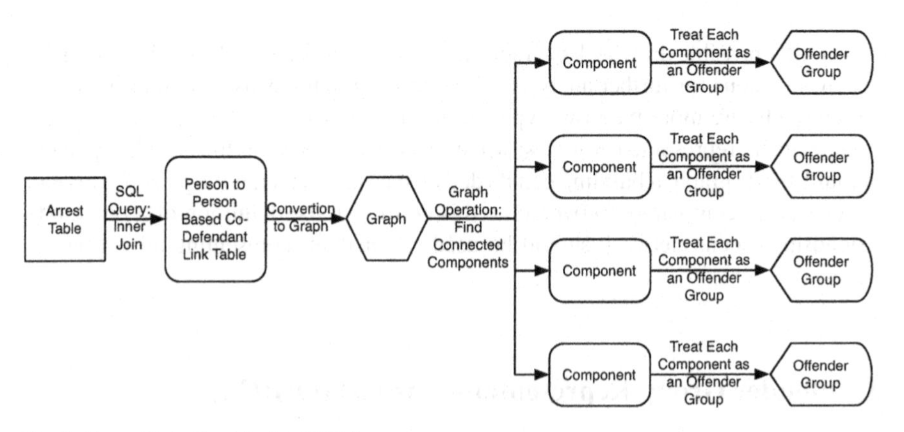

Fig. 8 Group Detection Model (GDM)

This ensures that the information doesn't include duplicate arrest records and a person only has one unique number or reference. The source of link information is gathered from police arrest records using an inner join query (Figure 8). Inner join query result, which we call co-defendant link table; consisting of From Offender, To Offender, and W (how many times this offender pair caught together by the police) is produced with inner join SQL query. Then this link table is converted to graph where nodes represent offenders, edges represent crimes using OGRM representation model (Figure 7). Number of times caught together is counted to be used for edge weight (W). At this point a subgraph detection operation is needed; we used strongly connected components (SCC) algorithm because it is scalable and gives concrete results. A directed graph [12] is called strongly connected if for every pair of vertices has a path towards each other. The strongly connected components of a directed graph are its maximal strongly connected subgraphs like in Figure 9.

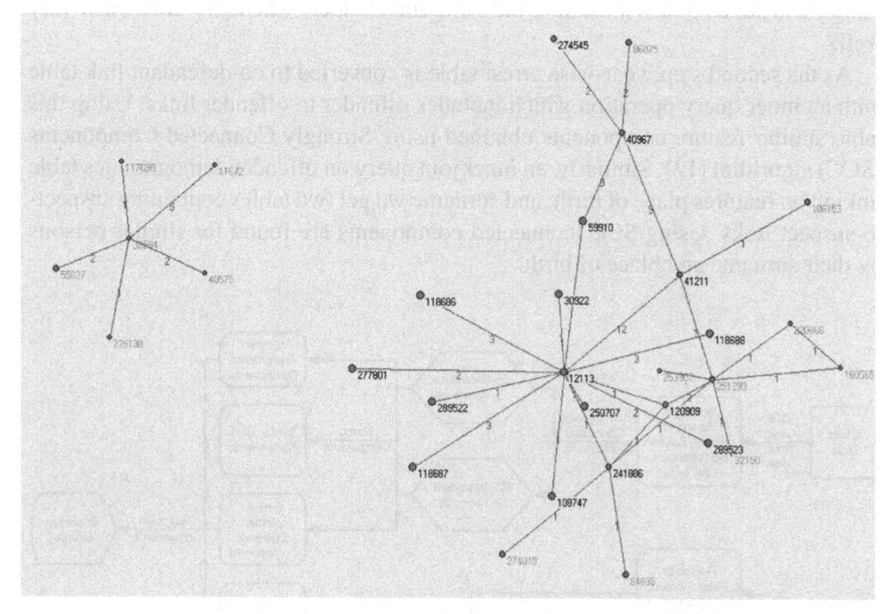

Fig. 9 This graph shows two theft groups generated from police arrest table using GDM, consisting of 31 offenders with their person_id numbers (nodes), 30 undirected "co-defendant" links (edges)

In GDM, every component represents a unique offender group because one offender can only belong to one group thereby concrete a result of group membership is obtained.

5 Offender Group Detection Model (OGDM)

Due to lack of enclosure of crime specific attributes in GDM, we needed to develop our model for better results on offender group detection. OGDM is developed to add more functionality by including time, location, modus operandi and demography similarity features for offender group detection. As exhibited in Figure 10, following steps are taken in OGDM. As well as the arrest table, which is used in GDM, we also need some demographic knowledge features about offenders; literally offender surname and offender birthplace.

The first step is taken on arrest table by dividing all arrest records in year-wise segments. This prevents linking too many relationships between crimes and offenders like a snowball which gets bigger and bigger without control. During testing GDM and OGDM, it is understood that only a bunch of highly skilled criminals generally operates most of criminal activities. From this point of view, offender networks in a city would draw a typical small-world-graph, like the graph represents WWW. This can be because of the underground economy generated by crimes, money and stolen goods must have the same stakeholders who know each other very well.

As the second step, year-wise arrest table is converted to co-defendant link table with an inner query operation which includes offender to offender links. Using this table similar feature components obtained using Strongly Connected Components (SCC) algorithm [12]. Similarly, an inner join query on offender demographics table linking on features place of birth, and surname we get two tables containing suspect-to-suspect links. Using SCC, connected components are found for similar persons by their surname and place of birth.

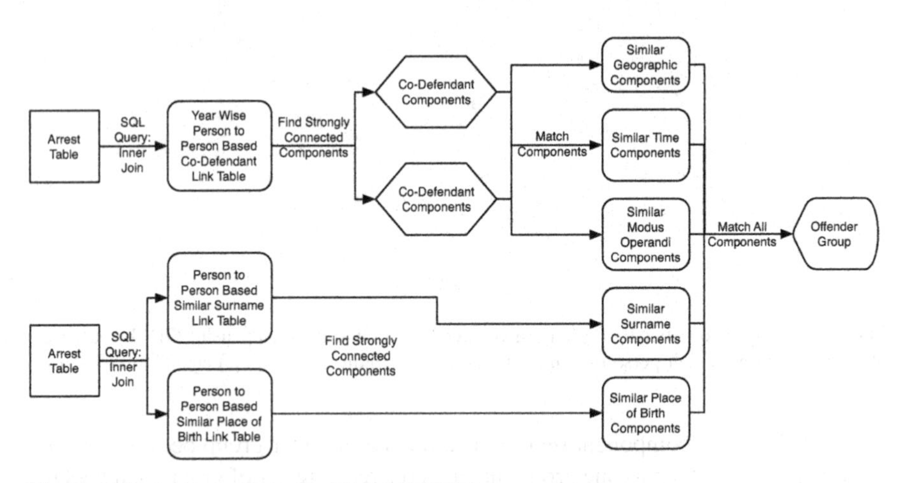

Fig. 10 Offender Group Detection Model (OGDM)

As the third step, all components obtained until this step are matched for their similarities. Similarities between component pairs with their similarity scores are put in a rank-ordered list of possible relations of persons. From top to bottom, each link represents from most to least important links between offender pairs with a jaccard coefficient similarity score [23].

5.1 Computing Similarity Score

The Jaccard coefficient is the size of the intersection divided by the size of the union of the sets. So for given two components, for each component containing relationship between person x and person y, Jaccard coefficient similarity score is:

$$\mathscr{J}(x,y) = \frac{|x \cap y|}{|x \cup y|} \tag{1}$$

So for person (offender or suspect) x, (s)he can be linked to person (offender or suspect) y, using their crimes and demographics matched. In some cases, x and y persons have more than one similarity of features. To represent this similarity score as total $\Delta\mathscr{J}(x,y)$, we need to use Jaccard coefficient similarity scores for every feature in crimes that person x and y committed and person x and y demographics. Thereby using $\mathscr{J}$BirthPlace Sim(x,y), $\mathscr{J}$Time Sim(x,y), $\mathscr{J}$ModusOperandi Sim(x,y), total similarity $\mathscr{J}(x,y)$ is reached by:

$$\Delta\mathscr{J}(x,y) = \sqrt{\begin{array}{l} \mathscr{J}\text{BirthPlace Sim}(x,y)^2 \\ +\ \mathscr{J}\text{Geographic Sim}(x,y)^2 \\ +\ \mathscr{J}\text{ModusOperandi Sim}(x,y)^2 \\ +\ \mathscr{J}\text{Time Sim}(x,y)^2 \\ +\ \mathscr{J}\text{Surname Sim}(x,y)^2 \end{array}} \tag{2}$$

$\Delta\mathscr{J}(x,y)$ is measured by euclidean distance of these scores. Getting euclidean distance provides a total score $\Delta\mathscr{J}(x,y)$ which varies between 0 and 1. In practise, before calculating this score, noise should be removed. Noise level is considered below 0.1 or 0.2 depending on how many crimes and offenders to be predicted. If noise value is decided lower, size of groups will rise with less number of offender groups obtained, elsewhere noise value decided higher, size of groups will be more accurate but too many small sized offender groups will be obtained. After removing noise (the least important links) and deciding a threshold cut value for similarity score, links constitute groups and each group is considered an independent offender group.

5.2 Using Terrorist Modus Operandi Ontology

Similar modus operandi components in OGDM are decided via terrorist modus operandi ontology, as seen in Figure 11.

Ontologies are beneficial for importing the domain knowledge from sources to reuse [21] so we decided to create a terrorist modus operandi system to use in OGDM. As shown in Figure 11, main node of ontology named "Terror Crime", one below second and third level there are 27 terror crimes such as "being member of terrorist organisation", "delivery of terrorist propagandize material".

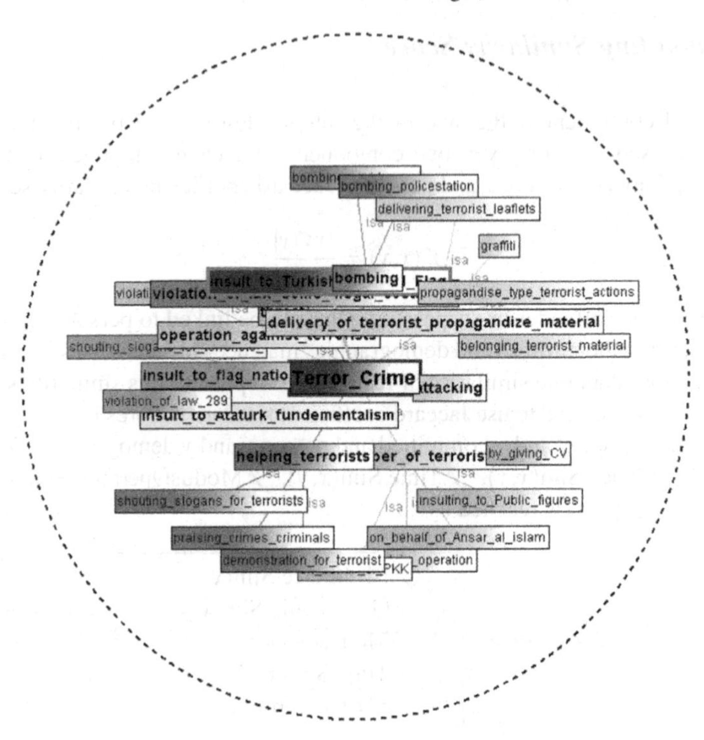

Fig. 11 Modus Operandi Ontology for Terror Crimes

5.3 Deciding Threshold

Deciding a threshold cut value for similarity score should be decided upon a domain expert's recommendation. After getting final table to decide offender groups, we need to try top-to-bottom approach to visualise links and nodes. Similarity scores as link weights decide which links should be considered as valid. It is not possible to get every links to be accepted. Higher numbers of links with low similarity values constitute more bigly and cluttered offender groups, whereas lower number of links with high similarity scores give the most accurate results. Reminding the fact that by the lower number of links and high similarity scores many offender groups would remain unprocessed and unpredicted.

In practice, $\Delta \mathscr{J}(x,y)$ threshold values ranking between 0.25 and 0.50 produced satisfying results. For example some similarities such as surname similarity proved useful then other similarity types such as geographical proximity similarity of crimes. As a result we developed metrics to decide feature selection, asked domain experts which results are promising. With the help domain expert's comments on visualised offender groups and reduction of least important features, ideal offender groups are detected (Figure 12).

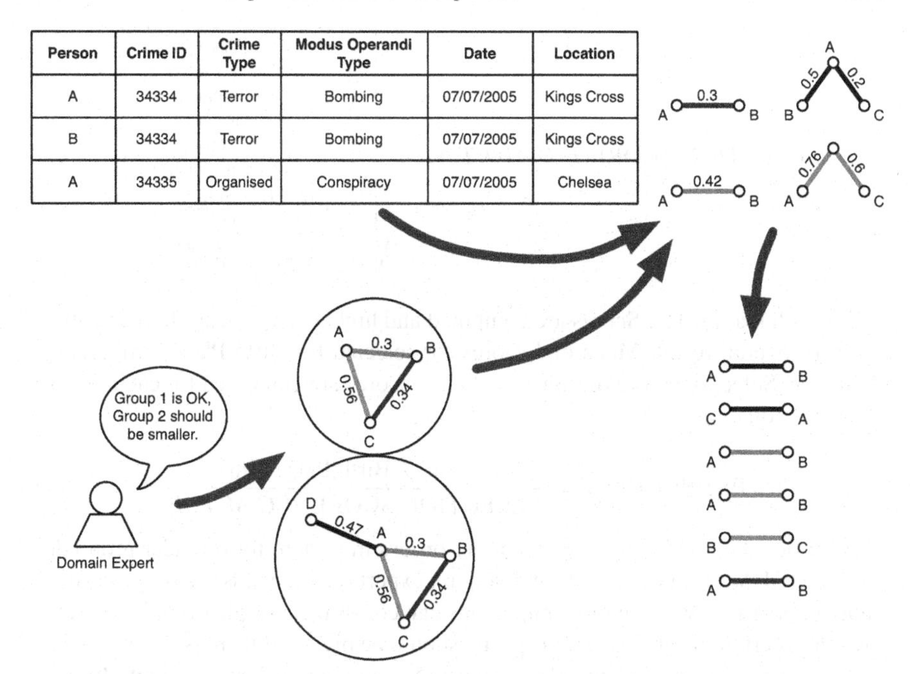

Person | Crime ID | Crime Type | Modus Operandi Type | Date | Location
A | 34334 | Terror | Bombing | 07/07/2005 | Kings Cross
B | 34334 | Terror | Bombing | 07/07/2005 | Kings Cross
A | 34335 | Organised | Conspiracy | 07/07/2005 | Chelsea
0.3
0.42
0.5
0.2
0.76
0.6
Group 1 is OK, Group 2 should be smaller.
Domain Expert
0.56
0.34
0.47

Fig. 12 Domain expert makes comments on how size of groups and members should be

5.4 Feature Selection

We need to help domain experts for finding out which feature in our data is the most promising; so that domain experts can decide the best threshold can be used to reduction of unnecessary data. One possible technique for feature selection is based on comparison of means and variances. To summarize the key characteristics of the distribution of values for a given feature, it is necessary to compute the mean value and the corresponding variance. Equation 3 and 4 formalizes a scoring test (TEST), where B (for birthplace similarity score), G (for geographic similarity score), MO (for modus operandi similarity score), T (for Time similarity score) and S (for Surname similarity score) are sets of feature values measured for different matching types and components total nums 1 … to total num N, are the corresponding number of samples:

$$\text{SELECTION SCORE}(B,G,MO,T,S) = \sqrt{\begin{aligned}&\frac{\text{var}(\mathcal{J}\,\text{BirthPlace Sim})}{\text{total nums for } B} \\ &+ \frac{\text{var}(\mathcal{J}\,\text{Geographic Sim})}{\text{total nums for } G} \\ &+ \frac{\text{var}(\mathcal{J}\,\text{ModusOperandi Sim})}{\text{total nums for } MO} \\ &+ \frac{\text{var}(\mathcal{J}\,\text{Time Sim})}{\text{total nums for } T} \\ &+ \frac{\text{var}(\mathcal{J}\,\text{Surname Sim})}{\text{total nums for } S}\end{aligned}} \tag{3}$$

For each feature Test Scores are computed and higher score means higher importance for overall result. Mean of J scores for selected for Birth Place Similarity is divided by Selection Score in Equation 4. Test scores are computed for each feature using this equation.

$$\text{TEST SCORE}(B) = \frac{\text{mean}(\mathcal{J}\,\text{Birth Place Sim})}{\text{SELECTION SCORE}(B,G,M,T,S)} \tag{4}$$

After deciding which features are more or less important for offender group detection, OGDM produces results and domain expert recommends the best threshold value. In practice, two or more components can constitute a single offender group if domain expert thinks the size of the group should be bigger, if there is no need to extend group size or add new members, selected component is kept as target offender group (Figure 10).

6 Experiments and Evaluation

We evaluated the performance of GDM and OGDM model. Experiments are conducted to answer the following questions;

- Will GDM and OGDM identify offenders as terrorists with high recall value? (Equation 6)
- Will GDM and OGDM identify offenders refer to one particular terrorist group with high precision value? (Equation 5)

6.1 Performance Matrix

We evaluated GDM and OGDM performance using precision and recall that are widely used in information retrieval. Precision, in this scenario, is defined as the percentage of correctly detected offenders within one particular terrorist group in cluster divided by correctly detected related offenders. Recall is the percentage of correctly detected terrorists belongs to one particular terrorist group in cluster divided by correctly detected terrorists.

$$\text{precision} = \frac{\text{True Positives (TP)}}{\text{True Positives (TP)} + \text{False Positives (FP)}} \tag{5}$$

$$\text{recall} = \frac{\text{True Positives (TP)}}{\text{True Positives (TP)} + \text{False Negatives (FN)}} \tag{6}$$

Table 1 Performance matrix for GDM and OGDM on terrorist groups

	Offenders considered as terrorists	Offenders considered as non-terrorists
Offenders considered refer to the same terrorist group	True Positive (TP)	False Positive (FP)
Offenders considered refer to other groups (not to the same terrorist group)	False Negative (FN)	True Negative (TN)

6.2 Testbed: Terrorist Groups Detected in Bursa

We conducted experiments on Bursa data set. Crime records in Bursa, Turkey crime and offender demographics data set are available from 1991 to August 2007 as 318 352 crimes committed by 199 428 offenders. Crime types are varying from organized crimes, narcotic gangs, theft or illegal enterprises to terrorist groups.

All experimental setup and operations are done using R and related R libraries [15, 16, 27, 40]. There were 9 terrorist groups and 78 other types of offender groups available such as theft groups, drug dealing groups, mafia type groups. When visualised links and groups presented to domain experts, they recommended setting our threshold value for jaccard similarity as ten per cent. Nine terrorist groups are selected as "golden standard" to measure recall and precision values for GDM and OGDM. Results for terrorist groups detected are presented in below Table 2.

Table 2 Recall and Precision Results for GDM and OGDM

Group Number and Name	GDM Recall	GDM Precision	OGDM Recall	OGDM Precision
(2) TDKP-Legal Cells	1	0.93	0.44	1
(3) TDKP Illegal Cell	1	0.93	0.44	1
(4) TKP/ML Cell	1	1	1	0.90
(5) PKK Cell	0.84	1	0.59	1
(6) Racist Terrorist Group	0.75	0.86	0.75	1
(7) Communist DHKPC Cell	0.70	1	0.53	1
(8) Communist TIKB Cell	0.94	1	0.94	1
(9) MKP Terrorist Cell	1	1	0.90	1
(10) PKK Cell	0.86	1	0.56	1
Overall Mean	0.809	0.872	0.615	0.99

Domain experts' feed back about GDM and OGDM results were promising [37, 38]. For OGDM, they commented positively when OGDM runs on theft, mafia and drug dealing groups. OGDM also performed apparently better then GDM on non-terrorist offender groups.

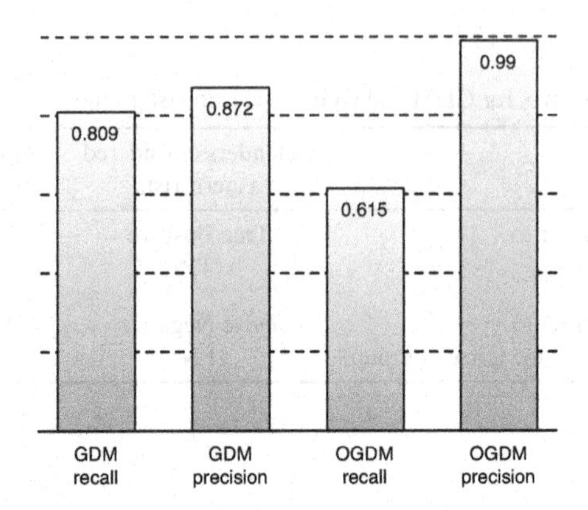

Fig. 13 Comparison of recall and precision values

When it comes to terrorist groups, GDM and OGDM had different outputs. For precision OGDM performed better than GDM which means OGDM detected terrorist offenders within their particular group with high accuracy. But GDM sometimes shows simple offenders within terrorist groups which was not desirable. On the other hand, OGDM performed worse than GDM for recall measure. As can be seen in figures below, GDM didn't produce bulky results whereas OGDM produced bulky results especially it linked unrelated terrorists to target terrorist groups which can be misleading.

7 Conclusion

Domain experts state that it is essential for terrorist groups to look for demographic similarity. Since terrorist cells are isolated from the outer world, they don't know other cell members quite well but for a terrorist cell it is likely that members within cell are coming from similar demographic origins like place of birth similarity and surname similarity. Terrorists care for not being arrested by the police because of minor crimes, so they only have terrorism related modus operandi skills. From this point of view OGDM performs better to link terrorist according to their demographic information. Another advantage of using OGDM on terrorist groups comes from location and time similarity aspect because many terrorist groups have memorial places and special anniversaries to be known and remembered. They do their attacks on these anniversary days and on places which makes terrorist organisation be remembered by public audience. Many detected links between terrorists by OGDM

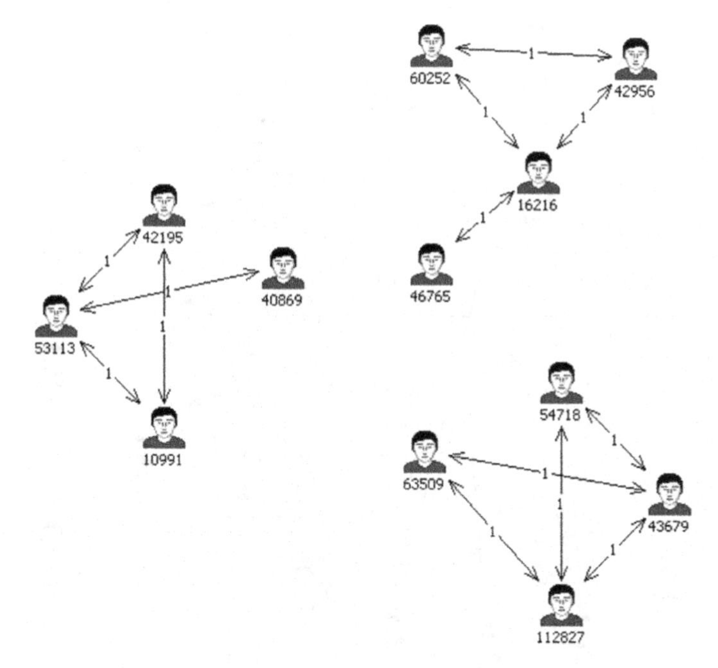

Fig. 14 (2) TDKP-Legal Cells according to domain experts

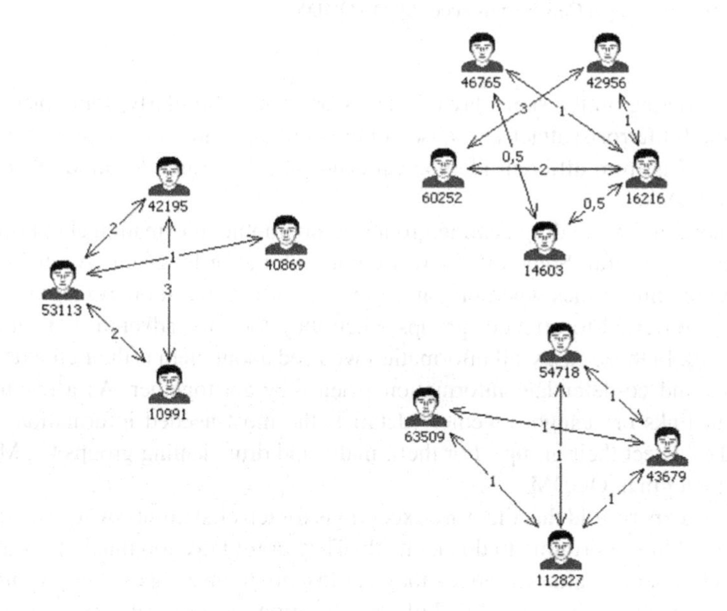

Fig. 15 (2) TDKP Legal Cells results according to GDM

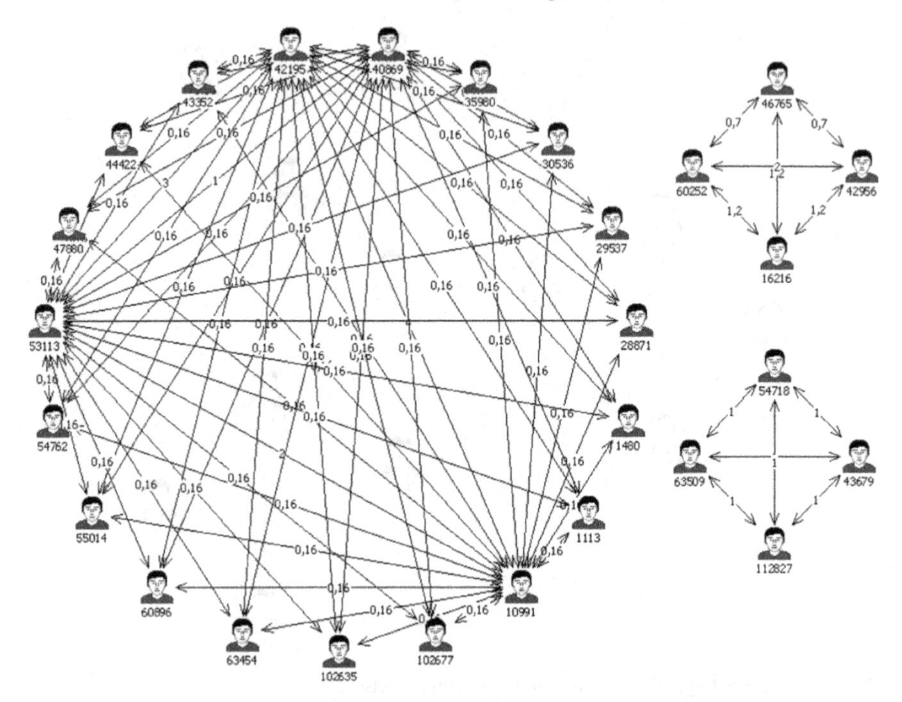

Fig. 16 (2) TDKP Legal Cells results according to OGDM

arise from acting on the same day of attacks or crimes. Similarly, some places are vulnerable for terrorist attacks because of high publicity and remembrance; interestingly terrorists from different-ideological origins choose same location when they decide to attack.

For theft, mafia and drug dealing groups demographic information about offenders is not very useful. Some offenders are not seeking for long-run criminal career and they commit crimes when they need money, drugs etc. Some peer groups can easily be converted to offender groups when they look for adventure or violence [13, 14, 20]. In those cases, all information we need about them is their co-existence in crimes and co-defendant information when they act together. As a result, co-defendant links or "caught-together" detail is the most needed information when we need to detect their groups. For theft, mafia and drug dealing groups GDM performed better than OGDM.

Domain experts told that there are exceptions for terrorist groups whose members are gathered in a short time to do an attack. They don't have too much information about each other and after the attack they got lost not to meet again. Only similarity for those is their modus operandi skills, not location or temporal similarity. When looking for links, terrorists with similar modus operandi skills should be more focused.

Both GDM and ODGM performed well for detecting terrorist groups but OGDM produced bulky results which can be misleading. Police arrest data, especially co-defendant information is very valuable for detecting non-terrorist offender groups. Demographic data can be more valuable for detecting terrorist groups.

When compared to COPLINK CrimeNet Explorer and TMODS, GDM and OGDM have some specific advantages. OGDM uses terror crimes modus operandi ontology to link terrorist whereas TMODS use terrorist attack case scenario ontology to match cases. OGDM uses one-mode, person-to-person graph representation whereas TMODS uses multi-mode entity-to-entity representation on graphs which can be confusing visually. COPLINK CrimeNet Explorer is not tested on terrorist groups, so its output on a terrorist group detection case should be examined to have a better comparison.

Finally it is very important for law enforcement agencies to keep their arrest data with unique crime reference numbers and person id numbers, thereby the police and other agencies can benefit from this for using this information deciding the links between terrorists.

References

1. Analyst Notebook, i2 Analyst Notebook i2 Ltd. (2007), http://www.i2.co.uk/
2. Chen, H., Atabakhsh, H., et al.: Visualisation in Law Enforcement. In: CHI 2005, Portland, Oregon, USA, April 2–7, ACM, New York (2005)
3. Chen, H., Xu, J.J.: Fighting Organised Crimes: using shortest-path algorithms to identify associations in criminal networks. Decision Support Systems 38(3), 473–487 (2003)
4. Chen, H., Chung, W., et al.: Crime data mining: a general framework and some examples. Computer 37(4), 50–56 (2004)
5. Chen, H., Schroeder, J., et al.: COPLINK Connect: information and knowledge management for law Enforcement. Decision Support Systems 34, 271–285 (2002)
6. Chen, H., et al.: Visualization in law enforcement in CHI'05 extended abstracts on Human factors in computing systems 2005 ACM: Portland, OR, USA p. 1268–1271
7. Chen, H., et al.: COPLINK Connect: information and knowledge management for law enforcement. Decision Support Systems, 2002. 34(2002): p. 271–285.
8. Coffman, T., Greenblatt, S. and Marcus, S.: Graph-based technologies for intelligence analysis. Communication of ACM, 2004. 47(3): p. 45–47.
9. Coffman, T.R. and Marcus, S.E.: Dynamic Classification of Suspicious Groups using social network analysis and HMMs. in 2004 IEEE Aerospace Conference, March 6–13, 2004.
10. Cook, D.J., Holder, L.B.: Mining Graph Data. 2007, Hoboken, New Jersey, US: John Wiley & Sons
11. Cook, D.J., Holder, L.B.: Graph-Based data mining. IEEE Intelligent Systems 15(2), 32 – 41 (2000)
12. Cormen, T.H., Leiserson, C.E., Rivest, R.L., Stein, C.: Introduction to Algorithms, 2nd edn. MIT Press and McGraw-Hill (2001)
13. Clarke, R.V. and Cornish, D.B.: Modeling Offenders' Decisions: A Framework for Research and Policy, in Crime and Justice: An Annual Review of Research, M. Tonry and N. Morris, Editors. 1985, University of Chicago Press: Chicago.
14. Dahbur, K. and Muscarello, T.: Classification system for serial criminal patterns. Artificial Intelligence and Law, 2003. 11(2003): p. 251–269.
15. Gabor, Csardi igraph: R package version 0.1.1 (2005), http://cneurocvs.rmki.kfki.hu/igraph
16. Gentleman, R., Whalen, E., Huber, W., Falcon, S.: Graph: A package to handle graph data structures. R package version 1.10.6 (2006)
17. Getoor, L., Diehl, C.P.: Link Mining: A Survey. SIGKDD Explorations 7(2), 3–12 (2005)

18. Getoor, L., et al.: Link Mining: a new data mining challenge. SIGKDD Explorations 5(1), 84–89 (2004)
19. Greenblatt, S., Coffman, T., Marcus, S.: Emerging Information Technologies and enabling policies for counter terrorism, in Behaivoural Network Analysis for Terrorist Detection. 2005, Wiley-IEEE Press: Hoboken, NJ.
20. Guest, S.D., Moody, J., Kelly, L., Rulison, K.L.: Density or Distinction? The Roles of Data Structure and Group Detection Methods in Describing Adolescent Peer Groups. Journal of Social Structure 8(1), http://www.cmu.edu/joss/content/articles/volindex.html (viewed at July 28, 2007)
21. Gomez-Perez, A.: Ontological Engineering with examples from the areas of knowledge management, e-commerce and the semantic web. Springer, London (2005)
22. Frank, O.: Statistical estimation of co-offending youth networks. Social Networks 23, 203–214 (2001)
23. Hauck, R.V., M. Chau, and Chen, H.: COPLINK: Arming Law Enforcement with New Knowledge Management Technologies, in Advances in Digital Government: Technology, Human Factors, and Policy, W. McIver and A. Elmagarmid, Editors. 2002, Kluwer Academic Publishers.
24. Hauck, R.V. and Chen, H.: COPLINK: A case of Intelligent analysis and knowledge management. 1999, University of Arizona: Tucson, Arizona.
25. Hunter, A.: Leninist Cell Data Analysis. 2002, 21st Century Technologies Inc.: Austin,TX.
26. Kantardzic, M.: Data Mining: Concepts, Models, Methods, and Algorithms. 2003, New York, US: John Wiley & Sons.
27. Lapsley, M. and from October 2002 Ripley, B.D.: RODBC: ODBC database access. R package version 1. 1–7 (2006)
28. Lin, S. and Brown, D.E.: An outlier-based data association method for linking criminal incidents. Decision Support Systems, 2004. 41(2006): p. 604–615.
29. Marcus, S.M., Moy, M., Coffman, T.: Social Network Analysis In Mining Graph Data. Cook, D.J., Holder, L.B. (eds.), John Wiley &Sons, Inc., Chichester (2007)
30. Marcus, S. and Coffman, T.: Terrorist Modus Operandi Discovery System 1.0: Functionality, Examples, and Value. 2002, 21st Century Technologies: Austin, TX.
31. Marcus, S.E., Moy, M. and Coffman,T.: Social Network Analysis, in Mining Graph Data, D.J. Cook and L.B. Holder, Editors. 2007, John Wiley & Sons, Inc: Hoboken, New Jersey, US.
32. McCue, C.: Data Mining and Predictive Analytics Intelligence Gathering and Crime Analysis. BH Press Elsevier Oxford, England (2007)
33. Mena, J.: Investigative Data Mining for security and criminal detection, Butterworth, US (2003)
34. Moy, M.: Using TMODS to run the best friends group detection algorithm. 21st Century Technologies Internal Publication (2005)
35. Nath, S.V.: Crime Pattern Detection Using Data Mining. in 2006 IEEE/WIC/ACM International Conference on Web Intelligence and Intelligent Agent Technology (WI-IAT 2006 Workshops)(WI-IATW'06). 2006: IEEE.
36. Nooy, W.D., Mrvar, A., et al.: Exploratory Social Network Analysis with Pajek. Cambridge University Press, New York (2005)
37. Ozgul, F., Bondy, J., Aksoy, H.: Mining for offender group detection and story of a police operation. in Sixth Australasian Data Mining Conference (AusDM 2007). 2007. Gold Coast, Australia: Australian Computer Society Conferences in Research and Practice in Information Technology.
38. Ozgul, F., Erdem, Z. and Aksoy, H.: Comparing Two Models for Terrorist Group Detection: GDM or OGDM? LNCS, 2008. 5075(ISI 2008 – PAISI 2008): p. 149–160.
39. Pioch, N.J. and Everett, J.O.: POLESTAR – Collaborative Knowledge Management and Sensemaking Tools for Intelligence Analysts. in CIKM '06 5–11 November 2006. 2006. Arlington, VA, US: ACM.
40. R Development Core Team: R: A language and environment for statistical computing. R Foundation for Statistical Computing, Vienna, Austria (2006),

41. Schild, U.J. and Kannai, R.: Intelligent computer evaluation of offender's previous record. Artificial Intelligence and Law, 2005. 2005(13): p. 373–405.
42. Scott, J.: Social Network Analysis. SAGE Publications, London (2005)
43. Senator, T.E.: Link Mining Applications: Progress and Challenges. SIGKDD Explorations 7(2), 76–83 (2005)
44. Sentient Data Detective, Sentient Information Systems (2007), http://www.www.sentient.nl/
45. Smith, M.N. and King, P.J.H.: Incrementally Visualising Criminal Networks. in Sixth International Conference on Information Visualisation(IV'02). 2002: IEEE.
46. Xu, J. and Chen, H.: Untangling Criminal Networks: A Case Study. ISI 2003 LNCS, 2003(2665): p. 232–248.
47. Xu, J., Chen, H.C.: CrimeNet Explorer: A Framework for Criminal Network Knowledge Discovery. ACM Transactions on Information Systems 23(2), 201–226 (2005)
48. Wasserman, S., Faust, K.: Social Network Analysis: Methods and Applications, p. 266 (1994)

Part IV
Behavior

CAPE: Automatically Predicting Changes in Group Behavior

Amy Sliva, V.S. Subrahmanian, Vanina Martinez, and Gerardo Simari

Abstract There is now intense interest in the problem of forecasting what a group will do in the future. Past work [1, 2, 3] has built complex models of a group's behavior and used this to predict what the group might do in the future. However, almost all past work assumes that the group will not change its past behavior. Whether the group is a group of investors, or a political party, or a terror group, there is much interest in when and how the group will change its behavior. In this paper, we develop an architecture and algorithms called *CAPE* to forecast the conditions under which a group will change its behavior. We have tested *CAPE* on social science data about the behaviors of seven terrorist groups and show that *CAPE* is highly accurate in its predictions—at least in this limited setting.

1 Introduction

Group behavior is a continuously evolving phenomenon. The way in which a group of investors behaves is very different from the way a tribe in Afghanistan might behave, which in turn, might be very different from how a political party in Zimbabwe might behave. Most past work [1, 4, 2, 3, 5] on modeling group behaviors focuses on learning a model of the behavior of the group, and using that to predict what the

Amy Sliva
University of Maryland College Park, College Park, MD 20742, USA, e-mail: asliva@cs.umd.edu

V.S. Subrahmanian
University of Maryland College Park, College Park, MD 20742, USA, e-mail: vs@cs.umd.edu

Vanina Martinez
University of Maryland College Park, College Park, MD 20742, USA, e-mail: mvm@cs.umd.edu

Gerardo Simari
University of Maryland College Park, College Park, MD 20742, USA, e-mail: gisimari@cs.umd.edu

N. Memon et al. (eds.), *Mathematical Methods in Counterterrorism*,
DOI 10.1007/978-3-211-09442-6_15,

group might do in the future. In contrast, in this paper, we develop algorithms to learn when a given group will *change* its behaviors.

As an example, we note that terrorist groups are constantly evolving. When a group establishes a standard operating procedure over an extended period of time, the problem of predicting what that group will do in a given situation (hypothetical or real) is easier than the problem of determining when, if, and how the group will exhibit a significant change in its behavior or standard operating procedure. Systems such as the CONVEX system [1] have developed highly accurate methods of determining what a given group will do in a given situation based on its past behaviors. However, their ability to predict when a group will change its behaviors is yet to be proven.

In this paper, we propose an architecture called *CAPE* that can be used to effectively predict when and how a terror group will *change* its behaviors. The *CAPE* methodology and algorithms have been tested out on about 10 years of real world data on 5 terror groups in two countries and—in those cases at least—have proven to be highly accurate.

The rest of this paper describes how this forecasting has been accomplished with the *CAPE* methodology. In Section 2, we describe the architecture of the *CAPE* system. Section3 gives details of an algorithm to estimate what the environmental variables will look like at a future point in time. In Section 4, we briefly describe an existing system called CONVEX [1] for predicting what a group will do in a given situation s and describe how to predict the actions that a group will take at a given time in the future. In each possible future situation (i. e. setting of environmental variables) predicted in Section 3, we show how CONVEX can be used to learn a probability distribution that a group will take an action a (or one of many actions $a_1,\dots,a_n$) at a given time in the future. This section shows how to use CONVEX to generate a forecast about what the group will do at a given point t in time, *assuming the group behaves normally*.

However, a group may deviate from its normal behavior. How should such changes in behavior be studied? Section 5 focuses on this important problem. Intuitively, we try to learn the conditions under which a group changed its behaviors, and use these conditions in order to predict the onset of behavioral changes. Unlike the focus of systems such as SOMA [2, 3, 6, 7, 8] and CONVEX [1], the major object of study is *change in behavior*, and not the *behavior* itself. We propose the concepts of a *change table* and *change analysis algorithms* that are used to study the change table, generating rules that determine the conditions under which various terror groups changed their behaviors. This section concludes with the definition of the *CAPE* algorithm. When this method predicts a change in behavior, this result is what *CAPE* predicts; however, if no change in behavior is predicted, then *CAPE* uses the predictions of normal behavior based on the methods described in Section 4.

In Section 6, we describe *preliminary* results of experiments to evaluate the accuracy of our algorithms by applying them to the behaviors of seven terror groups in two countries. Depending upon the granularity with which actions are modeled, our algorithms are either 80 % accurate, or about 69.32 % accurate. This suggests that the granularity at which predictions are made is a key problem.

2 *CAPE* Architecture

Figure 1 shows the architecture of the *CAPE* framework. *CAPE* consists of two major parts—one focusing on *learning* the conditions under which a group changed its behavior, and another focusing on *forecasting* what the group will do.

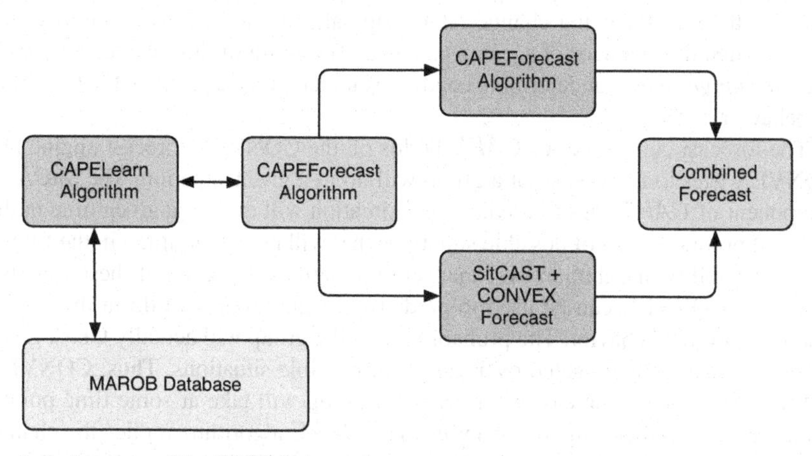

Fig. 1 Architecture of the *CAPE* framework.

Figure 1 shows the learning architecture underlying *CAPE*. The data that *CAPE* currently processes is the Minorities at Risk Organizational Behavior (MAROB) data set [9] that contains information about various terrorist groups. MAROB data exists for over 117 groups from Morocco to Afghanistan, as well as for several other countries such as Bangladesh, the Philippines, and Russia. MAROB contains a relational table for each terror group. The rows in the relational table correspond to years (in a few cases, semi-annual periods of time). The columns correspond to various MAROB variables that fall broadly into two categories—*environmental variables* describing the environment in which a group functioned during a given time frame, and *action variables* describing actions taken by the group during a given time frame. Note that environmental variables can include actions taken by other groups. Each variable v in MAROB has a domain $dom(v)$ consisting of *integers* from 0 to some value between 1 and 7—thus, all MAROB variables can assume somewhere between two and eight possible values. Examples of environmental variables include the level of diaspora support for a group, the level of foreign state financial support for a group, the level of repression of the group by the state in which the group is based, and so forth. Examples of action variables include whether the group used suicide bombings as a strategy, whether they used attacks against domestic security organizations as a strategy, whether they mounted transnational attacks on foreign groups, etc. [9] is an excellent introduction to MAROB.

Though CAPE has currently only been tested on MAROB data, it can be used on any other data set that has a similar form, i. e. where certain attributes are designated as independent attributes (generalizing the environmental variables in MAROB), and certain other attributes are designated dependent attributes (generalizing the action variables in MAROB) and where the goal is to learn conditions on the independent attributes that are predictors of changes in the dependent attributes. The algorithms are in no way tied to MAROB data.

The learning architecture of *CAPE* is based on the concept of a *change table*. The change table captures that part of a MAROB table that changed. A *change analyzer* then analyzes the change table automatically and learns conditions that specify when the behavior of a group changed. The result of the change analyzer is a set of *change rules* that determine conditions under which a group is likely to alter its behavior.

The forecast component of *CAPE* builds on the CONVEX forecast engine [1]. CONVEX tries to forecast what a group will do in a given situation. The SitCAST component of *CAPE* tries to predict what situation will exist at a given time in the future. It produces a set of possible situations that will be true at time t in the future, and a probability distribution over that set of situations. For each of these possible situations, CONVEX can be used to predict what the group will do in the future, assuming normal behavior. The probability that the group will actually take a given action can then be aggregated over the set of possible situations. Thus, CONVEX and SitCAST jointly forecast what actions a group will take at some time point t in the future, together with a probability. The *CAPE* algorithm applies the change rules to a given situation to predict the onset of behavioral changes. If the change rules predict that a behavior will change, then this is what *CAPE* returns. Otherwise, *CAPE* returns what the CONVEX-SitCAST combination predicts.

3 SitCAST Predictions

Suppose we have a MAROB table T_G for some terror group G. We use the notation $E_1,\dots,E_m$ to denote environmental variables, and $A_1,\dots,A_n$ to denote action variables. For a given time period y in the table T, $y.E_i$ denotes the value of environmental variable E_i during time period y and $y.A_j$ denotes the value of action variable A_j during time period y. Let us suppose that we have k rows $y_1,\dots,y_k$ in the table G corresponding to *equally spaced* time periods $y_1 < y_2 < \dots < y_k$. Our goal is to try and predict what the situation will be like during time periods y_{k+s} for $s \geq 1$.

To achieve this, we start by building on top of any time series prediction algorithm (e. g., a linear regression, quadratic regression, or logistic regression program) ts [10]. Let $y_{k+s}.E_i$ denote the value of E_i predicted by ts for the time period y_{k+s}.

Definition 1. A *possible situation* at time y_{k+s} according to prediction algorithm ts is a mapping ps from $\{E_1, \ldots, E_m\}$ to the domain of each E_i such that for all $1 \leq i \leq m$, $\lfloor y_{k+s}.E_i \rfloor \leq ps(E_i) \leq \lceil y_{k+s}.E_i \rceil$.

In other words, the value of y_{k+s} must be set to either $\lfloor y_{k+s}.E_i \rfloor$ or $\lceil y_{k+s}.E_i \rceil$. Let $\mathscr{PS}(ts, s)$ denote the set of all possible situations at time y_{k+s} using time series predictor ts.

Example 1. Consider a simple example shown below where only four environmental variables are considered.

Time	E_1	E_2	E_3	E_4
1	2	4	1	1
2	2	3	1	2
3	1	3	1	2
4	1	3	1	2
5	1	2	1	2

Suppose we wish to predict what the situation will be like at time 6 and suppose the time series prediction function, ts, used is simple linear regression. In this case, $y_6.E_1 = 0.5$, $y_6(E_2) = 1.8$, $y_6.E_3 = 1$ and $y_6.E_4 = 2.4$.

As all our variables (action and environment) have integer domains, it follows that $y_6.E_1$ must be either 0 or 1, $y_6.E_2$ must be either 1 or 2, $y_6.E_3 = 1$ and $y_6.E_4$ is either 2 or 3. This leads to *eight* possible situations listed below.

Situation	E_1	E_2	E_3	E_4
S_1	0	1	1	2
S_2	0	1	1	3
S_3	0	2	1	2
S_4	0	2	1	3
S_5	1	1	1	2
S_6	1	1	1	3
S_7	1	2	1	2
S_8	1	2	1	3

SitCAST can induce a probability distribution on the set of situations as follows. Let us consider variable E_2 whose predicted value at time 6 is 1.8. One can view this as a probability statement saying that the value of E_2 is 2 with 80 % probability and the value of E_2 is 1 with 20 % probability. If we were to assume that the E_i's values are independent of one another, then we can associate the following probabilities with each situation listed in the example above.

Thus, we see that the most probable situations at time 6 are S_3 and S_7, each with 24 % probability of occurring.

Situation	Prob. Calculation	Probability
S_1	$0.5 \times 0.2 \times 1 \times 0.6$	0.06
S_2	$0.5 \times 0.2 \times 1 \times 0.4$	0.04
S_3	$0.5 \times 0.8 \times 1 \times 0.6$	0.24
S_4	$0.5 \times 0.8 \times 1 \times 0.4$	0.16
S_5	$0.5 \times 0.2 \times 1 \times 0.6$	0.06
S_6	$0.5 \times 0.2 \times 1 \times 0.4$	0.04
S_7	$0.5 \times 0.8 \times 1 \times 0.6$	0.24
S_8	$0.5 \times 0.8 \times 1 \times 0.4$	0.16

4 CONVEX and SitCAST

The CONVEX system [1] develops methods to automatically predict what actions a given group will take in a given situation (real or hypothetical) by examining the similarity between the hypothesized situation and past situations in which the group has been. Once SitCAST has been used to forecast what a situation might be in the future (e. g. at time 6 in the running example given above), we can use CONVEX to predict what actions the group will take in each of the possible situations.

Suppose we consider a single action A_i whose intensity value we want to predict. That is, we want to predict the value of $y_6.A_i$ in the example above. In this case, we can apply CONVEX to each possible situation S_i predicted to occur at a future time t. The probability that A_i will have the value v at time y_{k+s} will then be given by the following formula.

$$\mathbf{P}(y_{k+s}.A_i = v) = \Sigma_{S \in \mathscr{P}\mathscr{S}(ts,s) \wedge CONVEX(S,A_i)=v} prob(S).$$

In other words, to compute the probability of $y_{k+s}.A_i$ having the value v, we perform the following steps:

1. Find all possible situations S that can arise at time $k+s$ such that CONVEX predicts that those situations will cause action variable A_i to have value v.
2. Add up the probability of all such situations. This is the probability that $y_{k+s}.A_i = v$.

Let us apply this to the running example mentioned above.

Example 2. Suppose we consider an action A with three possible values: 0,1,2. Let us suppose that CONVEX predicts that:

1. $A = 0$ in possible situations S_1, S_4, S_6.
2. $A = 1$ in possible situations S_2, S_7.
3. $A = 2$ in possible situations S_3, S_5, S_8.

The probability that $A = 0$ above is the sum of the probabilities that either situation S_1, S_4 or S_6 will occur. These probabilities are 0.06, 0.16 and 0.04 respectively, yielding a 26 % probability that $A = 0$ at time 6. This table summarizes the probabilities involved.

Action value	Prob. Summation	Total Prob.
$A = 0$	$0.06 + 0.16 + 0.04$	0.26
$A = 1$	$0.04 + 0.24$	0.28
$A = 2$	$0.24 + 0.06 + 0.16$	0.46

This tells us that the *most likely* situation to arise is that the group in question will engage in action A with level 2 intensity with 46 % probability.

Note that though CONVEX itself runs in polynomial time, the combination of CONVEX and SitCAST can take exponential time because the number of possible situations to consider might be exponential in the number of environmental variables. One way around this problem is the following. Let w be a real number between 0 and 0.5 inclusive. Whenever an environmental variable is predicted to be a real number r between $\lfloor r \rfloor$ and $\lceil r \rceil$, we do the following.

- Check if $\lfloor r \rfloor + 0.5 - w \leq r \leq \lfloor r \rfloor + 0.5 + w$. If so, then generate two possible situations.
- Otherwise, if $r \leq \lfloor r \rfloor + 0.5 - w$, then reset r's value to $\lfloor r \rfloor$.
- Otherwise, if $r \geq \lfloor r \rfloor + 0.5 + w$, then reset r's value to $\lceil r \rceil$.

To see how this works, suppose $w = 0.1$. This means, for instance, that if the predicted value $y_{k+s}.E_i$ is in the range $j.4$ to$j.6$ for an integer j, then two possible situations are generated for $y_{k+s}.E_i$. Otherwise, if the predicted value is between j and $j.4$, then the value is modified to j. If the predicted value is between $j.6$ and $(j+1)$, then the value is modified to $j+1$.

Example 3. Let us return to our running example where the predicted values are as follows. Let $w = 0.1$.

1. $y_6.E_1 = 0.5$. In this case, no change is made to $y_6.E_1$ because $y_6.E_1$ has value 0.5 which is between 0.4 and 0.6.
2. $y_6.E_2 = 1.8$. In this case, $y_6.E_2$ is reset to 2 because $y_6.E_2$ has a value greater than 1.6.
3. $y_6.E_3 = 1$. In this case, $y_6.E_2$ stays 1 because $y_6.E_3$ has a value less than 1.4.
4. $y_6.E_3 = 2.4$. In this case, no change is made to $y_6.E_4$ because $y_6.E_4$ has value 2.4 which is in the interval $[2.4, 2.6]$.

It can be seen that by using this method, we will only generate the four possible situations, shown as situation S'_1, S'_2, S'_3, S'_4 in the first table that follows.

Situation	E_1	E_2	E_3	E_4
S'_1	0	2	1	2
S'_2	0	2	1	3
S'_3	1	2	1	2
S'_4	1	2	1	3

This result shows a 50 % reduction in the set of generated possible situations. If, on the other hand, we had set $w = 0.31$, then we would only have two situations to consider—see situations $S_1^\star, S_2^\star$ in the second table below.

Situation	E_1	E_2	E_3	E_4
$S_1^\star$	0	2	1	2
$S_2^\star$	1	2	1	2

5 The *CAPE* Algorithm

The combination of CONVEX and SitCAST does very well in predicting what a group will do; however, both CONVEX and SitCAST focus on how a group will behave based on *past behaviors* of the group. Groups occasionally change their behaviors, and predicting such *changes* in behavior is often much more important than predicting that the group's behavior will conform to what is normal for the group. In this section, we will present the *CAPE* algorithm for forecasting group behaviors. Suppose A_i is an action variable. We are interested in learning the conditions under which A_i changes for a given group G.

5.1 The Change Table

The first concept we define when learning the conditions under which a group changes its behavior is a *change table*.

The *change table* $CH(G,A_i)$ for group G w.r.t. A_i is a table derived from the table for G as follows.

(Step 1) The set of rows in $CH(G,A_i)$ is given by $\{y_j \mid y_j.A_i \neq y_{j-1}.A_i\}$.
(Step 2) The set of columns in $CH(G,A_i)$ consists of the column associated with A_i and the set of all E_r's such that $y_{j-1}.E_r \neq y_{j-2}.E_r$.
(Step 3) If $y_j.E_r$ is not eliminated in the previous two steps, then its value is set to the pair $(y_{j-2}.E_r, y_{j-1}.E_r)$.
(Step 4) If $y_j.A_i$ is not eliminated in the first two steps above, then its value is set to the pair $(y_{j-1}.A_i, y_j.A_i)$.

In other words, the change table $CH(G,A_i)$ for group G w.r.t. A_i eliminates all rows in the original table where the action A_i did not exhibit a change in value from the previous time period y_{j-2}. In addition, it eliminates all columns except for those environmental variables which changed from $y_{j-2}.E_r$ to $y_{j-1}.E_r$. The change table *documents the changes that have occurred in the original table for group G, based on the assumption that changes in an environmental variable from time period*

$(j-2)$ to $(j-1)$ are potentially responsible for a change in the action variable one time period later, i. e. from time period $(j-1)$ to j.[1]

Example 4. Let us expand the example table we have been using with an action attribute.

Time	E_1	E_2	E_3	E_4	A
1	2	4	1	1	1
2	2	3	1	2	1
3	1	3	1	2	2
4	1	3	1	2	2
5	1	2	1	2	1

This table shows that the value of the action attribute A changed twice—once in time period 3 and once in time period 5. This means that the change table will have just two rows after the execution of Step 1 of the change table construction algorithm.

Time	E_1	E_2	E_3	E_4	A
3	1	3	1	2	2
5	1	2	1	2	1

In Step 2, all attributes that did not change in the preceding time interval get eliminated. Thus, attributes E_1 and E_3 get eliminated because they did not change their values from either time periods 1 to 2 or from the time period 3 to 4, yielding the following table.

Time	E_2	E_4	A
3	3	2	2
5	2	2	1

In Step 3, we replace the values of the environmental variables in the above table by the changes exhibited.

Time	E_2	E_4	A
3	(4,3)	(1,2)	2
5	(3,3)	(2,2)	1

What the first row of this table means is that the value of E_2 changed from 4 in time 1 to 3 in time 2 and that the value of E_4 changed from 1 in time 1 to 2 in time 2. We now execute Step 4 to obtain the following table.

[1] It is easy to account for other lags in a similar manner.

Time	E_2	E_4	A
3	(4,3)	(1,2)	(1,2)
5	(3,3)	(2,2)	(2,1)

This result is the final change table. The row associated with time 3 says that there was a change in the action variable A from time 2 (value 1) to time 3 (value 2). The fact that E_2 changed from time 1 (value 4) to time 2 (value 3) and the fact that E_4 changed from time 1 (value 1) to time 2 (value 2) are potential causes of the change in attribute A one time period later.

The final change table, as shown above, encapsulates all changes in the action A (corresponding to a behavior) that we are trying to model. More importantly, the final change table dramatically reduces the number of possible environmental attributes to consider when modeling the conditions under which changes occur in action A.

Note that in the form presented above, changes in the environmental variables from time y_{j-2} to y_{j-1} are used to learn changes in the action variables associated with a group from time period y_{j-1} to y_j. In other words, these change rules can be used to make predictions "one period ahead." It is easy to learn rules that apply multiple periods ahead via the following simple changes to the change table construction. The following algorithm looks h time periods ahead.

(Step 1') The set of rows in $CH(G,A_i)$ is given by $\{y_j \mid y_j.A_i \neq y_{j-1}.A_i\}$.
(Step 2') The set of columns in $CH(G,A_i)$ consists of the column associated with A_i and the set of all E_r's such that $y_{j-h}.E_r \neq y_{j-h+1}.E_r$.
(Step 3') If $y_j.E_r$ is not eliminated in the previous two steps, then its value is set to the pair $(y_{j-h}.E_r, y_{j-h+1}.E_r)$.
(Step 4') If $y_j.A_i$ is not eliminated in the first two steps above, then its value is set to the pair $(y_{j-1}.A_i, y_j.A_i)$.

This algorithm merely changes the parameters associated with E_r to look $h-1$ time periods back instead of just one time period back. We call this the *h-change table* and denote it by $CH^h(G,A_i)$.

5.2 Learning Predictive Conditions from the Change Table

When learning conditions that predict change in an action variable, we only consider *conjunctive conditions* on environmental attributes. If E_i is an environmental variable and c is a value in its domain, then $E_i = c$ is an *environmental atom*. If $EA_1, \ldots, EA_n$ are environmental atoms, then $(EA_1 \wedge \ldots \wedge EA_n)$ is an *environmental condition of size n*. There is no loss of generality in the assumption that only conjunctions are considered because we are learning *sets* of conditions and hence disjuncts can be easily accounted for.

Our *CAPE-Learn* algorithm given below learns sets of conditions from both the original table and the change table for a given group G. It uses the usual concept of satisfaction of a condition by a tuple [11]. The *CAPE-Learn* algorithm also takes as input, a special boolean function called *STAT-TESTS* which takes an environmental condition EC as input and tests whether the environmental condition satisfies various statistical conditions that a user may wish to impose. We place no restriction whatsoever on how *STAT-TESTS* may be implemented—it could compute p-values and ensure that they fall within a given bound, or it might involve a t-test or it might involve no tests at all. The person invoking the algorithm can decide what statistical tests, if any, he wants to use for his needs.

algorithm *CAPE-Learn*$(G, T_G, CH(G,A_i), v_1, v_2, sz, \tau_1, \tau_2, STAT\text{-}TESTS)$
% We are interested in changes of A_i from value v_1 to v_2;
% We are only interested in conjunctive conditions of size sz or less;
% τ_1, τ_2 are real numbers in the $[0,1]$ interval;
$Ans = \emptyset$;
$\mathscr{C} = \{EC \mid EC$ is an environmental condition of size sz or less and each attribute in EC occurs in $CH(G,A_i)\}$;
for each $EC \in \mathscr{C}$ **do**

$$precision(EC) = \frac{card(\{y_j \mid y_j \in CH(G,A_i) \wedge y_j \models EC \wedge y_j.A_i = (v_1,v_2)\})}{card(\{y_j \mid y_j.A_i = (v_1,v_2)\})};$$

$$recall(EC) = \frac{card(\{y_j \mid y_j \in T_G \wedge y_j \models EC \wedge y_j.A_i = (v_1,v_2)\})}{card(\{y_j \mid y_j \in T_G \wedge y_j \models EC\})};$$

if $precision(EC) \geq \tau_1 \wedge recall(EC) \geq \tau_2 \wedge STAT\text{-}TESTS(EC)$ **then**
$Ans = Ans \cup \{EC\}$.
end for
Return *Ans*.
end algorithm

The *CAPE-Learn* algorithm cycles through the set of all environmental conditions that are of size sz or less and that are only composed of environmental conditions defined over the environmental attributes and values that occur in the change table. The change table is usually substantially smaller than the original table about a given group. For instance, we have run experiments with a total of 5 terrorist groups and the size of the change table, as a proportion of the size of the original table is given by the following table:

Group	MAROB Table Size	Change Table Size	Ratio of Change Table to MAROB Table
1	1224	40	3.27 %
2	1224	37	3.02 %
3	1176	33	2.81 %
4	1176	51	4.34 %
5	1176	35	2.98 %

In other words, the size of the change table is *very small* compared to the size of the original MAROB table—typically around 3 to 4 % of the size of the original MAROB table. Once the set $\mathscr{C}$ has been generated, we compute the precision and recall of each environmental condition EC in $\mathscr{C}$.

The precision of an environmental condition EC, denoted $precision(EC)$ is the ratio of two quantities:

- The number of time periods y_j in the change table which satisfy EC where action variable A_i changed value from v_1 to v_2.
- The number of time periods y_j in the change table where action variable A_i changed value from v_1 to v_2.

Thus, the precision computes the conditional probability of EC being true, given that the action variable A_i changed value from v_1 to v_2. We clearly want this conditional probability to be high for any EC that is returned to the user. Note that the precision is computed without ever looking at the original MAROB table—only the change table is used.

In contrast, the *recall* of an environmental condition EC, denoted $recall(EC)$, connects the change table with the MAROB table. It tries to compute the conditional probability of action variable A_i changing its value from v_1 to v_2, given that environmental condition EC is true. Thus, this is computed over the *original MAROB table* (not the change table) as follows.

- Find the number of time periods y_j in the original MAROB table which satisfy EC and where action variable A_i changed value from v_1 to v_2,
- Find the number of time periods y_j in the original MAROB table which satisfy EC, and
- Take the ratio of the first quantity to the second.

We obviously want environmental conditions returned to the user to have a high recall as well. Thus, the only environmental conditions returned by the *CAPE-Learn* algorithm are those that have a sufficiently high recall (exceeding a given threshold) and have a sufficiently high precision (exceeding a threshold as well), and that satisfy various statistical tests that the user designates.

The complexity of the *CAPE-Learn* algorithm is dominated by the computation of the set $\mathscr{C}$. A few tricks can be used to reduce the size of $\mathscr{C}$.

Definition 2. Suppose T_G is the MAROB table associated with group G, and $CH(G,A_i)$ is the change table associated with changes in the value of action variable A_i from a value v_1 to a value v_2. An environmental atom $E_j = c$ is said to be *coherent* iff E_j occurs as a column in $CH(G,A_i)$ and either $(c,-)$ or $(-,c)$ is a value in column E_j of $CH(G,A_i)$.

If $EA_1,\ldots,EA_k$ are *coherent* environmental atoms, then $(EA_1 \wedge \ldots \wedge EA_k)$ is a *coherent* environmental condition of size k.

In our *CAPE-Learn* algorithm we generate only *coherent* environmental conditions when computing $\mathscr{C}$. It is easy to see that the number of coherent environmental conditions is relatively small.

Example 5. Let us consider the change table below that was generated in Section 5.1.

Time	E_2	E_4	A
3	(4,3)	(1,2)	(1,2)
5	(3,3)	(2,2)	(2,1)

There are only four coherent environmental atoms w.r.t. this change table: $E_2 = 3, E_2 = 4, E_4 = 1$ and $E_4 = 2$. Thus, if sz is set to 3, then $\mathscr{C}$ can only consist of $4 \times 3 \times 2 = 24$ possible values.

The complexity of the *CAPE-Learn* algorithm is

$$\mathbf{O}(card(\mathscr{C}) \times card(CH(G,A_i))).$$

If the change table consists of c columns and each column has at most b values in it, then this complexity boils down to

$$\mathbf{O}(b^c \times card(CH(G,A_i))).$$

Though there is an exponential factor in this computation, c has been quite small in the real world MAROB data that we have looked at.

5.3 The *CAPE-Forecast* Algorithm

The *CAPE* system uses the rules learned via the *CAPE-Learn* algorithm to come up with forecasts s periods ahead of schedule. The *CAPE-Forecast* algorithm is quite simple. It is given a MAROB table T_G associated with a group containing information on the group's activities during time periods $y_1, \ldots, y_k$, and it is asked to forecast the value of an action variable A_i, from its domain $\mathscr{V}_{A_i}$, during time period y_{k+s}. Informally speaking, it achieves this as follows.

- It applies every single CAPE rule learned by the *CAPE-Learn* algorithm for action A_i to forecast the value of variable A_i at time y_{k+s}. In the event of conflict in what the rules predict, a rule with the highest recall is chosen (recall is defined in the *CAPE-Learn* algorithm). If a conflict still exists, the first conflicting value is chosen.
- If the preceding step causes a forecast of A_i which is different from the most recent value $y_k.A_i$ that is known to hold at time period y_k, then *CAPE-Forecast* returns the forecasted value.
- Otherwise *CAPE-Forecast* returns the value forecast by CONVEX and SitCAST.

Algorithm *CAPE-Forecast* is presented below.
algorithm *CAPE-Forecast*$(G, T_G, CH(G,A_i), \mathscr{V}_{\mathscr{A}_i}, sz, \tau_1, \tau_2, STAT\text{-}TESTS, y_{k+s})$
% We first generate the set of all *CAPE* rules using the *CAPE-Learn* algorithm
% for all values of A_i changing from $y_k.A_i$ to some value v_j.

% Each rule is examined to determine if it applies to an environmental change
% from times y_{k-1} to y_k in T_G. Of these, we apply the rule with
% the highest precision to make the change prediction
$Rules = \{\}$;
$prediction = null$;
$probability = 0$;
for each $v_j \in \mathcal{V}_{\mathcal{A}_i}$ s.t. $y_k.A_i \neq v_j$ **do**
    $Rules =$ *CAPE-Learn*$(G, T_G, CH^s(G, A_i), y_k.A_i, v_j, sz, \tau_1, \tau_2, STAT\text{-}TESTS)$.
    for each $EC \in Rules$ **do**
        if $y_k \models EC \wedge recall(EC) > probability$ **then**
            $probability = recall(EC), prediction = v_j$;
    end for
end for

if $prediction = null$ **then**
    $prediction = SitCAST + CONVEX.prediction$;
    $probability = SitCAST + CONVEX.probability$;

Return $(prediction, probability)$.
end algorithm

The algorithm initially starts with no prediction and with a probability of 0 (corresponding to the recall of the best rule found so far). It uses *CAPE-Learn* to learn all rules for predicting action A_i s time periods in the future. It then considers each environmental condition *EC* learned in this manner. If the current year (y_k) satisfies the condition *EC* and the recall of *EC* exceeds the best recall found so far for any rule, then it sets *probability* to this recall and stores *EC* as the best rule (environmental condition) found so far. This process is repeated until all learned *EC*s are considered.

6 Experimental Results

We applied the *CAPE* architecture to study the behavior of seven terror groups using data manually collected by the MAROB effort [9]. We only looked at one action that these seven groups took - their propensity to engage in rebellion against the state. We tried to predict whether the group would engage in rebellion at all and the intensity of that rebellion by predicting rebellion one time period (half year) ahead by using *CAPE-Learn* on past data to learn the rules.

We conducted two experiments. In the first experiment, the MAROB data for rebellion by these seven groups merely indicated if the groups engaged in rebellion or not. In this case, *CAPE-Forecast* accurately predicted the rebellion status of these groups with 80 % accuracy.

In the second experiment, the intensity of rebellion was described on a scale of 0 (zero rebellion) to 7 (denoting full fledged civil war). *CAPE-Forecast* accurately predicted the exact intensity of rebellion 69.32 % of the time.

What these two experiments suggest is that *CAPE-Forecast* does a very good job of making predictions, and that these predictions are significantly more accurate than an arbitrary guess would be (simple guessing would give 50 % accuracy in the first experiment above, and 12.5 % accuracy in the second experiment). However, this also suggests that *CAPE*'s accuracy decreases when highly fine grained forecasts are required. Further experiments are needed in order to assess *CAPE*'s accuracy with changes in the granularity of forecasts desired. Moreover, the above experiments only apply to situations one time period ahead—how *CAPE*'s accuracy changes with longer look-aheads needs to be further studied.

7 Related Work

There has been intense interest in the last few years in the problem of learning models of the behaviors of groups and applying them to forecast what the group might do in the future [4, 5, 2, 3]. Such works focus on what the group would *normally* do, not how it would change its behaviors. In contrast, this paper focuses on learning conditions under which a group will change its behavior—to date, we are aware of no work on forecasting *changes* in behavior of terror groups.

Specifically, Schrodt [4] and Bond et al. [5] have come up with methods to build Hidden Markov Models to describe how a country might evolve over time. However, these HMMs are painstakingly constructed over time using data gathered about the country. This is very time-consuming and needs to be repeated for each country or each group, and has a degree of subjectivity to it. Nevertheless, HMMs and their variants (e. g. stochastic automata, stochastic Petri nets, etc.) can form a very valuable modeling tool and mechanism and deserves continued study. Their main advantage is that they are not necessarily based on past data.

Another method to forecast how a group might behave in the future is our group's Stochastic Opponent Modeling Agents (SOMA) system developed in [2, 3, 6, 7, 8]. SOMA looks at the data collected about a given group and then *automatically* extracts stochastic behavioral rules. This has an advantage over [4, 5] in that our method to build a behavioral model is fully automated and fast, as shown in [6]. These behavioral rules are then used for predictive purposes using a linear programming based algorithm. No detailed analysis of the accuracy of this forecast algorithm is available (the SOMA Forecast Engine is still under development) at this time and the compute time for generating forecasts can be expensive [7, 8].

In contrast to all these methods, the CONVEX [1] system does *not* build a model of a group. It uses data about a group in order to assess the similarity between a given situation (expressed as the query vector) and information about the group's behavior in the past when confronted with similar situations. Similar situations that the group has encountered in the past are then used directly for predictions—the "model building phase" present in related works is therefore completely skipped. Hence, CONVEX predictions can be made solely by examining the data and directly predicting outcomes without a country or group specific model that takes time to build and is often built using subjective methods. However, as in the case of the other work mentioned above, CONVEX, by virtue of examining similarities between a current situation and the past, predicts what is expected behavior of a group, and not how the group changes behaviors. In this paper, we build upon CONVEX and show how it is one component of a larger system to predict behavioral changes.

8 Conclusions

There are numerous applications where we wish to predict not only the ordinary behavior of a group, but when the group will change its behavior. Financial organizations are interested in tracking the behavior of major investors and political organizations. Political parties are interested in tracking the behavior of various special interest groups. Governments are interested in the behaviors of foreign political and military organizations.

In this paper, we have made two important contributions. First, we developed the *CAPE-Learn* algorithm to learn conditions under which a group changes a certain type of behavior of interest. Second, we developed the *CAPE-Forecast* algorithm to use the rules and conditions learned by *CAPE-Learn* to forecast what a group will do in future time periods.

We tested *CAPE* on one behavior (propensity to engage in rebellion) engaged in by seven terrorist groups outside the USA, and found that *CAPE* exhibited high predictive power in these limited experiments.

Much work remains to be done. Despite the promising results above, we believe *CAPE* needs to be tested out on a much wider range of groups, and a much wider set of actions. Second, experiments are needed to study the accuracy of *CAPE* in forecasting behavior change multiple time periods into the future. Third, further experiments are needed to study the accuracy of *CAPE*, as more fine grained, pinpoint forecasts are needed.

Acknowledgements

The work reported in this paper was funded in part by AFOSR grants FA95500610405 and FA95500510298.

References

1. Martinez, V., Simari, G.I., Sliva, A., Subrahmanian, V.S.: Convex: Similarity-based algorithms for forecasting group behavior. IEEE Intelligent Systems **23**(4) (2008) 51–57
2. Subrahmanian, V.S., Albanese, M., Martinez, V., Nau, D., Reforgiato, D., Simari, G.I., Sliva, A., Wilkenfeld, J.: Cara: A cultural adversarial reasoning architecture. IEEE Intelligent Systems **22**(2) (2007) 12–16
3. Subrahmanian, V.S.: Cultural modeling in real-time. Science **317**(5844) (2007) 1509–1510
4. Schrodt, P.: Forecasting conflict in the balkans using hidden markov models. In: Proc. American Political Science Association meetings. (2000)
5. Bond, J., Petroff, V., O'Brien, S., Bond, D.: Forecasting turmoil in indonesia: An application of hidden markov models. In: International Studies Association Convention, Montreal. (2004) 17–21
6. Martinez, V., Simari, G.I., Sliva, A., Subrahmanian, V.S.: The SOMA terror organization portal (STOP): Social network and analytic tools for the real-time analysis of terror groups. In Liu, H., Salerno, J., Young, M., eds.: Social Computing, Behavioral Modeling and Prediction, Spring Verlag (2008) 9–18
7. Khuller, S., Martinez, V., Nau, D., Simari, G.I., Sliva, A., Subrahmanian, V.S.: Finding most probable worlds of probabilistic logic programs. In: Proc. 2007 International Conference on Scalable Uncertainty Management. Volume 4772., Springer Verlag Lecture Notes in Computer Science (2007) 45–59
8. Khuller, S., Martinez, V., Nau, D., Simari, G., Sliva, A., Subrahmanian, V.S.: Computing most probable worlds of action probabilistic logic programs: Scalable estimation for $10^{30,000}$ worlds. Annals of Mathematics and Artificial Intelligence **51**(2-4) (2007) 295–331
9. Wilkenfeld, J., Asal, V., Johnson, C., Pate, A., Michael, M.: The use of violence by ethnopolitical organizations in the middle east. Technical report, National Consortium for the Study of Terrorism and Responses to Terrorism (2007)
10. Bowerman, B., O'Connell, R., Koehler, A.: Forecasting, Time Series and Regression. Southwestern College Publ (2004)
11. Ullman, J.: Principles of Database and Knowledge Base Systems. Volume 2. Computer Science Press, Maryland (1989)

Interrogation Methods and Terror Networks

Mariagiovanna Baccara and Heski Bar-Isaac

Abstract We examine how the structure of terror networks varies with legal limits on interrogation and the ability of authorities to extract information from detainees. We assume that terrorist networks are designed to respond optimally to a trade-off caused by information exchange: Diffusing information widely leads to greater internal efficiency, but it leaves the organization more vulnerable to law enforcement. The extent of this vulnerability depends on the law enforcement authority's resources, strategy and interrogation methods. Recognizing that the structure of a terrorist network responds to the policies of law enforcement authorities allows us to begin to explore the most effective policies from the authorities' point of view.

1 Introduction

Apprehending terrorists and extracting information from them and, in particular, the different investigation and interrogation methods used are the topics of an ongoing debate which once again took center stage upon the start of the U.S. war on terror.[1] In this article, we do not aim to examine the ethical and legal aspects of the debate, but, highlight the fact that the efficacy of different methods may vary to the extent

Mariagiovanna Baccara
Department of Economics, Stern School of Business, New York University, 44 W Fourth Street, New York, NY 10012, USA, e-mail: mbaccara@stern.nyu.edu

Heski Bar-Isaac
Department of Economics, Stern School of Business, New York University, 44 W Fourth Street, New York, NY 10012, USA, e-mail: heski@nyu.edu.

[1] A similar debate took place, for example, at the beginning of the 1900s, during the engagement of the US against the Spanish colony in the Philippines. The media reported cruelties against Filipino prisoners to gather information on the counter-insurgency (see "The Water Cure" by P. Kramer, *The New Yorker*, 2/25/2008).

N. Memon et al. (eds.), *Mathematical Methods in Counterterrorism*,
DOI 10.1007/978-3-211-09442-6_16,

that the structure of terror networks strategically adapts to different law enforcement policies.

The starting point for our work is the observation that the way terrorist networks are designed is the result of a trade-off: Diffusing information throughout the organization allows it to operate more successfully, but it leaves the organization more vulnerable to law enforcement detection. In choosing their design, terror organizations act as strategic entities that respond optimally to the policies implemented by the authorities. Empirically, it certainly appears that protecting information is a crucial concern of terror networks since their survival depends on how effective they are in preventing information leakage. The available evidence suggests that terror network are characterized by a fairly decentralized "cell" structure; however, there are some agents that appear to hold more information than others (which we term an "informational hub").[2]

In this article, we first aim to describe how authorities should expect the organization to respond strategically to alternative policies. In particular, we highlight the role and the consequences of *investigation and interrogation methods* on the information structure of the terror organization. Given our findings, we take the authorities' point of view and discuss the problem of choosing the most effective policy to fight terror networks.

It is worth highlighting from the outset, what we mean by the "network" and the costs and benefits of "informational links."footnote In particular, within any organization (and terrorist organizations are no exception) many networks coexist simultaneously and often interact: ranging from the network of decision-making authority, to communication networks, "productive networks" who act and work together, and more informal networks based on social interactions. Moreover networks can vary in whether they are directed (for example, an authority network where Ann may be able to veto Bob's decisions) or undirected (for example if Ann is Bob's cousin, then it's a safe bet that Bob is Ann's cousin). We focus on a directed network and, specifically, if Bob is linked to Ann, this should be understood as Ann having some information about Bob. This information could, for example, allow Ann to ensure that Bob acts in the interests of the group. Examples of this kind of information include identity, whereabouts, or incriminating evidence about a person.[3] We view the network of who holds such information as independent of the other structures that coexist within the organization. For example, planning, decision-making, coordinating activity and communication can take place under code-names.[4] In that case, coordinating actions can be quite independent of who within the organization

[2] See, for instance, Arquilla and Ronfeldt [2].

[3] Though these examples are drawn from gangs and crime rather than from terrorist structures, Thompson [11], for example, describes that in his role as a journalist reporting on organized crime, he had to divulge his address and that of his close family members. Charlie [5] describes how committing a murder in front of peers is often an initiation ritual used by U.S. gang members.

[4] Indeed, this was exactly what was done, for example, by the Italian resistance in WWII (see, for instance, Oliva [9]) . Moreover, the investigations on Al-Qaeda suggest that the same Al-Qaeda member often uses different names depending on whom he is interacting with, as described, for instance, in the account of the investigation following Daniel Pearl's murder in Pearl [10].

knows the real name or whereabouts of other members of the organization. However, the diffusion of this kind of information clearly has implications both for the extent of trust (and the efficient functioning of the organization) and the organization's vulnerability with respect to its enemies.

In the model, we simply assume a fixed benefit of linking one member of the organization to another. This benefit can be viewed as arising from the enhanced ability of the organization to disciple such a member, leading to better outcomes from the group's perspective.[5] Here, we devote more attention to fleshing out the costs of the network, and how they respond to law enforcement policies. In particular, we suppose that individuals within the organization are directly vulnerable according to the authorities' investigation policies. Further, if caught, then a member of the network will jeopardize other member of the network depending on the extent to which he has information about them (that is, depending on whether they are linked to him), and depending on the law enforcement's ability to extract such information through interrogations.

Since we view the structure of the network as designed in response to the costs (of vulnerability to the authorities) and benefits of information links, we can characterize the network that arises in any set of circumstances, and show how different law enforcement capabilities, resources or policies affect the structure of the terrorist network. In particular, we show that both centralized and decentralized networks can arise as optimal responses to investigation policies. On the one hand, given a certain investigation budget, the optimal network tends to be more decentralized when the authorities invest similar resources to seek each agent. On the other hand, when the budget is allocated asymmetrically, the optimal organization is a *mixed structure* in which one cell acts as "information hub" and holds information about a certain number of members of the organization. However, there are other members of the organization that remain independent from the hub and are organized in binary cells. We also find that sometimes, as the interrogation techniques become harsher, the terror network responds by increasing the number of cells in the mixed structure. Therefore, our results suggest that there are circumstances in which harsher interrogation methods do not change the number of links in the organization (and, so, do not affect its operation), but just have the effect of increasing the degree of decentralization in the way information is shared in the network.

After we characterize the effect of different policies on network structure, we can turn to the assessment of the authorities' policies and, therefore, to the characterization of the optimal policies. Our results suggest that the three instruments available to the authorities (investigation methods, interrogation methods and severity of punishments) are not independent but are *strategically interrelated.* If interrogation methods and severity of punishments are bounded by legal limits, the investigation methods should be tailored to the legal environment. We show the circumstances under which a symmetric investigation strategy (i. e., allocating the same budget to detect all potential members of the organization) is optimal, and we also look at how optimal investigation strategies vary with changes in the legal environment within

[5] Such a benefit is derived endogenously in Baccara and Bar-Isaac [3].

which the authorities act. In particular, we find that as legal limits broaden, the cost of a link in a binary cell increases. This makes it more likely that a symmetric allocation of the investigation budget can prevent any links from arising. However, as legal boundaries become narrower, it is more likely that a symmetric budget allocation leads to a fully efficient allocation. In this case, we show that the authorities can do better by pursuing an asymmetrical investigation strategy.

1.1 Related Literature

This article builds on Baccara and Bar-Isaac [3] which introduced a model that is more explicit on how information links lead to trust within the organization, and allows for the probability of detection to vary with the extent of a terrorist's cooperation with the organization. Baccara and Bar-Isaac [3] also provides some additional discussion of assumptions, and considers a number of extensions. Here, we simplify the model in a number of respects – in particular, in presenting a reduced form benefit for an informational link – and extend the model to consider variety in interrogation methods. To our knowledge, this work is the first to address the optimal information structure in organizations subject to an external threat.

To our knowledge, this is the first article that addresses a terror network's strategic response to changes in the investigation and interrogation methods. However, several papers have some elements that are related to our work. Farley ([6], [7]) considers the robustness of a terrorist cell. In that work, robustness is with regard to maintaining a chain of command in a hierarchy. Garoupa [8] looks at the organizational problem of an illegal activity and at the trade-off between enhancing internal productivity and leaving members of the organization more exposed to detection. He takes a different approach, focusing on the optimal size of the criminal organization and taking its internal structure as given.

This paper is also related to the literature on social networks. In particular, Ballester et al. [4], under the assumptions that the network structure is exogenously given and observed, characterize the "key player" – the player who, once removed, leads to the optimal change in aggregate activity. In this paper, instead, we start by asking how a network can be (endogenously) built to make criminal activity as efficient as possible.

2 Model

Suppose that there are $N > 2$ members in a terrorist organization, with N an even number and a single law enforcement agency.[6] The law enforcement authority acts

[6] Allowing N to be an odd number presents no conceptual difficulties, but adds to the number of cases that need be considered in some of the results with regard to how to treat the last odd agent, with no real gain in insight.

first and sets a given detection strategy as specified in Sect. 2.1. The N terrorists have the possibility of forming an information structure by exchanging information among themselves as specified below in Sect. 2.2.[7] After forming an information structure, the N terrorists generate the benefits described in Sect. 2.2. Finally, detection takes place.

The assumption that the law enforcement authority chooses its policies before the terror structure is formed can be justified on the grounds that law enforcement policies and investigating budgets are broadly laid out and are hard to fine-tune once a certain policy is in place.

2.1 Law Enforcement Agency

At the end of the game, each terrorist can be detected by the enforcement agency. If the terrorist is detected, this imposes a direct cost to the organization of $k > 0$. This cost may include a punishment for the individual, such as time in prison, and a cost to the organization of recruiting a new member.

There are two ways for a terrorist to be detected, a *direct* way and an *indirect* way. In particular, an independent Bernoulli random draw determines whether a particular terrorist is detected *directly*. The direct detection of a particular terrorist at each period is *independent* of other terrorists' detection. Thus, a terrorist i can be detected directly by the authority according to some probability α_i. This probability depends on the extent of the enforcement agency's scrutiny of terrorist i.

Second, the law enforcement authority might also detect terrorists indirectly. Indeed, we assume that when the agency detects a terrorist who has information about other members of the organization, the agency also detects each of these with probability $\gamma \in (0,1]$. First, note that this implies that the enforcement agency's ability to detect terrorists indirectly depends on the structure of the terror group. Second, γ is a parameter that depends on the ability of the law enforcement agency to extract information from detected terrorists. For instance, the parameter γ is determined by the interrogation methods and the ability to strike deals with prisoners in exchange for information.

The law enforcement agency has a budget $B \in [0,N]$ to allocate for investigating the N members of the organization and devotes $\alpha_i \in [0,1]$ to investigate member i where $\sum_{i=1}^{N} \alpha_i \leq B$. Without loss of generality, we label terrorists so that $\alpha_1 \leq \alpha_2 \leq \cdots \leq \alpha_N$. We refer to α_i as the enforcement agency's level of scrutiny (or investigation) of terrorist i.

[7] We assume that the N agents constitute an organization through some other production structure that is independent of the information structure. Although we do not explicitly model the formation process, one could assume that the information structure is determined by a "benevolent" third party. Indeed, this is essentially the approach advocated by Mustafa Setmariam Nasar, an Al-Qaeda strategist who suggested that cell-builders be from outside the locale or immediately go on suicide missions after building cells.

2.2 Information Structure

We assume that each of the terrorists has a piece of *private* and *verifiable* information about himself and can decide to disclose this information to any of the other terrorists. Examples of such information could be the identity of the player, his whereabouts, some incriminating evidence, etc. We formalize the fact that i holds such information about j by an indicator variable μ_{ij}, such that $\mu_{ij} = 1$ if and only if i knows the information regarding j ($\mu_{ij} = 0$ otherwise). We use the notation $j \to i$ to represent $\mu_{ij} = 1$ (and, similarly, for instance $i, j, k \to l$ to represent $\mu_{li} = \mu_{lj} = \mu_{lk} = 1$).[8] The set $\mathscr{I}$ of all the possible organization (or "information") structures among N people is a subset of the set $\{0,1\}^{N^2}$ of values of the indicator variables, and we denote by μ its generic element.

We suppose that if $\mu_{ij} = 1$ for some $i \neq j$ this yields a benefit t for the organization. The idea here is that when someone in the organization holds information about j, it leads to more trust within the organization and, in particular, induces terrorist j to cooperate with the organization's goals, and yielding a benefit t to the organization. "Trust" and a benefit of a link in ensuring trust can be formalized in the context of cooperation in the infinitely-repeated Prisoners' dilemma – a standard approach in economics toolkit – and under the assumption that when i has information about j he can more easily punish him for not cooperating. A full discussion can be found in Baccara and Bar-Isaac [3].

A terrorist i is said to be indirectly linked to a terrorist j if there is a path of direct links that connect i to j. That is, if there is a set of terrorists $\{h_1, .., h_n\}$ such that $i \to h_1, h_1 \to h_2, ..., h_n \to j$.

The information structure affects the terrorists' probabilities of getting caught by the enforcement agency. Specifically, if i has information about another member of the organization j and i is detected (either directly or indirectly), player j is also detected *with probability* γ. As discussed in Sect. 2.1, if a terrorist i is detected and there is a single path of length d from terrorist i to terrorist j, j's probability of indirect detection as a result of i's capture is decaying in d and, more precisely, is γ^d. Note, however, that there could be multiple paths from i to j, and the total probability of detection of i depends on all of them.

Recall from Sect. 2.1 that each terrorist i is detected by the enforcement agency directly with probability α_i and indirectly as just discussed above. Overall, given an information structure μ, let $\beta_i(\mu)$ be the *total probability of detection* of terrorist i.

In Fig. 1, we give three examples of information structures. Panel A represents a decentralized structure in which terrorists are arranged in disconnected "cells" within which terrorists are linked to each other. Panel B represents a centralized structure (or "hierarchy"), in which terrorist 1 holds information on all other terrorists. Panel C represents a structure in which terrorists 1 and 2 form a cell, and the other two terrorists are connected to terrorist 1 in a hierarchical fashion.

If we assume that $\alpha_i = \alpha$ for $i = 1, \ldots, 4$, in the information structure represented by Panel A in Fig. 1, the probability of detection for each terrorist is $\beta_i(\mu) = \alpha +$

[8] Note that $\mu_{ii} = 1$ for all i.

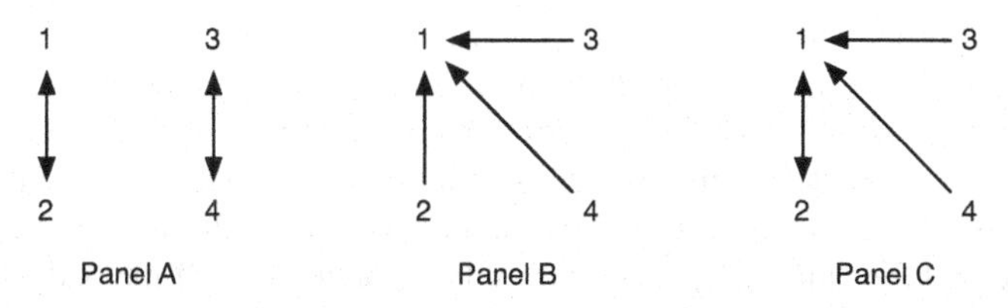

Fig. 1 Examples of information structurs

$\alpha(1-\alpha)\gamma$ for $i = 1,\ldots,4$. The first term is the probability of direct detection, while the second term is the probability of indirect detection (note that indirect detection occurs if the terrorist is not detected directly–with probability $(1-\alpha)$, the other member of the cell is caught directly (with probability α) and the authority is able to extract information (with probability γ).

In the information structure represented by Panel B in Fig. 1, the probability of detection for terrorist 1 is α, since terrorist 1 cannot be detected indirectly. However, $\beta_i(\mu) = \alpha + \alpha(1-\alpha)\gamma$ for $i = 2,3,4$. Again, the first term is the probability of direct detection, while the second term is the probability of indirect detection.

For the structure in Panel C, terrorists 1 and 2 form a cell, and their probability of detection is $\beta_i(\mu) = \alpha + \alpha(1-\alpha)\gamma$, while terrorists 3 and 4 can be detected if either terrorist 1 or 2 is detected. In particular, for $i = 3,4$ we have

$$\beta_i(\mu) = \alpha + (1-\alpha)\alpha\gamma + (1-\alpha)^2\alpha\gamma^2 \tag{1}$$

The first term of (1) is the probability of direct detection of terrorist i. The second term refers to the event in which terrorist 1 is detected directly and terrorist i is not. In this case, terrorist i is detected with probability γ. Finally, the third term represents the event in which terrorist 2 is detected directly, while terrorists 1 and i are not. In this case, terrorist i is detected with probability γ^2.(since terrorist 1 would first need to be detected as a result of terrorist 2's capture and then the authority would need to extract the information from terrorist 1).

2.3 Payoffs

The payoff for the terror organization is given by the sum of the benefits generated by the links (t for every linked member), minus the expected cost of direct and indirect detection. Bringing together benefits and costs, we can write down the net payoffs to the organization of an information structure μ as:

$$\Pi(\mu) = tn(\mu) - \sum_{i=1}^{N} k\beta_i(\mu), \tag{2}$$

where $n(\mu)$ denotes the number of members of the organization who are linked to some other member of the organization under the information structure μ. Note that the total probability of detection $\beta_i(\mu)$ depends on the probabilities of direct detections $\{\alpha_j\}$, the information structure μ, and the decay parameter γ.

We assume that the goal of the enforcement agency is to minimize the trust among the N terrorists. In other words, to minimize $n(\mu)$ and the benefits that the terror organization obtains from linking its members to each other. For simplicity, we assume that the authority gets no utility from saving part of the budget B. Also, the enforcement agency does not benefit from the the payments k incurred by the detected terrorists.[9]

Baccara and Bar-Isaac [3] provides some further justification and extensive discussion on the benefits of links and on potential extensions and developments of the model.

3 The Optimal Network

In this section, we study how a strategic terror organization chooses its structure, given that a specific enforcement policy is in place. In particular, we take the authority's scrutiny $\{\alpha_1,\ldots,\alpha_N\}$ as given, and we study the most efficient information structure that the N terrorists can form. These characterizations will allow us to tackle the problem of the government's optimal behavior.[10]

In characterizing the optimal information structures, it is useful to begin by focusing on optimal structures *given a fixed number of terrorists "linked" to other terrorists* – that is, a fixed number of terrorists that disclose their information to at least one other terrorist. Note that the benefits of having terrorists linked depend only on their number rather than on the structure of the organization. In particular, the potential benefit that the organization can yield from the links is constant with respect to all the information structures with the same number of terrorists linked to someone else. Thus, if the number of linked terrorists is fixed, an *optimal organization minimizes the cost of information leakage.*

We begin by characterizing the optimal information structure when the number of linked terrorists n is strictly less than N in the next Proposition.

Proposition 1. *The optimal structure to link $n < N$ terrorists is a hierarchy with the terrorist with the lowest probability of detection at the top, and the n terrorists with the highest probabilities of detection linked to him (i. e., $N, N-1, \ldots, N-n+1 \longrightarrow 1$).*

If the number of linked terrorists is less than N, the optimal structure is simply a hierarchy in which the top (the terrorist who receives the information from

[9] Indeed, these payments may be costly for the enforcement agency. For example, they may consist of detention in prison facilities.

[10] Note that even though we assume that the enforcement agency determines the level of scrutiny, these probabilities could also be exogenously given and due to some intrinsic characteristics of the agents. For example, some members of the organization may be more talented in evading detection (some may have a cleaner criminal record or simply might be able to run faster). If this is the case, the optimal organization characterizations we provide in this section can be seen as self-contained.

the others) is the member with the lowest probability of detection and the $n < N$ "inked" terrorists are those with the n highest probability of detection. The proof of Proposition 1 is very simple. Recall that, without loss of generality, we have $\alpha_1 \leq \alpha_2 \leq \cdots \leq \alpha_N$. Suppose, first, that $n = 1$, so we need to find the way to generate the "cheapest" possible link in terms of information leakage costs. The only event in which this link becomes costly is the case in which member i is independently detected and member j is not. This event has probability $\alpha_i(1-\alpha_j)$. Then, the cost of the link is minimized when α_i is as small as possible and α_j is as large as possible. It follows that the "cheapest" possible link is the one that requires member N to disclose his information to member 1 (the link $N \to 1$). If $n = 2$, the second cheapest link one can generate after $N \to 1$ is $N-1 \to 1$, and so on. Notice that Proposition 1 implies that the information leakage cost under an optimal structure in which there are $n < N$ links is simply $k\alpha_1\gamma\sum_{i=1}^{n}(1-\alpha_{N-i+1}) + k\sum_{i=1}^{N}\alpha_i$.

Next, consider the case where all N terrorists are linked to someone else. To start the characterization, consider a cell $\{i,j\}$. Let $\rho(i,j) \equiv \frac{\alpha_i+\alpha_j-2\alpha_i\alpha_j}{2-\alpha_i-\alpha_j}$. This is a useful ratio in understanding the organization members' proclivity to be linked as a binary cell rather than as subordinates to another terrorist. If two members $\{i,j\}$ are in a cell, terrorist i will get caught and the organization suffers the cost k with probability $\alpha_i + (1-\alpha_i)\alpha_j\gamma$. However, if each of them is independently linked to a third terrorist (the same for both, and who may be linked to others) with *overall probability of detection* η, terrorist i will get caught and the organization will suffer a cost k with probability $\alpha_i + \eta\gamma(1-\alpha_i)$, and terrorist j will get caught with probability $\alpha_j + \eta\gamma(1-\alpha_j)$. Then, having the terrorists $\{i,j\}$ forming an independent cell rather than linking each of them to the third terrorist minimizes the cost of information leakage if and only if

$$\alpha_i + (1-\alpha_i)\alpha_j\gamma + \alpha_j + (1-\alpha_j)\alpha_i\gamma < \alpha_i + \eta\gamma(1-\alpha_i) + \alpha_j + \eta\gamma(1-\alpha_j) \quad (3)$$

or, equivalently,

$$\rho(i,j) = \frac{\alpha_i + \alpha_j - 2\alpha_i\alpha_j}{2-\alpha_i-\alpha_j} < \eta. \quad (4)$$

Thus, for any couple of terrorists, the higher is $\rho(i,j)$, the greater is the advantage of forming a cell rather than being linked to a third terrorist. *Note that $\rho(i,j)$ is decreasing in both α_i and α_j* – that is, the higher the probability of detection of a terrorist, the lower $\rho(i,j)$ of any cell to which he might belong. Note, also, that $\rho(i,j)$ does not depend on γ. This is because decaying probabilities of detection affect the optimality of independence for a cell only if linking themselves to a third terrorist creates indirect links in the organization – that is, if it affects the probability η.

We now characterize the optimal information structure with N linked terrorists in the following proposition.

Proposition 2. *The optimal information structure with N linked terrorists is described as follows. Let $i^* \in \{2,..,N\}$ be the largest even integer such that $\rho(i-1,i) > \alpha_1 + (1-\alpha_1)\alpha_2\gamma$ (if no such integer exists, set $i^* = 1$): All the terrorists $i = 1,..,i^*$ are arranged in binary cells as $1 \leftrightarrow 2$, $3 \longleftrightarrow 4,..,i^*-1 \longleftrightarrow i^*$, and the terrorists $i = i^*+1,..,N$ all reveal their information to terrorist 1, that is, $i^*+1,..,N \rightarrow 1$.*

Proposition 2 states that the optimal way to link N members in a terror organization is to divide the members into two groups according to their probabilities of detection: a group comprising the i^* members with the lowest probabilities of detection, and another group with the $N-i^*$ members with the highest probability of detection. The members belonging to the first group are arranged in binary cells formed by members with adjacent probability of detection (i. e. $1 \leftrightarrow 2$, $3 \longleftrightarrow 4,..,i^*-1 \longleftrightarrow i^*$). All the members belonging to the second group reveal their information to terrorist 2 ($i^*+1,..,N \rightarrow 1 \leftrightarrow 2$).

The number of terrorists i^* belonging to the independent cell component depends on how steeply the ratio $\rho(i,i+1)$ of each couple grows. If α_1 and α_2 are very low relative to the other terrorists' probabilities of detection, it could be the case that $\rho(i-1,i) > \alpha_1 + (1-\alpha_1)\alpha_2\gamma$ for all $i = 4,..,N$. In this case, Proposition 2 requires that an optimizing organization links all the members $3,..,N$ to terrorist 1 (who remains linked in a cell with 2).[11] However, if α_3 and α_4 are close enough to α_2, then $\rho(3,4) > \alpha_2 + (1-\alpha_2)\alpha_1\gamma$, and Proposition 2 prescribes 3 and 4 to form a cell rather than being linked to 1, and so on.

The optimal information structure described in Proposition 2 is illustrated in Fig. 2, when there are $N = 8$ terrorists and $i^* = 6$.

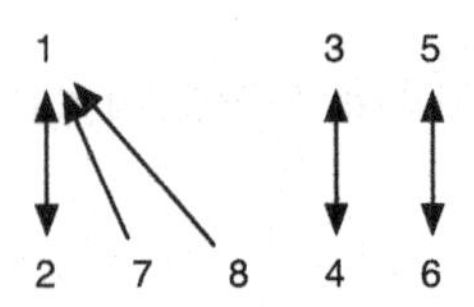

Fig. 2 Optimal information structure described in Proposition 2

Note that with a full characterization of the form of optimal structures, we can easily calculate the information leakage of linking n terrorists and inducing them to cooperate. Comparing costs and benefits of links then allows us to characterize the optimal information structure.

Lemma 1 and Proposition 2 allow us to define the net benefit to the organization of having n links. In particular, if the organization has N links, this yields a direct "trust" benefit of Nt, a direct cost of $k\sum_{i=1}^{N}\alpha_i$ of detection, and a cost of indirect

[11] In particular, if α_1 and α_2 approach zero, all these links have an arbitrarily small information leakage cost, so the organization's information leakage cost is the same as in the structure with no links.

detection: For a terrorist i (from i^*+1 to N) who is subordinate to the $1 \longleftrightarrow 2$ informational hub, the cost of indirect detection is $k(\alpha_1\gamma + \alpha_2(1-\alpha_1)\gamma^2)(1-\alpha_i)$, whereas for a terrorist $2i$ in a cell with terrorist $2i-1$, the cost of indirect detection is $k\gamma(1-\alpha_{2i})\alpha_{2i-1}$. Summing all benefits and costs over all terrorists, we can obtain that the value to an organization of having all N terrorists linked (in the most efficient way) is:

$$V(N) = Nt - k\sum_{i=1}^{N}\alpha_i - k(\alpha_1\gamma + \alpha_2(1-\alpha_1)\gamma^2)\sum_{i=i^*+1}^{N}(1-\alpha_i)$$
$$-k\gamma\sum_{i=1}^{\frac{i^*}{2}}[(1-\alpha_{2i-1})\alpha_{2i} + (1-\alpha_{2i})\alpha_{2i-1}] \quad . \tag{5}$$

Similarly we can write $V(n) = nt - k\sum_{i=1}^{N}\alpha_i - k\gamma\alpha_1\sum_{j=N-n+1}^{N}(1-\alpha_j)$ for $n \in \{1,\ldots,N-1\}$ and $V(0) = -k\sum_{i=1}^{N}\alpha_i$.

With expressions for the value of having n links, we can, therefore, find the number of links n^* that maximizes the value of the organization, $V(n)$. The following proposition follows trivially from the earlier results and characterizes the optimal information structure.

Proposition 3. *If $\gamma = 1$, the optimal information structure is as follows: (i) If $n^* = 0$, the optimal information structure is an anarchy. (ii) If $0 < n^* < N$, the optimal structure is an individual-dominated hierarchy where the hub is terrorist 1 and the subordinates are terrorists $N,\ldots N-n+1$. (iii) Finally, if $n^* = N$, the optimal structure is the mixed structure described in Proposition 2.*

Note that when $\alpha_i = \alpha$ for all i, the additional cost of each link is constant and equal to $k\gamma\alpha(1-\alpha)$. It follows that, in this symmetric case, the optimal structure either consists of no links or all terrorists are in binary cells.

Corollary 1. *Let $\alpha_i = \alpha$ for all i. If $t > k\gamma\alpha(1-\alpha)$, then the optimal information structure is a binary cell structure. Otherwise, the optimal information structure has no links.*

This concludes the characterization of the optimal information structure for a given scrutiny distribution $\{\alpha_1,..,\alpha_N\}$. Next, we endogenize scrutiny and discuss the strategy of the enforcement agency.

4 The Enforcement Agency

The enforcement agency aims to minimize production within the criminal organization. To do so, the agency has several instruments to affect how criminal organizations operate. First, as discussed in Sect. 2.1, the enforcement agency has a budget B to allocate to detect terrorists.

Second, two further parameters in our model affect the enforcement agency's ability to minimize cooperation within the organization. These parameters are γ, which measures how effective the interrogation techniques are in extracting information from detected terrorists, and k, which measures the severity of the punishment for detected terrorists. Note that both these parameters are typically constrained by legal limits. Thus, the parameters (γ, k) can be viewed as describing the legal environment in which authorities act and the effectiveness of their interrogation methods.

In this section, first we study how these three instruments should be used and how they are interrelated. In particular, we first show how, for any fixed legal environment (γ, k), an authority should optimally allocate its investigation budget to minimize cooperation within the organization – that is, to minimize the number of linked terrorists. Next, we look at how a criminal organization is likely to strategically respond to changes of γ and k – that is, changes in the legal environment – and, in turn, how the optimal investigation budget allocation should be optimally altered in response to such legal changes.

4.1 Investigation Budget Allocation

As discussed in Sect. 2.3, we assume that the enforcement agency's objective is to minimize the number of terrorists who cooperate – that is, the organization's production level n^*t, where n^* maximizes the value of the organization, $V(n)$.

Recall that, for a given legal environment (γ, k), the problem of the law enforcement agency is to allocate the investigation budget $B \in [0, N]$ to determine the scrutiny $\alpha_i \in [0, 1]$ of each terrorist i such that $\sum_{i=1}^{N} \alpha_i \leq B$. The enforcement agency acts first and chooses these scrutiny levels before the organization forms.

In the next result, we characterize the (weakly) optimal strategy for the enforcement agency in allocating its resources. Note that, if the authority allocates the same budget α to each terrorist, the cost of each link becomes $k\gamma\alpha(1-\alpha)$. Since this cost is maximized at $\alpha = \frac{1}{2}$, it is never optimal to set $\alpha > \frac{1}{2}$ in a symmetric allocation. Let, then, $\widehat{\alpha} = \min\{\frac{B}{N}, \frac{1}{2}\}$ be the optimal symmetric allocation.

Proposition 4. *A weakly optimal strategy for the enforcement agency is to set scrutiny symmetrically if* $k\gamma\widehat{\alpha}(1-\widehat{\alpha}) > t$ *and to not investigate one terrorist and investigate all others symmetrically (set* $\alpha_1 = 0$ *and* $\alpha_2 = \cdots = \alpha_N = \min\{\frac{B}{N-1}, 1\}$*) otherwise.*

A symmetric allocation of B can prevent the formation of any link if the cost of each link $k\gamma\widehat{\alpha}(1-\widehat{\alpha})$ is greater than the benefit of inducing an individual to cooperate. This is the case when $k\gamma\widehat{\alpha}(1-\widehat{\alpha}) > t$, and, in these circumstances, a symmetric allocation is optimal as it deters any cooperation.

However, if $k\gamma\widehat{\alpha}(1-\widehat{\alpha}) < t$, by Lemma 1, a symmetric allocation would imply the formation of a binary cell structure that reaches full cooperation within the organization. The question is whether, in these situations, the enforcement agency can do something else to prevent full cooperation. Proposition 4 addresses this question and suggests that, in this case, an allocation in which one terrorist remains undetected and the budget is equally divided among the other $N-1$ terrorists is optimal. Under this allocation, sometimes the organization still reaches full efficiency (in this case, we can conclude that the enforcement agency cannot prevent full efficiency for the terror network), but in some cases, a hierarchy with $N-1$ links arises. Since the hierarchy is strictly less efficient than a binary cell structure, this allocation strictly dominates the symmetric one.

If $k\gamma\widehat{\alpha}(1-\widehat{\alpha}) > t$, we show that there is no other allocation that strictly dominates $\alpha_1 = 0$ and $\alpha_2 = \cdots = \alpha_N = \min\{\frac{B}{N-1}, 1\}$.[12] The intuition for this part of Proposition 4 is the following. First of all, note that if two terrorists remain undetected ($\alpha_1 = \alpha_2 = 0$), the organization can form N links without incurring any additional information leakage costs with respect to the cost they would incur with no links (this is because the two terrorists can reveal information to each other at no cost and costlessly act as a hub for the $N-2$ terrorists). So, to deter full efficiency, the enforcement agency can leave, at most, one terrorist undetected. Suppose now that some cooperation is deterred by an allocation in which all terrorists are detected with some probability ($\alpha_1 > 0$). Then, the terrorist with the lowest allocation will act as a hub in a hierarchy, as described in Proposition 1. In the Appendix, we prove that under our assumption, there are exactly $N-1$ links in such a hierarchy. Then, moving all the resources from the hub to the other terrorists, as suggested in Proposition 4, is equivalent to the original allocation.

Let us turn to comment on the cooperation or trust outcomes in equilibrium. Proposition 4 states that in some circumstances (i. e., if $k\gamma\widehat{\alpha}(1-\widehat{\alpha}) \geq t$), the enforcement agency can prevent any cooperation in the organization by allocating its budget symmetrically. Note that these circumstances are more likely to occur when k is higher, or the benefit from trust t is lower. However, a higher budget B is beneficial as long as it is below the threshold $\frac{N}{2}$. Increases in B beyond that threshold would have no further effect on the cooperation level in the organization. On the other hand, when $k\gamma\widehat{\alpha}(1-\widehat{\alpha}) < t$, the optimal detection strategy for the authority deters, *at most*, one link and the trust in or cooperation of one terrorist in the organization.

4.2 Legal Environment and Interrogation Methods

Two parameters in our model directly describe the legal environment within which the authorities act. In particular, the parameter γ measures the probability with which an agent in the hands of the authority reveals the information he knows, and the

[12] Note that considering $\min\{\frac{B}{N-1}, 1\}$ guarantees that the resulting allocation on each agent is in the interval $[0,1]$.

parameter k measures the severity of the punishment inflicted on the detected agents. Thus, a broadening of the admissible measures (such as, for example, the Patriot Act) can be captured in this model as an increase of γ or k, or both.

In this section, we look at how a criminal organization is likely to strategically respond to changes of γ and k and, in turn, how effective such changes are and which investigation budget allocation should be optimally associated with them.

First, let us look at how, given a budget detection allocation $\{\alpha_1, \ldots, \alpha_N\}$, a change in the legal environment affects the optimal information structure. Examples of change in legal environments include banning the use of torture, and allowing reduced sentences and rewards for whistle-blowers.

From the results in Sect. 3, the effect of changes in γ on the optimal information structure is described in the following Proposition.

Proposition 5. *1. If $\gamma < \underline{\gamma}$, the optimal organization structure is a mixed strategy as described in Proposition 2, where the number of binary cells in the structure is increasing in γ.*

2. If $\underline{\gamma} < \gamma \leq \overline{\gamma}$, a simple hierarchy as described in Proposition 1 is optimal. The number of linked agent is decreasing in γ.

3. If $\overline{\gamma} > \gamma$, the optimal information structure has no links.

Proposition 5 suggests that for low levels of γ, an increase in γ does not have the effect of decreasing cooperation within the organization, but has the effect of increasing decentralization. However, sufficiently high increases in γ reduce the number of linked terrorists and make a simple hierarchy optimal.

To understand the argument behind point (1) of Proposition 5, recall from Proposition 2 that the optimal number of terrorists organized in cells is found by defining $i^* \in \{2,..,N\}$ as the largest even integer such that $\rho(i-1,i) > \alpha_1 + (1-\alpha_1)\alpha_2\gamma$. Since $\rho(i-1,i)$ is independent of γ, an increase in γ increases i^* and, thus, the number of resulting cells.

Proposition 5 has significant policy implications. Suppose that an asymmetric detection strategy is in place, and the criminal organization responds with a partially centralized structure, as described in Proposition 2. In such a structure, there are two agents (i. e., agents 1 and 2) who, directly or indirectly, hold a significant amount of information about other members of the organization. Thus, authorities may expect that toughening the interrogation methods (as, for example, by allowing the use of torture) will lead to more information being revealed following capture of such a critical member of the terror organization. However, our approach views the terror network as a strategic agent that will optimally respond to such environmental change. In particular, Proposition 5 suggests that there are circumstances under which the toughening of interrogation methods will fail to achieve less cooperation within the organization, and it will simply lead to *more decentralization* in the way information is shared among the members.

Next, we consider how an increase in the severity of the punishment (measured in our model by k) affects the optimal structure of the organization.

Proposition 6. *1. If $k < \underline{k}$, a mixed structure as described in Proposition 2 is optimal. The number of binary cells in the structure is constant in k.*

2. *If $\underline{k} \leq k < \overline{k}$, a simple hierarchy as described in Proposition 1 is optimal. The number of linked agents is decreasing in k.*
3. *If $k > \overline{k}$, the optimal information structure has no links.*

Propositions 5 and 6 suggest that, while a change in the interrogation methods (i. e., a change in γ) can lead to a more decentralized organization, a change in the severity of the punishment (i. e., a change in k) never has this effect.

Next, we look at how the enforcement agency should optimally adjust the optimal detection strategy $\{\alpha_1,..,\alpha_N\}$ after a change in the legal environment (γ,k).

Proposition 7. *Following an increase of γ or k, a symmetric investigation budget allocation is more likely to be optimal.*

The intuition behind Proposition 7 is simple. Given an investigation budget B, by broadening the legal limits and increasing γ or k (or both), the cost to the terror organization of a link in the binary cell structure increases. This makes it more likely that any cooperation within the organization can be prevented by a symmetric allocation of the budget B.

Although Proposition 7 is a direct consequence of Proposition 4, it offers interesting insights from a policy perspective. Indeed, this result suggests that the three instruments available to the authorities (investigation budget allocation, interrogation methods and severity of punishments) are not independent but are *strategically interrelated*. Thus, any change in the legal environment should be associated with a revised investigation budget allocation. In particular, if the legal environment broadens the legal limits within which the authorities can act (i. e., an increase in γ and k), a symmetric budget allocation is more likely to be optimal than before. However, if the legal limits for the authorities become narrower, it is more likely that a symmetric allocation will lead to a fully linked binary cell structure. Thus, the authorities could do better by implementing an asymmetric detection strategy as described in Proposition 4.

5 Extensions and Conclusions

In the framework outlined above, we assumed that the law enforcement authority acts first, and the terrorist organization chooses a structure in response. Another possibility is that the authority might choose its strategy first, the organization then forms, and the authority's strategy is then realized. If the authority is restriced to choose a pure strategy, this timing assumption leads to results equivalent to this paper's. However, if the authority can choose mixed strategies then this timing assumption is substantive and can lead to more effective strategies for the enforcement agency. For example, suppose that there are only two agents in the organization. Randomizing between choosing scrutinies of $(0,1)$ and $(1,0)$ is more likely to the deter creation of a binary cell than a symmetric allocation of $(\frac{1}{2},\frac{1}{2})$. However, in the case where there are many agents, then a symmetric and deterministic allocation might be more likely to deter the creation of N or $N-1$ links than mixing asymmetric ones.

Numerous extensions to framework above might be considered. Baccara and Bar-Isaac [3] provides a more careful consideration of the benefits of informational links and it considers a model where the a terrorist's behavior can depend on his position in the structure, and the probability of getting caught depends on behavior. Other potentially interesting extensions of this framework include exploring the process of the formation of the structure, and a more detailed and careful consideration of the costs to the organization of having members detected. In all these cases, the fundamental points will still apply: The structure of a terrorist network responds (in a way that can be characterized) to law enforcement policies, and these policies should be tailored to the legal environment.

Appendix

Proof of Proposition 2 *First step.* Recall that $\rho(j,i)$ is decreasing in both α_j and α_i. This follows easily by taking the derivative of $\rho(i.j)$ with respect to α_i and α_j and noting that $0 \leq \alpha_i, \alpha_j \leq 1$.

Second step Let us prove that among all possible binary cell information structures that pair N terrorists to each other $\{\mu \in I \text{ s.t. if } \mu_{ij} = 1 \text{ for some } i \neq j \text{ then } \mu_{ji} = 1 \text{ and } \mu_{ik} = 0\ \forall k \neq j\}$, the one which minimizes information leakage costs is $1 \longleftrightarrow 2$, $3 \longleftrightarrow 4, \ldots, N-1 \longleftrightarrow N$. To see this, let us first show that this result holds for $N = 4$. The claim is true if $1 \longleftrightarrow 2$, $3 \longleftrightarrow 4$ is better than either of the alternatives $1 \longleftrightarrow 4$, $2 \longleftrightarrow 3$ and $1 \longleftrightarrow 3$, $2 \longleftrightarrow 4$. This requires that:

$$\begin{aligned}&\alpha_1 + (1-\alpha_1)\alpha_2\gamma + \alpha_2 + (1-\alpha_2)\alpha_1\gamma + \alpha_3 + (1-\alpha_3)\alpha_4\gamma + \alpha_4 + (1-\alpha_4)\alpha_3\gamma \leq \\ &\alpha_1 + (1-\alpha_1)\alpha_4\gamma + \alpha_4 + (1-\alpha_4)\alpha_1\gamma + \alpha_2 + (1-\alpha_2)\alpha_3\gamma + \alpha_3 + (1-\alpha_3)\alpha_2\gamma\end{aligned} \tag{6}$$

and,

$$\begin{aligned}&\alpha_1 + (1-\alpha_1)\alpha_2\gamma + \alpha_2 + (1-\alpha_2)\alpha_1\gamma + \alpha_3 + (1-\alpha_3)\alpha_4\gamma + \alpha_4 + (1-\alpha_4)\alpha_3\gamma \leq \\ &\alpha_1 + (1-\alpha_1)\alpha_3\gamma + \alpha_4 + (1-\alpha_3)\alpha_1\gamma + \alpha_2 + (1-\alpha_2)\alpha_4\gamma + \alpha_4 + (1-\alpha_4)\alpha_2\gamma\end{aligned}. \tag{7}$$

Inequality (6) holds if $\alpha_1\alpha_2 + \alpha_3\alpha_4 \geq \alpha_1\alpha_4 + \alpha_2\alpha_3$ or if $(\alpha_4 - \alpha_2)(\alpha_3 - \alpha_1) \geq 0$, which is always the case. Inequality (7) also always holds.

Now, suppose that for a general even N the claim is not true. Then, there is an optimal structure in which it is possible to find 2 pairs $\{i_1, i_2\}$, $\{i_3, i_4\}$ such that $\alpha_{i_1} \leq \alpha_{i_2} \leq \alpha_{i_3} \leq \alpha_{i_4}$ is violated. Then, since that is the optimal structure, rearranging the terrorists in these pairs leaving all other pairs unchanged cannot reduce information leakage costs. However, this contradicts the result for $N = 4$.

Third step. It is clear that the best way to link terrorists 1 and 2 is to link them to each other since they are the two lowest-probability terrorists. Now, for any couple $\{N-1, N\}, \ldots, \{3, 4\}$ let us compare whether it is better from an information leakage point of view to link the pair to each other and independently from the others, or to have them linked to terrorist 1 (and 2) instead. If the terrorists N and $N-1$ are linked to each other, the cost of information leakage corresponding to the couple is $k[\alpha_{N-1}+(1-\alpha_{N-1})\alpha_N\gamma+\alpha_N+(1-\alpha_N)\alpha_{N-1}\gamma]$. If they are linked to terrorists 1 and 2, the cost of information leakage is

$k[\alpha_{N-1}+(1-\alpha_{N-1})(\alpha_1+\alpha_2(1-\alpha_1)\gamma)\gamma+\alpha_N+(1-\alpha_N)(\alpha_1+\alpha_2(1-\alpha_1)\gamma)\gamma]$.

Then, the couple $\{N-1,N\}$ should be linked to terrorist 1 (and then, since we have $1\leftrightarrow 2$, to the couple $\{1,2\}$) if and only if

$$\rho(N-1,N)<(\alpha_1+\alpha_2(1-\alpha_1)\gamma)\gamma \tag{8}$$

If condition (8) fails, by the first step of this proof we know that the condition will fail for any subsequent couple. Then, the optimal way to link the N terrorists to each other is to create a pairwise structure and, by the second step of this proof, we know that the optimal way to do this is to set $1\leftrightarrow 2$, $3\longleftrightarrow 4,..$ and $N\longleftrightarrow N-1$. If condition (8) is satisfied, we can link terrorists N and $N-1$ to the couple $\{1,2\}$, and we can repeat this check for the couple $\{N-2,N-3\}$. We repeat this process until we find a couple $\{i-1,i\}$ for which the condition

$$\rho(i-1,i)<(\alpha_1+\alpha_2(1-\alpha_1)\gamma)\gamma \tag{9}$$

fails. If we find such a couple, by the first step of this proof we know that the condition will fail for any subsequent couple, and, by the second step of the proof, we can arrange any subsequent couple in a pairwise fashion. □

Proof of Proposition 4 In order to prove this result, we prove the following Lemma first. Let $\widetilde{\alpha}\equiv\left\{0,\frac{B}{N-1},..,\frac{B}{N-1}\right\}$.

Lemma The allocation $\widetilde{\alpha}$ minimizes the net benefit of the Nth link (linking terrorist 1 to terrorist 2) compared to any other allocation α which generates exactly N-1 links.

Proof: Consider any allocation α that generates exactly $N-1$ links. Since $\alpha_1\leq\alpha_2\leq\cdots\leq\alpha_N$ and $\alpha_1\geq 0$, it follows that $\alpha_2\leq\frac{B}{N-1}$. We can compare the additional information leakage costs from the $N-th$ link, $V(N)=V(N)-V(N-1)$ and $\widetilde{V}(N)=\widetilde{V}(N-1)-\widetilde{V}(N-1)$ associated with each terrorist i under allocations α and $\widetilde{\alpha}$. In order to do that, let us consider the allocation $\widehat{\alpha}\equiv\{0,\alpha_2,..,\alpha_N\}$ and first compare α with $\widehat{\alpha}$. Under the optimal information structures with N links described in Proposition 2, given allocation α, either (a) terrorist i remains linked to terrorist 1 or (b) terrorist i is in a binary cell with some other terrorist j in the organization (which will be $i+1$ or $i-1$ depending on whether i is even or odd). In case (a), the incremental leakage cost for terrorist i is $k\gamma^2(1-\alpha_i)(1-\alpha_1)\alpha_2$, while under allocation $\widehat{\alpha}$ is going to be $k\gamma^2(1-\alpha_i)\alpha_2$. Trivially, $k\gamma^2(1-\alpha_i)(1-\alpha_1)\alpha_2<k\gamma^2(1-\alpha_i)\alpha_2$. In case (b), since the incremental information leakage cost for terrorists i and $i+1$ of the $N-th$ link under allocation α is $k\gamma(1-\alpha_i)\alpha_{i+1}+k\gamma(1-\alpha_{i+1})\alpha_i-k\gamma(1-\alpha_1)\alpha_i-k\gamma(1-\alpha_1)\alpha_{i+1}$ (where the first positive terms denote the new information leakage costs associated with these terrorists and the negative terms the old information leakage costs when they were subordinates in the $N-1$ hierarchy). Since the

cell is preferred to making i and $i+1$ subordinates to terrorists 1 and 2, it follows that

$$\begin{aligned}
& k\gamma(1-\alpha_i)\alpha_{i+1}+k\gamma(1-\alpha_{i+1})\alpha_i-k\gamma\alpha_1(1-\alpha_i)-k\gamma\alpha_1(1-\alpha_{i+1}) \\
< & \; k\gamma(\alpha_1+\alpha_2(1-\alpha_1)\gamma)(1-\alpha_{i+1})+k\gamma(\alpha_1+\alpha_2(1-\alpha_1)\gamma)(1-\alpha_i) \\
& -k\gamma\alpha_1(1-\alpha_i)-k\gamma\alpha_1(1-\alpha_{i+1}) \\
= & \; k\gamma^2(1-\alpha_i)(1-\alpha_1)\alpha_2+k\gamma^2(1-\alpha_{i+1})(1-\alpha_1)\alpha_2 \\
< & \; k\gamma^2(1-\alpha_i)\alpha_2+k\gamma^2(1-\alpha_{i+1})\alpha_2
\end{aligned}$$

The last expression is the information leakage cost associated with the allocation $\widehat{\alpha}$ (that is, the information leakage costs beyond those incurred in anarchy).

Next, we show that the allocation $\widetilde{\alpha}$ has a lower net benefit for the $N-th$ link $\widetilde{V}(N)-\widetilde{V}(N-1)$ than the allocation $\widehat{\alpha}$, that is, $\widetilde{V}(N)-\widetilde{V}(N-1)\geq\widehat{V}(N)-\widehat{V}(N-1)$. These two incremental values can be written down trivially:

$$\widetilde{V}(N)-\widetilde{V}(N-1)=t-k\gamma^2\sum_{i=3}^{N}\frac{B}{N-1}(1-\frac{B}{N-1})=t-k\gamma(N-2)\frac{B}{N-1}(1-\frac{B}{N-1})$$

and

$$\widehat{c}(N)=t-k\gamma^2\sum_{i=3}^{N}\alpha_2(1-\alpha_i)=t-k\gamma^2(N-2)\alpha_2+k\gamma^2\alpha_2\sum_{i=3}^{N}\alpha_i$$

Since $\sum_{i=3}^{N}\alpha_i<B<N-2$, it follows that information leakage costs under $\widehat{\alpha}$ are increasing in α_2, whose highest value is $\frac{B}{N-1}$ and when it takes this value the information leakage costs are equal to those under $\widetilde{\alpha}$. Thus, $\widetilde{c}(N)\geq\widehat{c}(N)\geq c(N)$. This concludes the proof of Lemma 5□

Let us now proceed to the proof of Proposition **4**.

Suppose now that $t<k\gamma\frac{B}{N}(1-\frac{B}{N})$. By Corollary 1, in this case, the symmetric allocation deters the organization from establishing any link, so this will be the optimal strategy for the enforcement agency. *In the rest of the proof we will then assume that* $t>k\gamma\frac{B}{N}(1-\frac{B}{N})$. In points (1)-(3), we go over all the possible budget allocations and show that the allocation $\widetilde{\alpha}=\{0,\frac{B}{N-1},..,\frac{B}{N-1}\}$ is optimal.

(1) Consider any allocation such that $\alpha_1=\alpha_2=0$. Then, the organization can reach full efficiency with zero additional information leakage cost with respect to anarchy. To see this, suppose that $\alpha_1=\alpha_2=0$; then, an organization with the links $\mu_{1i}=1$ for all $i\in\{2,..,N\}$, $\mu_{21}=1$ and $\mu_{ij}=0$ otherwise engenders full trust in the organization. Thus, it must be the case that, in order to prevent links between terrorist and deter efficiency, *at most one terrorist can be left with zero probability of detection.*

(2) Consider any allocation such that $\alpha_1>0$–that is, *all* the probabilities of detections are set to be positive. Since we are under the assumption that $t>k\gamma\frac{B}{N}(1-\frac{B}{N})$, if these probabilities are symmetric, full cooperation will ensue, and the allocation

$\widetilde{\alpha}$ cannot do worse than that. Suppose, then, that the allocation is asymmetric – that is, $\alpha_1 < \frac{B}{N}$. Following the characterization in Proposition 3, the terrorists will then form an optimal organization.

First, suppose the parameters are such that the organization has N links. Then, the allocation we are considering reaches full efficiency, and the allocation $\widetilde{\alpha}$ cannot do worse than that.

Suppose, instead, that the optimal organization given the allocation α we are considering generates $N-1$ links. Then, by the Lemma 5, allocation $\widetilde{\alpha}$ performs at least as well.

Finally, suppose that under the allocation α there are $n < N-1$ linked terrorists. We argue that such a structure is impossible. In such organizations, according to Proposition 1, there are three types of terrorists to consider: the top of the hierarchy terrorist 1, the $N-n-1$ independent terrorists $2,..N-n$, and the n terrorists who reveal their information to terrorist 1–that is, $N-n+1,..N$. Without loss of generality, we will restrict our attention to the allocations that give the same probability of detection to each terrorist in the same category (if the probability is not the same, it is easy to see that it is possible to substitute such probabilities with the average in each category and still obtain the same structure of organization). Let's name such probabilities α_1, α_2 and α_N respectively. The probability allocations we are restricting our attention to have to satisfy the following constraints:

(i) $0 < \alpha_1 \leq \alpha_2 \leq \alpha_N \leq 1$ (by feasibility and by Proposition 1);

(ii) $k\gamma\alpha_1(1-\alpha_2) \geq t$ (it is not optimal for the organization to link the $N-n-1$ independent to terrorist 1);

(iii) $t \geq k\gamma\alpha_1(1-\alpha_N)$ (it is optimal for the organization to link the n terrorists to terrorist 1);

(iv) $\alpha_1 + (N-n-1)\alpha_2 + n\alpha_N \leq B$ (the resource constraint).

Note that $k\gamma\alpha_1(1-\alpha_2) \leq k\gamma\alpha_2(1-\alpha_2) \leq k\gamma\frac{B}{N}(1-\frac{B}{N})$ since $\alpha_2 \leq \frac{B}{N} < \frac{1}{2}$ (otherwise either the (iv) or is violated or it cannot be that $\alpha_1 \leq \alpha_2 \leq \alpha_N$), but then (ii) cannot hold since $t > k\gamma\frac{B}{N}(1-\frac{B}{N})$. If follows that such a structure is impossible.

(3) In points (1)-(2) we showed that if $t > k\gamma\frac{B}{N}(1-\frac{B}{N})$, all the allocations such that $\alpha_1 = \alpha_2 = 0$ or $\alpha_1 > 0$ are (weakly) dominated by allocation $\widetilde{\alpha}$. Finally, let us consider an allocation such that $\alpha_1 = 0$ and $\alpha_2 > 0$. Under this allocation, it is clear that an organization with $N-1$ linked terrorists can arise costlessly. Thus, the best the enforcement agency can do is to try to prevent the $N-th$ link from arising. Observe that, if $\alpha_1 = 0$, the characterization in Proposition 2 yields, for each $i \in \{4,..,N\}$, $\rho(i-1.i) \leq 1-\alpha_2$ (easy to check since $\alpha_2 \leq \alpha_j$ for all $j \in \{3,..,N\}$). Then, in the optimal organization, all the terrorists are linked to terrorist 1, without binary cells (besides the cell $\{1,2\}$). Then, the cost of the $N-th$ link for the organization is $k\gamma\alpha_2\sum_{i=3}^{N}(1-\alpha_i)$, and it is maximized (under the constraints $\alpha_2 \leq \alpha_i$ for all i and $\sum_{i=2}^{N}\alpha_1 = B$) by $\alpha_i = \frac{B}{N-1}$ for all $i \in \{2,..,N\}$, which is allocation $\widetilde{\alpha}$. □

References

1. Anselmi, T.: La testimonianza di Tina Anselmi sulla Resistenza raccolta da Alessandra Chiappano. (2003) http://www.novecento.org.
2. Arquilla J., Ronfeldt, D. F.: Networks and Netwars: The Future of Terror, Crime, and Militancy. (2002) *Rand Corporation.*
3. Baccara, M., Bar-Isaac, H.: How to Organize Crime. The Review of Economic Studies **75** (2008) 1039-1067
4. Ballester, C., Calvó-Armengolz A., Zenou, Y.: Who's Who in Networks. Wanted: The Key Player. Econometrica **74** (2006) 1403-1418
5. Charlie M.:Into the Abyss: A Personal Journey Into the World of Street Gangs. Manuscript (2002)
6. Farley, J. D.: Breaking Al Qaeda Cells: A Mathematical Analysis of Counterterrorism Operations (A Guide for Risk Assessment and Decision Making). Studies in Conflict and Terrorism **26** (2006) 399-411
7. Farley, J. D.: Building the perfect terrorist cell. conference presentation (2006)
8. Garoupa, N.: Optimal law enforcement and criminal organization. Journal of Economic Behavior & Organization **63** (2007) 461-474
9. Oliva A.: Esercito and Democrazia. Vangelista Editore (1976)
10. Pearl M.: A Mighty Heart: The Inside Story of the Al Qaeda Kidnapping of Danny Pearl. Scribner Publishing (2003)
11. Thompson, T.: Gangs: A Journey into the Heart of the British Underworld. Hodder & Stoughton (2005)

Terrorists and Sponsors. An Inquiry into Trust and Double-Crossing

Gordon H. McCormick and Guillermo Owen

Abstract We consider the conditions that lead to the dissolution of state-terrorist coalitions. While such coalitions have well known advantages, they also have structural weaknesses that are largely ignored in the literature on the state sponsorship of terrorism. Each player in the coalition has interests that are only partially shared and, in some cases, at odds with those of its partner. Long term cooperation must be based on mutual advantage and mutual trust, both of which are subject to change over time. We examine the conditions that are needed to begin and maintain a cooperative strategy and the circumstances that lead a state and a terrorist group to leave the coalition and double-cross its partner. Equilibrium strategies for both players are defined and interpreted.

1 State-Terrorist Coalitions

Every significant terrorist organization since the late 1960s has depended, to one degree or another, on the active or passive support of one or more state sponsors. The benefits of these coalitions are well understood. For terrorist groups, state sponsorship provides a much needed source of financial, material and diplomatic aid, and often most importantly, a safe base of operations from which they can plan and launch their attacks and then withdraw. For the sponsor, this relationship offers the opportunity to pursue an aggressive and otherwise risky foreign policy through the agency of a surrogate actor, shielded by the fact that the nature and sometimes even the existence of its ties to the group is unknown. It is difficult to hold a state account-

Gordon H. McCormick
Department of Defense Analysis, Naval Postgraduate School, Monterey, CA 93943, USA, e-mail: gmccormick@nps.edu

Guillermo Owen
Department of Applied Mathematics, Naval Postgraduate School, Monterey, CA 93943, USA, e-mail: gowen@nps.edu

N. Memon et al. (eds.), *Mathematical Methods in Counterterrorism*,
DOI 10.1007/978-3-211-09442-6_17, © Springer-Verlag/Wien 2009

able for the actions of a sub-state actor over which it has little or no control, even if that group is based within its borders. The political cost of retaliating, in such cases, is much greater than it would be if the same sponsor carried out the same actions on its own. To the degree that an effective response, furthermore, depends on the cooperation of other governments, as it does in the case of most political and economic sanctions and even many military options, the probability of success is reduced. The cost of retaliation, in these circumstances, is generally high and the likelihood of success is low, making a response less likely.[1]

As important as state sponsorship has been to the rise of international terrorism over the past forty years, however, there is often less to these coalitions than meets the eye. Concern over the apparent ideological, religious, or communal ties that appear to bind sponsors and terrorist groups together in a united front often disguises the fact that these relationships are generally weak, temporary bargains struck between independent actors. The average life-span of a state-terrorist coalition is less than eight years, with a median life cycle of six years. Those coalitions that have managed to endure for notably longer periods of time, such as the alliances forged over the years between Iran and Hizbollah (26 years), Syria and the Kurdistan Worker's Party (19 years), and Syria and Hamas (15 years), divert attention from the fact that most such partnerships fail relatively early. While these coalitions can be deadly as long as they last, they often do not last long, forcing some groups, such as al Qaeda, the PLO, and the now defunct Abu Nidal Organization, to move through a succession of sponsors to stay in the game.

The underlying instability of state-terrorist coalitions is due to several factors. First, these relationships are usually narrowly defined, "negative" alliances. With few exceptions, they are not grounded in a complex web of shared interests, as we see in the case of enduring inter-state alliances, but tend to be issue specific or, at best, based on a relatively small set of correlated objectives. This is due, at least in part, to the fact that most terrorist groups are themselves effectively single-issue actors. While a state may share all or some of a sponsored group's objectives, it will also have other interests, in addition to these, that are different and sometimes even at odds with the goals of its partner. The thing that typically holds this relationship together, furthermore, is not a common vision of the future, but a common opponent. As circumstances change, the advantage conferred by such tactical alliances changes, in turn. While the state and the sponsored group frequently share a social, religious, or ideological identity, once their core interests begin to diverge this will not be sufficient to maintain the coalition.

Second, the two parties in this relationship do not share the same risks. For a terrorist group, state sponsorship is a source of protection. For the state, by contrast, it is a potential source of jeopardy. A terrorist group will not come under any additional pressure, above and beyond what it might otherwise experience, due to the fact that it has entered into a tactical alliance with its sponsor. The opposite, of course, is true for its benefactor. While there may be offsetting gains that continue to make a coalition attractive, these benefits are gained at an expected cost. The

[1] For a good overview of the relationship between terrorists and their sponsors, see Byman [2].

fact that the risks it assumes are lower than they would be if the sponsoring state carried out the same actions directly, does not mean that they are not greater than they would be if it did not enter into such a covert alliance in the first place. For the state, then, the benefits of the partnership must be sufficiently large to compensate for the associated cost. Where the risk is high, the marginal advantage for the state is likely to be relatively low. Even small changes in the terms or circumstances of the coalition, in such cases, can end the partnership.

Finally, even assuming there are no changes in the underlying circumstances that brought the terrorist group and sponsor together in the first place, the expected payoffs enjoyed by the two players in this arrangement are subject to change, and this at an uneven rate. This is, again, most apparent in the different risk dynamics the two players face over time. The risk of retaliation faced by a terrorist group is relatively stable, due in large measure to the inherent difficulty of identifying and localizing a sufficiently large segment of the organization within a sufficient time frame to put it out of action. While most targets of terrorist attack may be able to hit what they can see, in the case of underground opponents, they are much less capable of seeing what they wish to hit. The objective weakness of terrorist groups, in this respect, is offset by the fact that they are notoriously difficult to attack. Apart from any protections offered by its sponsor, the terrorist group's risk of retaliation is mitigated by its own relative invisibility.

This is not the case for the sponsoring state which can generally expect to assume a greater risk from this partnership as time goes on. What protects the state, as we have suggested, is not its invisibility – it is an open target – but the plausible deniability that surrounds the agency relationship it has formed with its client. As long as this can be maintained, a sponsor is likely to enjoy some protection from retaliation from the targets of terrorist attack. The problem it must confront, however, is that the ambiguity that surrounds the existence and nature of the coalition will tend to erode the longer the association continues. As time goes on, more evidence of the relationship and the sponsor's own responsibility for whatever has occurred will begin to come to light. As it does, the risks of retaliation and, therefore, the risks associated with maintaining the coalition will increase. Put another way, the longer an active partnership continues, the more liability the sponsoring state can expect to assume for the actions of its ally.

One factor that determines how much risk the two players in this relationship will assume is the nature and consequences of the actions that result from the alliance. All things being equal, high profile, high casualty attacks carry a higher risk of retaliation and, therefore, a higher expected cost than low profile actions or even high profile actions that result in a small number of casualties. One way in which sponsors can attempt to offset any increase in liability, in this respect, is by forcing or persuading its partner to reduce the severity and/or frequency of its attacks. If successful, the risk that is assumed by a sponsor can be reduced or at least managed over time as the nature of its complicity in the actions of its client becomes an open secret. As we have discussed elsewhere, state-terrorist coalitions often evolve through a natural life-cycle due to the growing risk a sponsor state can face when it backs the same group over an extended period of time. A secret alliance that may begin with a series of high profile actions during the opening phase of the cycle, will often evolve into something less provocative as the sponsor's involvement is exposed and his risks increase (Manuel, McCormick, Owen, [5]).

This introduces another element of instability into state-terrorist coalitions. All other things being equal, the benefits that accrue to the sponsor from the coalition, just as the risks it assumes, are tied directly to the level of activity carried out by its client. It follows from this that the less it can get away with over time, the less it can expect to achieve. This can make maintaining the alliance less attractive, even at a lower expected cost. The same is true for the terrorist group. In this case, the implications for the future of the coalition can be even more decisive. As noted above, terrorists tend to be less concerned with the prospect of retaliation than their sponsors in the first place. Nor do their risks increase in the same way with time. What they are concerned about is staying in the game, and their ability to do so is a function of their ability to hold the headlines through the audacity of their attacks. Any reduction in its attack profile, in this respect, poses a risk to a group's political existence, a position which can put it directly at odds with the interests of its sponsor. With this in mind, it is clear that state-terrorist coalitions can come under significant strain once a sponsor is forced to begin to rein in its client. Dissolving the coalition can become more attractive than continued cooperation.

The stress this can place on state-terrorist coalitions is reinforced by the fact that each side can receive a payoff by being the first to double-cross the other, apart from any long-run advantages it might enjoy by dissolving the relationship. To make matters worse, there is also a corresponding cost associated with being double-crossed. For the sponsor, there is often a political and even an economic premium to be gained by renouncing a terrorist ally and rejoining the mainstream international community. If the group is caught off guard and unprepared by this move, this can leave it vulnerable to retaliation. Terrorist groups, for their part, are likely to be tempted to use the advantages of their position one last time in the waning days of an alliance to carry out a final spectacular attack, unconstrained by any concern its sponsor might have over the growing risk of retaliation. Such an attack can subject its sponsor to a much higher risk of retribution at the very moment the partnership is ending, and for the same reason. This can be an attractive choice for a sponsored group if the long-run advantages of cooperation have declined significantly and the short-term gains of a double-cross are high. By choosing the time to defect, furthermore, the group can avoid being double-crossed first.

The structural tension this creates in state-terrorist coalitions is similar to what we see in a temporal repetition of the Prisoner's Dilemma Game (Luce and Raiffa [4], Axelrod [1]).[2] To overcome this tension, each side must retain a minimal level of trust in the other. There must be sufficient trust to form a coalition and then to continue to cooperate over time as the internal and external circumstances of the original alliance evolve. How much trust is enough depends on the value each side assigns to a cooperative strategy. As the latter changes, so will the former. The higher

[2] There is a difference between our treatment and that of Axelrod [1], namely that we assume that the game will end with the first double-cross. Axelrod, by contrast, assumes that the game will continue.

this value is believed to be, the more risk each side is willing to accept that they will be double-crossed first. As the payoff from cooperation declines, as it does in most state-terrorist coalitions, the level of trust each player must have in the other to continue to cooperate will increase. While it is easy to imagine that the two players could come to trust each other more in the early stages of their association, as they prove themselves to be reliable partners, this pattern can begin to reverse itself over time should each come to believe that the other has less and less reason to cooperate as the advantages of doing so decrease. Should each player conclude, furthermore, that the other is thinking the same way, it can result in a mutual, self-reinforcing decline in confidence and the preemptive end of the partnership.

The discussion that follows examines the implications of these stability factors in state-terrorist coalitions. We examine the conditions under which a cooperative strategy can continue, and the conditions that will result in a double-cross and the dissolution of the alliance. Equilibrium strategies are defined and interpreted for both players.

2 The Mathematical Model

We consider a situation in which a sponsor state, H, and a terrorist organization, T, form a coalition to attack a third party target state, V. For as long as the relationship between T and H is ambiguous it is difficult for V to retaliate. Targeting T directly is not likely to be effective due to its high level of secrecy. Targeting T indirectly by attacking H, on the other hand, is politically costly due to the uncertainty that surrounds its relationship with T and the degree of responsibility it actually has for T's actions. The expected benefits of targeting H are low for the same reason. It is unclear whether H can control or otherwise influence T's behavior, even if it could be compelled to try to do so.

As time progresses, the ambiguity surrounding and protecting this coalition will gradually recede. The longer the relationship continues and the more attacks that are carried out, the more that V will learn about the nature of H's alliance with T. As this occurs, the expected benefits of targeting H to indirectly influence T's behavior will increase, and the expected costs of doing so decline. All other things being equal, retaliation by V against H will become an increasingly attractive option. Depending on the circumstances, this may occur even where the exact nature of H's relationship with T is inconclusive, but where it has become clear that a coalition does in fact exist. As retaliation becomes an increasingly viable possibility, it will eventually be in H's interests to pressure T to decrease the size and/or the frequency of its attacks. As this occurs, however, the advantages of cooperation will decline for both partners.

We note that it is possible that V will discount T's attacks if they are far enough in the past. In this case, the collaboration between T and H may continue indefinitely with a reduced risk of retaliation as long as the coalition exercises some restraint and the two coalition partners continue to find it beneficial to operate within this boundary.

One problem the coalition faces, however, as noted above, is that each player can gain a payoff by being the first to break out of their alliance and double-cross its partner. Its partner, in this instance, will incur a cost by being double-crossed. A very strong, last attack by T that violates the restraints imposed by H can generate a high one-time advantage for T, but can subject H to a much higher risk of retaliation by V. Similarly, H can often gain a payoff by essentially trading its relationship with T (and even T itself) for recognition and support from V and the larger international community. This option will be more attractive if the players have alternative potential partners waiting in the wings with whom they can form a new coalition, one that will again commence at a higher level of activity (and, therefore, a higher payoff) than the level at which the old alliance ends. Because of the risk posed by a double-cross, each player must trust the other. How much trust is required to form a coalition and then maintain it over time depends on the changing advantages each player assigns to a cooperative strategy.

There is an additional consideration, both for T and for H. Since some of the value of a double-cross today lies in the fact that alternative partners (alternative sponsors for T, or alternative proxies for H) exist, it is not advisable to double-cross too quickly. Doing so can negatively affect each player's "alliance worthiness" in the eyes of other potential partners. The direct gains from a double cross, in this respect, will increase with time. In a similar way, the cost of being double-crossed (especially for H) is likely to be lower if this happens early in the relationship, as the actual relationship between H and T is still unclear to outside observers. H's effective liability for T's actions, in this regard, is low. T's vulnerability to a double-cross, for its part, is a function of its dependency on H's continued support. This is likely to be lower during the opening stages of the alliance than it will be once the coalition has matured. T, at this point, is not only likely to be more closely tied to H, but H's knowledge of T will be more acute than it was at the beginning of their association, making the cost of a double cross much greater.

Let us look first at H's payoffs. The continuing gains from cooperation (per time period) would be of the form

$$v_H(t) = C_H + k_H e^{-\sigma t} \tag{1}$$

where the initial gain, $C+k$, decreases to a long-run benefit, C, with a dissipation rate corresponding to the coefficient σ in the exponent.

The immediate gain from double-crossing at time t would be of the form

$$D_H(t) = A_{1H} - A_{2H} e^{-\lambda t} \tag{2}$$

The immediate loss from being double-crossed at time t is of similar form:

$$E_H(t) = B_{1H} - B_{2H} e^{-\lambda t} \tag{3}$$

Apart from this, there is a time discount factor α, so that a loss or benefit of z units, t time units in the future, has a present value of $ze^{-\alpha t}$.

Suppose, then, that H double-crosses at time x. In that case, H gains the amount $D_H(x)$, but loses all future gains from cooperation, so that H's net, discounted gain is

$$K_H(x) = (A_{1H} - A_{2H}e^{-\lambda x})e^{-\alpha x} - \int_x^{\infty} (C_H + k + He^{-\sigma t})e^{-\alpha t}dt$$
$$= e^{-\alpha x}[A_{1H} - A_{2H}e^{-\lambda x} - C_H/\alpha - k_H e^{-\sigma x}/(\sigma + \alpha)] \quad (4)$$

Suppose, on the other hand, that T double-crosses at time y. In that case, H's discounted gain (a negative amount, of course) is

$$L_H(y) = (-B_{1H} + B_{2H}e^{-\lambda y})e^{-\alpha y} - \int_x^{\infty} (C_H + k_H e^{-\alpha t})e^{-\alpha t}dt$$
$$= e^{-\alpha y}[-B_{1H} + B_{2H}e^{-\lambda y} - C_H/\alpha - k_H e^{-\sigma y}/(\sigma + \alpha)] \quad (5)$$

If both H and T double-cross (simultaneously) at time y, then we assume that H neither gains nor loses directly from the double-cross, but he loses the gains from future collaboration, so the payoff to H is

$$M_H(y) = -e^{-\alpha y}[C_H/a + k_H e^{-\alpha y}/(\sigma + \alpha)] \quad (6)$$

Thus, if T chooses to double-cross at time y, and H, at time x, the payoff to H will be

$$\prod_H(x,y) = \begin{cases} K_H(x) & \text{if } x < y \\ L_H(y) & \text{if } x > y \\ M_H(y) & \text{if } x = y \end{cases} \quad (7)$$

The payoff to T will be of similar form, though of course with different parameters.

3 Equilibrium Strategies

Generally speaking, we can expect that there will be three types of equilibrium:

1. H and T value cooperation and trust each other, and will collaborate indefinitely.
2. They do not trust each other enough to continue to cooperate, given the gains associated with doing so, and this will lead to a double-cross.
3. Apart from this, there is a mixed strategy equilibrium, in which each of T and H chooses the moment to double-cross its partner probabilistically. Alternatively each player decides probabilistically, at each moment of time, whether to double-cross the other or continue to cooperate.

The first two equilibria are easy enough to understand. What is not clear is what conditions will lead to one or the other. In fact, this seems to depend on the extent of the possible payoffs, as well as each side's "mistrust factors", φ and ψ, (which we discuss below). This can best be understood by looking at the third possible equilibrium.

Let us, then, consider the mixed-strategy equilibrium. (We proceed as in Owen [6], or Shiffman [7], Karlin [3].)

Suppose that H and T use mixed strategies $F(x)$ and $G(y)$ respectively, continuous distributions, with densities $f(x)$ and $g(y)$ respectively, positive on an interval of the form (a,b) where b could possibly be ∞.

Note that, if T uses the mixed strategy G, while H uses the pure strategy x, then the expected payoff to H is

$$E(x,G) = \int_a^x L_H(y)g(y)dy + \int_x^b K_H(x)g(y)dy$$
$$= \int_a^x L_H(y)g(y)dy + K_H(x)[1-G(x)] \tag{8}$$

Now, for equilibrium, this expected payoff must be the same for all x in the interval (a,b). Thus, for all such x,

$$\partial E/\partial x = 0$$

and this reduces to

$$[K_H(x) - L_H(x)]G'(x) + K'_H(x)[G(x)-1] = 0. \tag{9}$$

This equation has the solution

$$G(x) = 1 - ce^{-\mu(x)}$$

where

$$\mu(x) = \int_0^x \frac{K'_H(t)dt}{K_H(t) - L_H(t)} \tag{10}$$

and c is an arbitrary constant. Since $G(x) < 1$, c is necessarily positive. The density is then

$$g(x) = c\mu'(x)e^{-\mu(x)} = \frac{cK'_H(x)e^{-\mu(x)}}{K_H(x) - L_H(x)} \tag{11}$$

Since g, as a density, must be positive, we see that K' must also be non-negative in the interval (a,b).

Because of this, we now distinguish two cases, depending on whether αA_{1H} is larger or smaller than C_H. Note that, asymptotically, $K_H(x)$ is equivalent to $(A_{1H} - C_H/\alpha)e^{-\alpha x}$. Note also that usually, $K_H(0) < 0$.

In the first case, where $\alpha A_{1H} > C_H$, K_H eventually becomes positive, and reaches a maximum, before decreasing to zero.

Example 1:

An example of this appears in Fig. 1, where we assume that $A_{1H} = 35$, $A_{2H} = 30$, $C_H = 10$, $k_H = 4$, $\lambda = 0.6$, $B_{1H} = 20$, $B_{2H} = 0$, $\sigma = 0.9$, and $\alpha = 0.5$.

In this case, the function K_H is negative from $x = 0$ to $x = 1.23$, and reaches its maximum value at $b = 2.6$. Letting a be some number in the interval $[0, b)$, then $G(a) = 0$, so

$$G(a) = 1 - ce^{-\mu(a)} = 0$$

Thus $c = e^{\mu(a)}$. Note that if $a = 0$, then $\mu(a) = 0$, so $c = 1$.

This gives us $G(b) = 1 - e^{\mu(a)-\mu(b)}$, which is not equal to 1. We conclude that there must be a jump at $x = b$, equal to $e^{\mu(a)-\mu(b)}$. In other words, according to the equilibrium strategy, T will double-cross during the interval (a, b) according to the density g. At time b, T will definitely double-cross if the game is not yet over. (Of course, H might himself double-cross before T does so.)

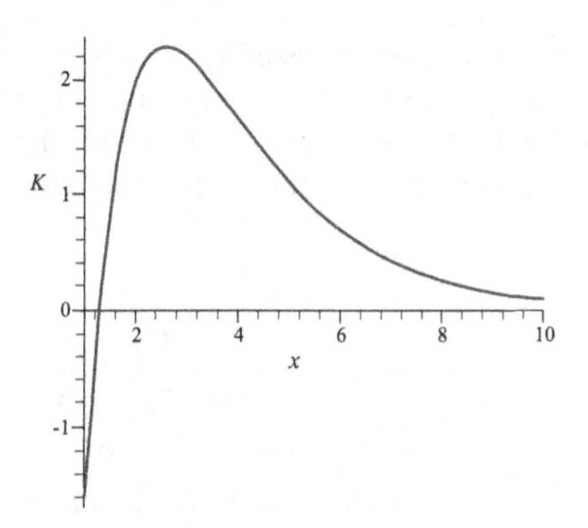

Fig. 1 Net gains, $K(x)$, from double-crossing at time x. The case $\alpha A_1 > C$.

In the second case, where $\alpha A_{1H} \leq C_H$, K_H will always be negative, but increase monotonically, so that $K'_H > 0$ for all x. In this case, T will double-cross according to the density g. If $\mu(\infty) = \infty$, i.e. if the corresponding improper integral diverges, then there will definitely be a double-cross. Note that, by (10), $\mu(\infty)$ is given by an improper integral. If, somehow, the improper integral converges, so that $\mu(\infty) < \infty$, then there is a positive probability that T will never double-cross. (Note that this can only happen if $\alpha A_{1H} = C_H$.) Then, if similar considerations hold for H, there is positive probability that the game will continue forever.

Example 2:

An example of this second case is illustrated in Fig. 2, where we assume that $A_{1H} = 15$, $A_{2H} = 10$, $C_H = 10$, $k_H = 4$, $\lambda = 0.6$, $\sigma = 0.9$, $B_{1H} = 20$, $B_{2H} = 0$, and $\alpha = 0.5$.

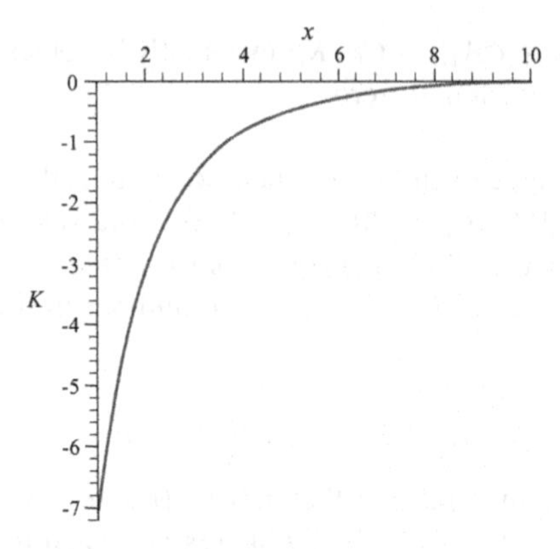

Fig. 2 Net gains, $K(x)$, from double-crossing at time x. The case $\alpha A_1 \leq C$.

Thus, if $\alpha A_{1H} \leq C_H$ (so that there is always some benefit from continuing the partnership), an equilibrium mixed strategy for T will be given by the density $g(x)$, for $a < x < \infty$.

Also of interest is the conditional probability density $q(x) = g(x|X \geq x)$, which gives us the probability that T will double-cross at time x or shortly thereafter, given that neither partner has double-crossed prior to that time. Essentially, for small h, the probability that T will double-cross during the time interval $(x, x+h)$ is $hq(x)$. This is given by

$$q(x) = g(x)/(1-G(x)) = \frac{cK'_H(x)e^{\mu(x)}}{[K_H(x) - L_H(x)]ce^{\mu(x)}}$$
$$= \frac{C_H + k_H e^{-\sigma x} - \alpha A_{1H} + (\lambda + \alpha) A_{2H} e^{-\lambda x}}{(A_{1H} + B_{1H}) - (A_{2H} + B_{2H}) e^{-\lambda x}} \tag{12}$$

For Example 1, the function q is given by Fig. 3. This function crosses the axis and becomes negative at approximately $x = 2.6$. In effect, the partnership will end at this moment as H will have too little to gain from continuing the collaboration.

For Example 2, on the other hand, we have Fig. 4. In contrast to Fig. 3 where q becomes negative, in this case q remains positive even as x goes to infinity.

4 Payoff to T

Of course, as mentioned above, it is necessary to look at the possible payoffs to both H and T in this situation. Up until this point, we have looked only at H's payoffs. The payoffs to T can be modeled in a similar way: an ongoing gain from cooperation,

$$u(t) = C_T + k_T e^{-\sigma t} \tag{13}$$

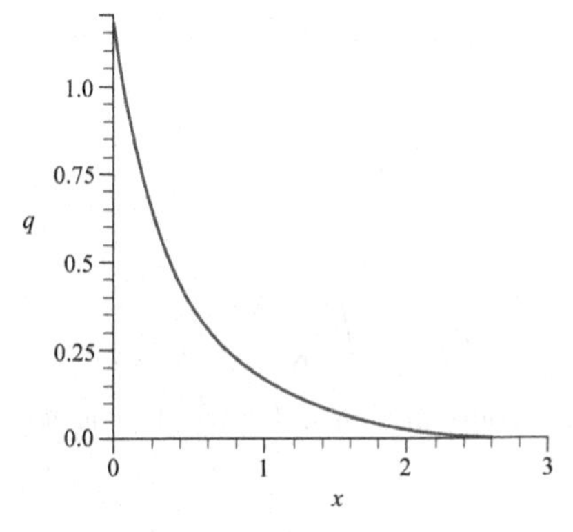

Fig. 3 Equilibrium probability $q(x)$ of a double-cross by H at time x. The case $\alpha A_1 > C$.

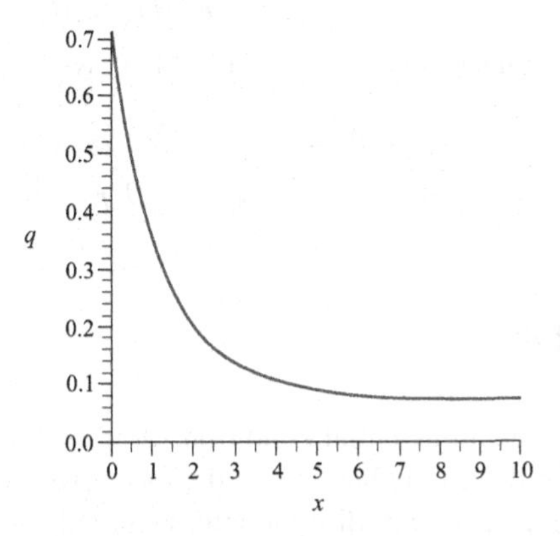

Fig. 4 Equilibrium probability $q(x)$ of a double-cross by H at time x. The case $\alpha A_1 \leq C$.

an immediate gain from a double-cross,

$$D_T(t) = A_{1T} - A_{2T}e^{-\lambda t} \tag{14}$$

and an immediate loss from being double-crossed

$$E_T(t) = B_{1T} - B_{2T}e^{-\lambda t} \tag{15}$$

Thus the net discounted gain to T, if he double-crosses at time y, will be

$$K_T(y) = e^{-\alpha y}[A_{1T} - A_{2T}e^{-\lambda y} - C_T/\alpha - k_T e^{-\sigma y}/(\sigma+\alpha)] \tag{16}$$

and the (negative) gain to T, if H double-crosses at time x, will be

$$L_T(x) = e^{-\alpha x}[-B_{1T} + B_{2T}e^{-\lambda x} - C_T/\alpha - k_T e^{-\sigma x}/(\sigma+\alpha)] \tag{17}$$

Thus (as mentioned above) the payoffs to T are similar to the payoffs to H, though with different parameters.

Proceeding as before, we obtain an equilibrium mixed strategy $F(y)$ for T, given by

$$F(y) = 1 - ce^{-\rho(y)} \tag{18}$$

where

$$\rho(y) = \int_0^y \frac{K_T'(t)dt}{K_T(t) - L_T(t)} \tag{19}$$

and c is an arbitrary constant. Since $F(y) < 1$, c is necessarily positive. The density is then

$$f(y) = c\rho'(y)e^{-\rho(y)} = \frac{cK_T'(t)e^{-\rho(y)}}{K_T(t) - L_t(t)} \tag{20}$$

and the conditional density, $r(y) = f(y)/[1 - F(y)]$, is given by

$$r(y) = \frac{C_T + k_T e^{-\sigma y} - \alpha A_{1T} + (\lambda + \alpha)A_{2T}e^{-\lambda y}}{(A_{1T} + B_{1T}) - (A_{2T} + B_{2T})e^{-\lambda y}} \tag{21}$$

5 The Trust Factor

Trust, as we have said, is an essential condition for a continuing alliance. It is of course very difficult to determine how much the two partners, T and H, trust each other. This is both subjective and idiosyncratic. Generally speaking, however, we might imagine that a rational H's willingness to continue to cooperate will depend on:

1. How probable H thinks it is that T will in fact double-cross him, and
2. How much H himself can gain by double-crossing T first, in relationship to the gains H can accrue by continuing an active partnership with T.

In turn, factor (1) depends on (i) how trustworthy T is in general, and (ii) how much T can gain by double-crossing, versus the gains T accrues by continuing to cooperate.

The coalitions formed between states and terrorist groups are mercenary pacts. Trust in such relationships is not based on shared principles but on the perception of shared interests. As long as each player believes that his partner has an interest in continuing to cooperate he will be "trusted" to do so. If, by contrast, one side or the other comes to believe that his partner will gain by defecting, the basis for continued cooperation will be eliminated. Note that this is true regardless of the accuracy of these beliefs.

In other words, we can imagine H saying to himself, "T is not an inherently trustworthy partner, but he has little to gain from double-crossing me and I have a lot to gain if we continue to cooperate. On the basis of this I will continue to trust him for the moment." Conversely he might say, "Even though T has no track record of double-crossing a previous partner, he has too much to gain and I have too much to lose if he double-crosses me, so I'm not willing to trust him. I will double-cross him first".

We model (1) by a mistrust factor, $\varphi(x)$, to represent how likely it is that T will double-cross at time x. Essentially, at time x, H has a (subjective) belief that T will double-cross between times x and $x+h$ (where h is small) with probability $h\varphi(x)$. Now, we will assume that there is an individual "unreliability" factor, of the form $\theta e^{-\gamma x}$, where θ is given exogenously. This decreases exponentially with time, as H gains more confidence in T's reliability. We multiply this by the quantity $D_T(x)$ which represents (H's belief) about T's gain from a double-cross, and divide by $-K_T(x)$ which represents the (perceived) advantage T can still hope to gain from cooperation. Thus we obtain

$$\varphi(x) = -\theta e^{-\gamma x} D_T(x)/K_T(x) \tag{22}$$

as the mistrust factor (how much H mistrusts T). Note that this is meaningful only so long as $K_T(x) < 0$; once $K_T(x)$ becomes positive, the game will effectively be over as T will gain more from double-crossing than he can expect to gain from cooperation, and so he can be expected to double-cross H, (assuming that H does not beat him to the punch).

In a similar way, we obtain the mistrust factor

$$\psi(y) = -\omega e^{-\gamma y} D_T(y)/K_T(y) \tag{23}$$

(where ω is also given exogenously) to measure the mistrust that T feels for H.

6 Interpretation

It is important to understand that we do not recommend that one follow the mixed-strategy equilibrium The point that needs to be made about this equilibrium is not so much that players will normally follow it, but, rather, that it is highly unstable, i.e., small changes in strategy by one of the players will lead to large changes by the other, etc. Specifically, if H's mistrust factor $\varphi(x)$ is greater than $q(x)$, then H will himself be more inclined to double-cross, which will cause T's mistrust factor, $\psi(x)$, to increase, thus making it more likely that T will double-cross, etc. Conversely, if $\varphi(x) \le q(x)$, then H will be more inclined to cooperate, etc. Thus the equilibrium pair $[q(x), r(y)]$ serves as a *threshold*: if each player feels that his partner is above this threshold, then they will soon double-cross; below it, they will be faithful and cooperate. In other words, a low value for q or r does not mean that things are going well; on the contrary, a low value for q means that things will go badly unless H has a sufficiently high offsetting confidence and trust in T. The reason for this is that, when q is small, there is not much to gain from cooperation, and much to be lost in case of a double-cross: thus, H will not continue to cooperate unless he is confident that T will continue to honor the partnership.

In the limit, at point b, where $K(x)$ reaches its maximum, we find that $q(b) = 0$; for $x > b, q(x) < 0$. But it is precisely here that H will certainly double-cross his partner, no matter how much he may trust T; i. e., even if $\varphi(x) = 0$, H will double-cross T. As another example of this phenomenon, note that, from equation (12), q increases with C. This does not mean (as a careless reading of the text might suggest) that T is more likely to double-cross when there is a greater long-term gain, but rather, that H will cooperate, even if he believes T is unreliable, if the prospects for a perceived cooperative gain are greater.[3]

To simplify notation, we classify H as **strong** if $\alpha A_{1H} \leq C_H$, and **weak** in the contrary instance, $\alpha A_{1H} > C_H$. In this latter case the maximizing point b (for $K_H(x)$) is important.

Similarly, we classify T as **strong** if $\alpha A_{1T} \leq C_T$, and **weak** otherwise.

We now distinguish three cases, depending on whether T and H are weak or strong.

Case I. Both T and H are strong:

In this case, a double-cross will never give greater benefits than the discounted long-term gains, and thus it is in each party's advantage to continue to cooperate. The only question is whether they will continue to trust each other sufficiently to do so.

The quantities $\varphi(0)$ and $\psi(0)$ are given (exogenously). Presumably, the partnership will not begin unless $\varphi(0) \leq q(0)$ and $\psi(0) \leq r(0)$. Assume then that this is so, and that the partnership indeed starts. Note, however, from (12) and (21) that q and r are decreasing functions. The question, then, is whether the players' mistrust factors will continue to be below the threshold values necessary to maintain their partnership over time.

Case I(a). $\varphi(0) \leq q(\infty)$ and $\psi(0) \leq r(\infty)$:

If this is so, then there should be no problem, as both H and T trust each other sufficiently. Of course some exogenous event may cause a sharp increase in either φ or ψ, but barring such an event, the partnership will continue indefinitely.

Case I(b). $\varphi(0) > q(\infty)$ or $\psi(0) > r(\infty)$ (or both):

[3] This idea of an unstable equilibrium as a threshold rather than as a recommendation can best be understood if we consider the case of two motorists, A and B, approaching each other from opposite directions along a road. Each can choose to drive on the left or on the right side of the road; they pass safely if both choose right, or if both choose left, but there will be an accident if one chooses right, and the other, left. There are of course two stable equilibria: the British equilibrium (both choose left) and the American (both choose right). There is also however a mixed-strategy equilibrium, in which each chooses right or left according to a (½, ½) probability distribution. This is clearly an equilibrium: since B is as likely to be on the right as on the left, A cannot gain by a unilateral change of strategy. But, if B is more likely to go on the right side than on the left, and A is aware of this, then A will be much more likely to drive on the right, which makes B more likely to drive on the right, etc. Thus the unstable equilibrium serves as a threshold: if they start just to the right of it, they will eventually move all the way to the right. Similarly if both of them are more likely to edge to the left than to the right, then eventually both will wind up on the left.

This is somewhat more problematic. The partnership, we assume, will start, but will be subject to stress as φ and ψ approach and exceed the threshold for cooperation. If this continues, one side or the other will defect and double-cross its partner. Note, this can occur even if the two partners are coming to trust each other more but not at a rate that keeps φ and ψ above the critical threshold needed to maintain the alliance.

Example 3 (Example 2 continued):

For the payoff to H, we use the same data as in Example 2. For T, the payoffs will be similar: $A_{1T} = 15, A_{2T} = 10, C_T = 10, k_T = 4, \lambda = 0.6, \sigma = 0.9, B_{1T} = 20$, $B_{2T} = 0$, and $\alpha = 0.5$. Finally, we use the unreliability factor $1.3e^{-x}$.

In Fig. 5, one of the curves is taken from Fig. 4, and represents $q(x)$. The other curve is the graph of $\varphi(x)$. As may be seen, $\varphi(x) \leq q(x)$ for $x < 0.7$ and again for $x > 2.8$. For $0.7 < x < 2.8$, however, φ is larger. Essentially, it seems that the partnership will start, but will be subject to serious stress at about time $t^* = 0.7$. H simply does not trust T enough. In these circumstances, H's possible gains from cooperation are not likely to be enough to keep him from defecting and double-crossing T. If somehow the partnership can survive those stresses until time $t^{**} = 2.8$, however, then it can continue indefinitely.

Example 4:

Suppose that everything is as in Example 3, except that the unreliability factor is $0.8e^{-x}$. In this case (see Fig. 6), the mistrust factor dissipates rather quickly (note γ here is larger than in Example 3) and we find that φ is always smaller than q. Thus in this case there should be no problem: we can expect the partnership will endure.

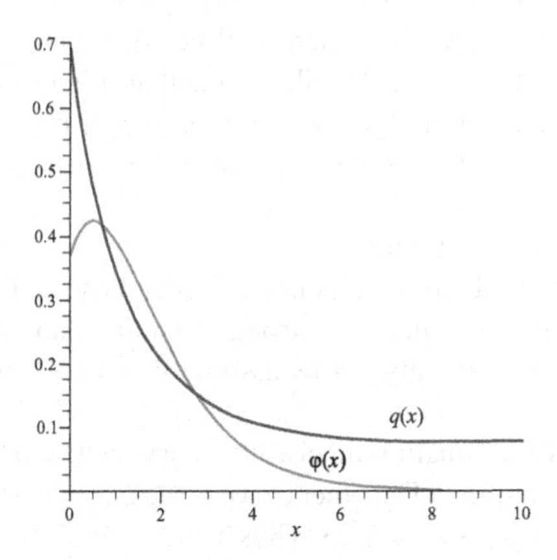

Fig. 5 Mistrust factor $\varphi(x)$ versus the equilibrium probability $q(x)$ resulting in a likely double-cross by H.

Case II. One of T or H is weak, the other is strong.

To simplify notation, assume that H is weak and T is strong. This means that $K_H(x)$ becomes positive at time t^*, and eventually reaches a maximum at $x = b$. Now, suppose T is aware of H's payoff function. In this case T realizes that it will be in

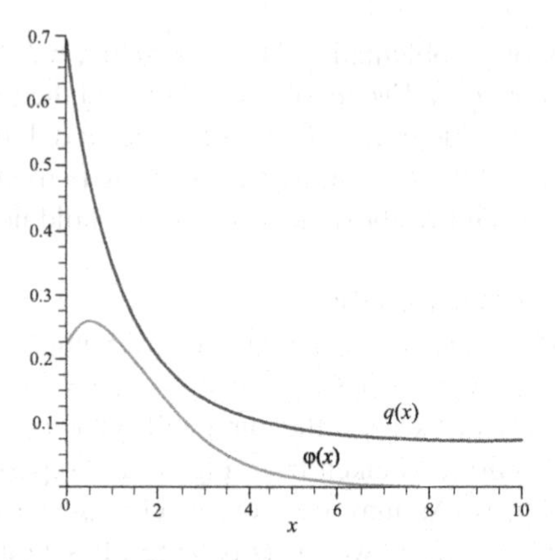

Fig. 6 Mistrust factor $\varphi(x)$ versus the equilibrium probability $q(x)$ resulting in continued cooperation.

H's interest to double-cross, certainly before time b, but more likely as early as time t^*. It will be in T's interest, under these circumstances, to preempt H's action and double-cross H first. This is true even though it would be in T's interest to continue to cooperate if were also in H's interest to do so. We can expect a breakdown in the coalition no later than time t^*.

Suppose, on the other hand, that T is unaware that H is weak and believes that it will continue to be in H's interests to maintain the coalition. In that case, T will not double-cross, but H will. In fact, we know $q(b) = 0$. Presumably $\varphi(0) \leq q(0)$, but $\varphi(b) > 0$, so (assuming continuity), there will be some value, t^{**} (smaller than b) with $\varphi(t^{**}) = q(t^{**})$. At this time, H will realize that it is in his interest to double-cross. Thus we can expect a double-cross *not later than* time t^*. An objective observer, who knows the payoff functions but not the mistrust function, would expect a double-cross *earlier than* time b.

Example 5 (Example 1 continued):

Assume the payoffs to H are as in Example 1, and those to T are as in Example 3. In this case, then, H is weak and T is strong. Assume also an unreliability factor (on both sides) of $0.1e^{-x}$. Finally, let us assume first that T is fully aware of H's weakness.

Figure 7 illustrates the situation nicely. One curve is taken from Fig. 3: it is the graph of $r(x)$, from Example 1. The other curve is the graph of $\varphi(x)$. As noted above, K_H becomes positive at about $t^* = 1.23$. Thus T must certainly double-cross no later than 1.23. In this case, however, because of his general mistrust of H, he is likely to double-cross H earlier, at or around $x = 1.08$ (where the two curves intersect).

Example 6. Example 5 continued:

Suppose, instead, that T is unaware that H is weak. Then H will double-cross T first. In fact, using the same parameters as in Example 5, suppose that H feels that T has an unreliability factor of $0.2e^{-1.35x}$. The figure below then shows $q(x)$ and $\psi(x)$. As we know, $K_H(x)$ reaches a maximum value at $b = 2.6$. Thus H will certainly double-cross T no later than time 2.6. However, because of the mistrust factor, his betrayal of T is more likely to come at or near time 2.45, which is again the point at which the two curves' intersect.

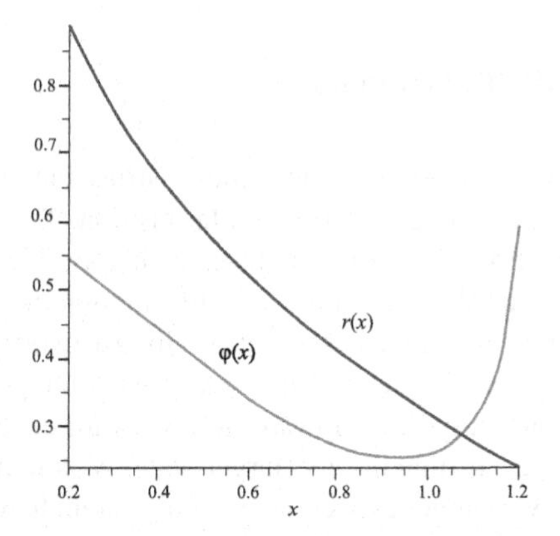

Fig. 7 Mistrust factor $\varphi(x)$ versus the equilibrium probability $q(x)$ resulting in a double-cross by T.

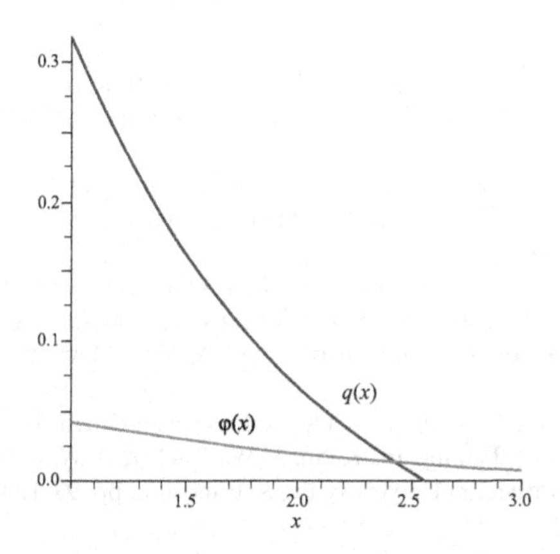

Fig. 8 Mistrust factor $\varphi(x)$ versus the equilibrium probability $q(x)$ resulting in a double-cross by H.

Case III. Both T and H are weak.

This would be similar to Case II. As discussed above, both $K_H(x)$ and $K_T(y)$ increase to reach their maxima at b and b' respectively. Assuming once again that $\varphi(0) \le q(0)$ and $\psi(0) \le r(0)$, the partnership can continue until some time t^* where either $\varphi(t^*) = q(t^*)$ or $\psi(t^*) = r(t^*)$. Clearly $t^* < \min\{b, b'\}$. At that time one or the other of the two partners (say H, in case $b < b'$) will find it in his interest to double-cross T, as no greater gains are possible, even if T remains faithful. T, of course, if he realizes that this is so, will himself want to double-cross first. Thus the partnership will end, one way or the other, no later than time t^*, and certainly some time before $\min\{b, b'\}$ is reached.

7 Conclusion. External Shocks

We note that we have assumed above that the mistrust factors φ and ψ both are continuous functions and evolve in a very regular fashion. This seems reasonable as long as the process is allowed to continue in its own way. The system, however, is always subject to external "shocks." Some new information may come to light which makes H, for example, believe that T is not generally trustworthy; this could cause a large increase in φ. Conversely, a new regime may come into power in H; this would certainly cause a discontinuity in ψ, upward or downward depending on the type of new regime. We may even imagine the target state, V, attempting to manipulate either φ or ψ. Clearly we cannot expect the general system here described to evolve in a mechanistic fashion.

References

1. Axelrod, R.: The Evolution of Cooperation, Basic Books, New York (1984)
2. Byman, D.: Deadly Connections: States That Sponsor Terrorism, Cambridge University Press, New York (2005)
3. Karlin, S.: Reduction of Certain Classes of Games to Integral Equations, In: Kuhn, H.W., Tucker, A.W. (eds.) Contributions to the Theory of Games, II, Princeton University Press, Princeton, pp. 125-158 (1953)
4. Luce, R. D., Raiffa, H.: Games and Decisions, John Wiley and Sons, New York (1957)
5. Manuel, C., McCormick, G.H., Owen, G.: Secret Coalitions and State Support for Terrorist Organizations, Technical Report, Department of Applied Mathematics, Naval Postgraduate School (2007)
6. Owen, G.: Game Theory, 3rd Edition, Academic Press, San Diego, 74-78 (1995)
7. Shiffman, M.: Games of Timing, In: Kuhn, H.W., Tucker, A.W. (eds.) Contributions to the Theory of Games, II, Princeton University Press, Princeton, pp. 97-124 (1953)

Simulating Terrorist Cells: Experiments and Mathematical Theory

Lauren McGough

Abstract How well do mathematical models of terrorist cells apply to the real-life struggle against terrorism? Certainly, mathematical models have been useful in the past for military planning and predicting the behavior of U.S. adversaries, but how well do mathematical projections of terrorist behavior actually hold up when tested on living people and real situations? This paper first presents a mathematical model of terrorist cells and their functionality, and then discusses the procedure and results of an experiment conducted to test this model's theoretical projections by comparing them with experimental results, thus confronting the question of theory versus reality.

1 Introduction

In the paper, "Breaking Al Qaeda Cells: A Mathematical Analysis of Counterterrorism Operations (A Guide for Risk Assessment and Decision Making)," Farley offers ideas on mathematically modeling terrorist cells (groups of terrorists who function as a unit) as "partially ordered sets" [2]. In a partially ordered set, or "poset," the people in a terrorist cell can be represented as nodes, and a connection between two people, such as a way of passing on commands or instructions, can be denoted as a line connecting the two nodes. This is very similar to the mathematical notion of graphs; however, unlike graphs, posets also account for the hierarchical structure of terrorist cells: that is, in a cell, there is at least one leader, who has subordinates, who then have subordinates underneath them, etc., down to the foot soldiers who carry out the attack. In other words, the direction in which commands are passed is captured. There is a partial order rather than a complete, or linear, order because oftentimes there are two or more people at the same hierarchical level within one

Lauren McGough
Phoenix Mathematics, Inc., 35 Northfield Gate, Pittsford, NY 14354, e-mail: lauren@phoenixmath.com

N. Memon et al. (eds.), *Mathematical Methods in Counterterrorism*,
DOI 10.1007/978-3-211-09442-6_18, © Springer-Verlag/Wien 2009

terrorist cell. A node at the top of the structure, known as a maximal node, may represent a leader of the cell, while the bottommost nodes, or minimal nodes, represent foot soldiers. A leader of the cell may pass commands to the foot soldiers through a series of middlemen in a hierarchical organization, and a terrorist attack may occur when a command has reached the foot soldiers of the cell.

As Farley suggests, using this model, it is possible to calculate the probable effect that a given action against a terrorist cell will have on its ability to pass on commands from the leaders to the followers. Then, using the assumption that a terrorist attack can only occur when a command has been passed from a leader of the cell down the various levels of the hierarchy to the foot soldiers who carry out the attack, one can estimate the probability that, by removing a certain number of people of unknown status in a known terrorist cell structure, one has broken all lines of command from the leaders of the cell to the foot soldiers.

Mathematically speaking, a chain from a maximal node of a poset to a minimal node is called a maximal chain. In modeling terrorist cells as posets, these maximal chains represent lines of command from leaders to foot soldiers. Therefore, one can infer that in order to prevent a terrorist attack from occurring, it is necessary to break all of the maximal chains by "removing" (killing or capturing) specific key people from the cell. Such collections of nodes that intersect all maximal chains in a poset are known as cutsets. Therefore, if the U.S. government has removed k people of unknown status from a terrorist cell Γ with a known structure containing n people, and there is a total number $Cut(\Gamma,k)$ of k-membered cutsets in this structure, the equation to find the probability, Pr, that the removed collection of people is a cutset (and therefore, that their removal has prevented a terrorist attack from occurring under these assumptions) is:

$$Pr = \frac{Cut(\Gamma,k)}{{}_nC_k}, \tag{1}$$

where
${}_nC_k = \frac{n!}{k!(n-k)!}$ and $r! = r(r-1)(r-2)\dots(2)(1)$ for all natural numbers r

This equation quantifies the effect of a given action of the U.S. government on a terrorist cell, finding the probability that this action has actually prevented a terrorist attack.

2 The Question of Theory versus Real-Life Applications

A reading of Farley's paper leads to the following question: How well does this mathematical theory of breaking all chains of command in a terrorist cell apply to real-life situations? While it is mathematically helpful to use nodes on a graph to represent people, the communication networks of real terrorist cells, as Farley observes, often have factors that come into play that the mathematical model does not take into account. Examples of these factors include: (a) the time that it takes for commands to be passed from person to person, and (b) the chances that the communication is disrupted for causes other than people in the cell being killed or captured, such as mechanical errors, or personal reliability. The following experiment was designed with the specific purpose of answering the question of how situational circumstances of real, operating terrorist cells can change the projected effects of actions against these cells predicted by this mathematical model.

3 Design

In designing an experiment to test how well this mathematical theory holds up in real life, it was logical to use a lifelike simulation of connections between people in a terrorist cell. This experiment simulated these connections using a chain of phone calls. To represent the way that a command travels from the leader(s) of the cell to the foot soldiers, people were assigned positions in a previously decided upon simulated cell structure. One person was placed at the maximal node in the structure, this person functioning as the leader of the cell. This "leader" was then assigned a message or "command" to pass on to his or her subordinates through phone calls. These subordinates then passed this message on to their direct subordinates. This pattern continued until the message reached the people representing the foot soldiers of the cell. Conceptually, this design was similar to the way that commands could be passed from person to person in a terrorist cell.

Further reasoning led to the conclusion that if a person in this simulation did not pass the message on to his or her subordinates, the chains of command that relied on that person were effectively broken. This was the equivalent of the U.S. government removing a terrorist from a terrorist cell. To simulate the removal of people from the cell, three people were chosen at random to be removed from the cell for each separate trial. Random selection improved the accuracy of the results by giving each combination an equal chance of being selected for testing where it was impossible to test each possible combination. This led to a higher likelihood that the results attained by this experiment were the same as if the experiment had tested all 455 possible combinations of ways to take three people from fifteen (the number of people in the simulations). Then, each person chosen was told not to pass on the message that was sent out for that specific trial. This had the same effect as removing these people from the simulated cell for that trial.

The structure for the simulated cell that worked best for the purposes of this experiment was a binary tree with fifteen members. There were several reasons for choosing this structure. Firstly, binary trees have one maximal node, corresponding to the fact that real terrorist cells might have only one leader. Also, a binary tree with fifteen elements has only four levels of hierarchy, corresponding to the fact that real terrorist cells most likely would not have more than four or five levels of hierarchy. Lastly, in a binary tree, each node is connected to two nodes below it, and therefore this structure evenly distributed the amount of work required for this experiment between all of the participants.

The last aspect of the design was measuring and tracking the experimental likelihood that removing three people from the simulated cell would break all of the chains of command. To accomplish this, it was necessary to track which people at the bottom level of hierarchy received each message or "command" and, in a real terrorist cell, would therefore have had the ability to carry out an attack. To implement this into the design, first, all of the messages sent out were recorded, and the people at the bottom level of the network who would receive the message when the message had gone through all lines of command were determined assuming the trial went according to the model's predictions. The people at the bottom level of the structure were assigned to call one person upon receiving the message. This person could then record which people at the bottom tier had received the message, making it possible to determine the path of successfully relayed messages. This showed what chains of command still functioned after the removal of those three people from the cell. Then, by determining which people at the bottom level had not received the message, it was possible to determine which chains of command had been broken with this removal of three people. Therefore, when a message was sent out but never returned, it followed that all of the chains of command had been broken with the removal of that specific combination of three people. By dividing the number of trials where the message did not reach the bottom level of the hierarchy by the total number of trials conducted, the data was used to find the experimental probability of cutting all lines of command by removing three people from this specific terrorist cell structure.

4 Procedure

The previously discussed design was used to conduct the experiment. First, fifteen volunteers were recruited to fill into the binary tree structure used for the design of the terrorist cell (see the appendix on page 316 for data). Each participant knew at least one other participant. Then, using connections that the participants had to one another in real life, each participant was assigned a role in the binary tree, and was given written instructions as to his or her role in the experiment (see the appendix for sample instructions).

Next, experimentation began. For each trial, the three people to be removed from the structure were chosen at random. To start the phone call chain, the "leader" was given a message to send out. Immediately, each of the three people chosen to be removed was notified not to pass on that specific message, in the order of hierarchical representation in the cell (notifying the person on the highest tier first, and the person on the lowest tier last). If any one was unavailable the first time of being called for notification, he or she was notified not to pass on the message as soon as he or she was available.

Every day, zero to ten trials were conducted, each with a new message. For every new message, a new combination of people was taken out (if a previously tested combination was chosen, it was not retested but the random selection was run until an untested combination was chosen). Fifty trials were conducted. After all of the trials were finished, the number of messages that did not come back was counted and used to calculate the experimental probability that no messages reached the bottom levels of the cell when removing three people. This data was compared with the theoretical probability that no messages would reach the bottom level of the cell, and conclusions were made based on this comparison.

5 Analysis and Conclusions

There are 455 possible ways to remove three nodes from a binary tree of 15 objects. Of these combinations, 105 are cutsets. Therefore, the theoretical probability that all of the chains of command are broken when three people are removed from a terrorist cell of fifteen people in a binary tree structure is about 23.0269 percent (105/455). However, using the data collected from this experiment, the probability that all chains of command were broken was 30.00 percent, as 15 out of the 50 messages that were sent out did not reach the bottom level of people. This means that the experimental probability of cutting the cell was actually about 7 percent higher than the theory suggested. Conducting a single proportion z-test to test the null hypothesis that this 30 percent generated from a random sample is not significantly different from the expected true proportion of 23 percent success rate, we find that the sample proportion is not significantly different from the predicted population proportion at the $\alpha = 0.05$ level, and thus we can not reject the null hypothesis that this sample proportion approximates a true proportion of 23 percent ($P = 0.245 > 0.05$).

These results have two potential interpretations. The first is that since the calculated sample proportion does not significantly differ from the expected proportion, real life situations are accurately represented in Farley's mathematical model, and any variation from the mathematical model to real-life situations in this experiment were merely matters of chance.

Another, perhaps more intriguing, interpretation, involves looking into the specific nature of the 7 percent difference between theoretical predictions and experimental findings in this case. Though according to the statistical test, the 7 percent difference could merely be a result of deviations from the expected proportion because of random sampling methods, the 7 percent difference could also derive from sources such as human error, mechanical error, and temporal effect on breaking terrorist cells. Indeed, this 7 percent difference in probability suggests that the real-life probability of breaking all chains of command by removing k people from a terrorist cell could actually be higher than the theoretical model suggests. Specifically, errors such as mechanical error and human error cause the operations of real networks to be less perfect than a mathematical model. As a result, the probability of breaking all chains of command with a given combination of three people might increase from the theoretical model. Therefore, it may be suggested that perhaps the mathematical model is conservative, in that the probability that it implies may actually be the lowest probability that taking out k people from the cell cuts all chains of command.

In particular, there were several interesting cases in the data that Farley's model does not take into account. These were cases where messages should have reached the bottom levels, but did not. For example, in two cases, no messages reached the bottom tier, but according to the mathematical model, at least one person at the bottom tier should have received the message. Interestingly, if these two cases had actually reached the bottom level of people as they were predicted to, the percentage of messages that would not have reached the "foot soldiers" would have been 26 percent, which is even closer than 30 percent to the theoretical prediction of about 23 percent. The other 3 percent difference would almost certainly be caused by error due to the randomness of the selection of combinations used. Therefore, 4 percent, more than half, of the 7 percent difference between the theoretical projection and the experimental results could be attributed to human and/or mechanical error in the execution of the experiment. These types of error could also occur in real terrorist cell networks. These cases further support the potential conclusion that the probability of breaking a terrorist cell by removing k people could be higher in real life than in theory.

There were other cases in the data where not all of the messages that should have reached the bottom level of the hierarchy did. It should be possible to calculate the "fractional expectation" of how many people will actually receive a message (as opposed to how many people theoretically "should" receive a message). This fractional expectation could be implemented into a mathematical model for projecting effects of actions on terrorist cells, and could be useful to military planners.

This experiment does have certain limitations. For example, terrorists are not necessarily motivated in the same way as the participants of this experiment, and people in a terrorist cell may be more or less reliable than the participants in this experiment. Also, real terrorists could have access to different ways of communicating with one another, and may have different schedules and life circumstances than the participants. These factors and others may lead to different results in a real terrorist cell. However, no matter how different terrorists' circumstances may be from the circumstances of the participants, it is still true that in any network involving humans there are built-in circumstances that mathematical models may not fully encompass, creating a potential for error in the models. Though the difference between a mathematical model and real-life probability may not have been exactly 7 percent if the experiment had used a real terrorist cell, it is important to remember that real terrorist cells also have other dangers and possible causes for error that are not built into this simulation. Therefore, the circumstantial differences do impose limitations on this experiment, but do not invalidate the conclusions.

Though this simulation's results are slightly different from the mathematical model's projections, the differences can be attributed to two sources: random sampling error, making the differences not statistically significant, and human error. One could either adopt the interpretation that the results imply that there is no difference between the mathematical model and true results, given that the difference between the calculated sample proportion and the true proportion was not statistically significant, or one could also, in looking at the sources of error in the data, conclude that some of the differences between the experimental results and the predicted results derive from human and mechanical error implicit to true systems. With either interpretation, we see that the mathematical model is still useful, for either interpretation implies that an action against a terrorist cell is at least as effective in real life as it is according to the calculation. If government officials plan strategies based on mathematical projections of effects of actions on terrorist cells, their actions may be just as effective or even more effective than they predict. Therefore, by using the mathematical model as a basis for decision-making, the government can ensure that its actions will most likely be at least as effective as the projections imply.

There are many possible expansions to this experiment. It would be useful to create a closer simulation – perhaps gather volunteers who are of closer age and personality type to real terrorists, and to more closely simulate a real terrorist cell: put volunteers through a simulated "training," put them in situations similar to the situations terrorists may be in, give them "missions" to complete, and instead of just seeing if the messages are passed on through the chains, see if they are able to complete the missions. Also, it could be beneficial to simulate terrorist cells more closely in terms of the amount of time it takes to transmit messages and the method of passing messages.

There are other factors that could be included into a simulation that we do not take into account. We should more closely simulate what strategies the U.S. employs in pursuing terrorist cells. For example, the following questions could serve as a template for more intricate expansions of this experiment:

- Does the U.S. government tend to have more knowledge about, and spend more energy going after, leaders, middlemen, or foot soldiers of terrorist cells?
- How many members of a terrorist cell is the U.S. government usually able to know about and remove from a terrorist cell, on average?
- Is it likely that the U.S. will be able to remove most or all of the members of a cell, half of the members, or less than half of the members? Is there an average range: say, does the U.S. usually have the ability to remove at least an eighth of a terrorist cell's members, and at most, for example, three-fourths of its members?

A simulation created using the answers to these questions could better model the way that the U.S. government fights terrorist cells, and it would then be possible to make more accurate conclusions about the effects of the government's actions against terrorist cells. The results of this experiment also suggest that the temporal component of terrorist operations increases the probability that an action against a terrorist cell disables the cell. Is it possible to analytically show the effect of this temporal component and estimate the mathematical effect of this temporal component on the U.S. government's ability to dismantle these cells?

Simulations and experiments like this could be beneficial in the war on terrorism. It is obviously impossible and impractical for the U.S. government to carry out tests on real terrorist cells. Simulations such as this one provide insight into the operations of networks like terrorist cells beyond the insight that mathematical models can offer, and can help the U.S. to better understand how mathematical models apply to real-life situations. In essence, realistic simulations can be important partners of theoretical models in helping the U.S. government to strategize and make better-informed decisions in order to more effectively fight the war on terrorism by saving resources and surviving the war with fewer casualties.

Acknowledgments

The author would like to acknowledge BaoTran, Brigitte, Caitlyn, Chelsea, Chris, Coral, Erica, Isabelle, Jake, Jennie, Kerri, Lisa D., Lisa S., Victoria and Winnie for all of their hard work and voluntary participation. Also, the author would like to thank G. Pelushi for reading this paper, and Professor Jonathan Farley for his help and guidance with research, along with proofreading this paper and providing comments.

Appendix

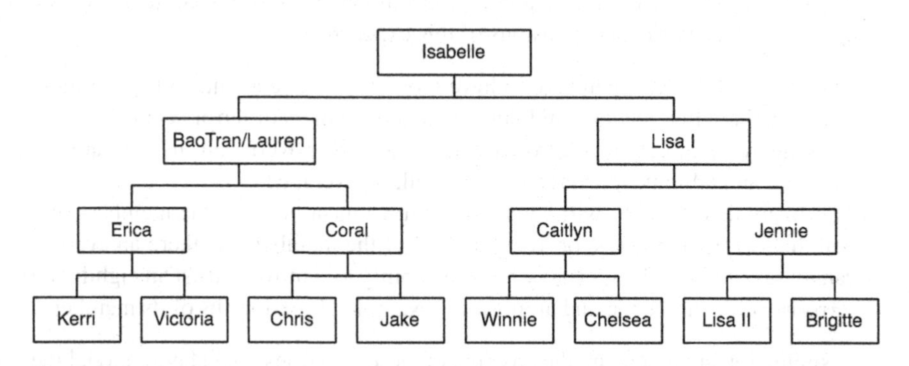

Fig. 1 Chart showing names of participants, and with whom each participant communicated in a binary tree structure.

References

1. Davey, B. A., Priestley, H. A.: Introduction to Lattices and Order, Second Edition. Cambridge University Press, Cambridge (2002)
2. Farley, J. D.: Breaking Al Qaeda Cells: A Mathematical Analysis of Counterterrorism Operations (A Guide for Risk Assessment and Decision Making). Studies in Conflict and Terrorism. 26, 399–411 (2003)

Part V
Game Theory

A Brinkmanship Game Theory Model of Terrorism

Francois Melese

Abstract This study reveals conditions under which a world leader might credibly issue a brinkmanship threat of preemptive action to deter sovereign states or transnational terrorist organizations from acquiring weapons of mass destruction (WMD). The model consists of two players: the United Nations (UN) "Principal," and a terrorist organization "Agent." The challenge in issuing a brinkmanship threat is that it needs to be sufficiently unpleasant to deter terrorists from acquiring WMD, while not being so repugnant to those that must carry it out that they would refuse to do so. Two "credibility constraints" are derived. The first relates to the unknown terrorist type (Hard or Soft), and the second to acceptable risks ("blowback") to the World community. Graphing the incentive-compatible Nash equilibrium solutions reveals when a brinkmanship threat is credible, and when it is not – either too weak to be effective, or unacceptably dangerous to the World community.

> Brinkmanship: "The practice, especially in international politics, of seeking advantage by creating the impression that one is willing and able to push a highly dangerous situation to the limit rather than concede." – American Heritage Dictionary (4th Ed. 2000)

1 Introduction

A nightmare scenario in the war on terror is that terrorists unleash chemical, biological or nuclear weapons against innocent civilians. This paper examines conditions under which a world leader such as the Secretary General (or Security Council) of the United Nations (UN) might credibly issue a brinkmanship threat of preemptive

Francois Melese
Defense Resources Management Institute (DRMI), Naval Postgraduate School, Monterey, CA, USA, e-mail: fmelese@nps.edu

N. Memon et al. (eds.), *Mathematical Methods in Counterterrorism*,
DOI 10.1007/978-3-211-09442-6_19,

action to deter sovereign states or transnational terrorist organizations from acquiring weapons of mass destruction (WMD).[1]

A historical precedent and classic illustration of a brinkmanship game is the Cuban Missile Crisis (see[2], [3], [5], [7], [8]). In 1962 the USSR began to install nuclear ballistic missiles in Cuba pointed at the U.S. Swift reaction by the U.S. nearly led to an all out nuclear war. The brinkmanship strategy played (explicitly or implicitly) by President Kennedy involved exposing the USSR, and the U.S., to a gradually increasing risk of mutual harm (Mutually Assured Destruction or "MAD"), partly outside the President's control.

As soon as the missile sites were discovered, one of Kennedy's top military advisors (Air Force General Curtis Lemay) recommended bombing Cuba. But Kennedy held back. Instead, the President decided to blockade Cuba to prevent any further shipments of arms and equipment from the USSR. He simultaneously issued a credible threat to Premier Krushev. If the USSR did not withdraw its missiles from Cuba, the President indicated there was a serious risk he would face such intense political pressure from his military advisors and the U.S. public that he would be forced to bomb the Cuban missile sites. Understanding the threat of nuclear war implied in the brinkmanship game being played by the U.S. – combined with a face-saving gesture from the U.S. to withdraw its missiles from Turkey – the USSR eventually backed down. Although risking a nuclear holocaust, Kennedy's brinkmanship strategy was ultimately successful in deterring the USSR.[2]

Today the world faces a different threat: that transnational terrorist organizations or state actors will acquire WMD and deploy them against innocent civilians and opportunistic military targets. Much like the brinkmanship game played by President Kennedy during the Cuban Missile Crisis, this paper considers the possibility the UN might issue a credible, probabilistic threat to deter the use of WMD by state and non-state (terrorist) actors.[3]

While President Kennedy leveraged the aggressive stance of General Curtis Lemay (his "loose cannon") to establish credibility in issuing his threat to the USSR, the UN Secretary General could point to the credible threat of U.S. retaliation and its policy of preemption in the war against terror. The UN Secretary General could issue the following warning: "Acquire WMD and all bets are off." Specifically, the UN Secretary General could warn that if it is discovered that a terrorist organization has acquired WMD the Security Council will pass a binding resolution condemning that action and open the way for a "coalition of the willing" to engage in swift and

[1] The generic term "terrorist" will be used for both state and non-state actors that acquire WMD. [9]

[2] It was later learned Cuba had an agreement with the USSR that provided for retaliation in the event the U.S. attacked Cuba.

[3] Suppose an outright threat to eliminate terrorists that acquire WMD is too dangerous to be tolerable to the world community, say due to the likelihood of significant collateral damage to innocent civilians. Then this threat can be reduced, yet still remain credible, by creating a probability rather than a certainty terrorists will suffer dire consequences if they do not comply.

harsh "preemptive retaliation" – regardless of how "justified" the terrorists' cause might be.[4]

What is clear is that any attempted preemptive move against elusive terrorist targets is risky (especially against those with newly acquired WMD capability). A credible threat of preemptive retaliation against terrorists having acquired WMD is likely to weigh heavily on terrorist organizations. Preemptive actions will have very bad consequences (negative payoffs) for the terrorist organization, but those retaliating are also clearly at risk, as are innocent civilians. In fact, in the event of substantial collateral damage to innocent civilians the blame could ultimately shift onto those engaged in retaliation (for example, onto the United States and its "coalition of the willing"). Nobody is likely to escape harm.

A Brinkmanship strategy always involves a probabilistic strategic action (preemptive retaliation in this case) that has a mutually harmful outcome. Retaliation might be bad for the terrorists, but it is also bad for those that carry out the threat. The objective is to make the brinkmanship threat a sufficiently credible and unpleasant option that it deters terrorists from using WMD, while not being so catastrophic and repugnant to those that would have to carry it out that they would refuse to do so. The key is to identify conditions under which issuing a brinkmanship threat is "incentive compatible" ... in the sense that it aligns interests of a Terrorist (Agent) with the UN (Principal) – namely to renounce any attempts to acquire and use WMD.

Given the probabilistic nature (say $0 \leq q \leq 1$) of the threat of retaliation, there is still a small probability $(1-q)$ the UN Security Council might not pass a resolution condemning the WMD attack and/or that the U.S. and its allies would not engage in a worldwide crackdown on the perpetrators.[5] To complete the brinkmanship strategy, the UN Secretary General needs to guarantee that if a terrorist organization complies and renounces the use of WMD, world condemnation of their actions would not be as severe, and (say for example in the case of Palestinian organizations) that the UN would continue to play a role in addressing and negotiating reasonable grievances that underlie their actions. According to Bruce Bueno de Mesquita of Stanford's Hoover Institute:

> [I]t is a mistake to view all terrorist organizations as being alike. I see two types of terrorists: One type I refer to as "true believers" **[HARD]** and the other as "relatively reasonable terrorists" **[SOFT]** ... The first steadfastly pursues a goal and has no interest in compromises or concessions ... Relatively reasonable terrorists, in contrast, are interested in bringing

---

[4] The term "preemptive retaliation" will occasionally be reduced to simply "preemption" or "retaliation" in what follows. The term is meant to indicate retaliation against the acquisition of WMD and preemption to prevent the use of WMD. A further refinement of the model might include all of the many steps involved in issuing and implementing a credible, internationally sanctioned threat of preemptive retaliation. For instance, faced with evidence a terrorist group has acquired WMD, what is the probability the Security Council passes a resolution (i. e., avoids a veto) that sanctions preemptive retaliation?

[5] This could be modeled more explicitly, but would add complexity and thus will be saved as a future refinement of the model. For example, including the probability the UN Security Council passes a resolution that sanctions preemption might be combined with the probability member countries can build a "coalition of the willing" to enforce that resolution.

their real or perceived plight to the attention of others; they seek greater understanding and hope for concessions but may accept a resolution ... that is far short of all-out victory.[6]

A key challenge in brinkmanship is to know whether we are engaging a HARD or SOFT terrorist type. The basic outline of the brinkmanship model is described in Table 1.

Table 1 The Brinkmanship Game

Sequential (Extensive Form) Two Player Game

1. Terrorist (agent)
   - Unknown Type
   - Nature determines: **Hard** (probability $= p$) or **Soft** (probability $= 1 - p$)
2. UN (principal)
   - Brinkmanship Decision: Threaten Preemption or Not Threaten
3. Terrorists (agent)
   - WMD Decision: Defy (Pursue WMD) or Comply (Not Pursue WMD)
4. UN (principal)
   - Brinkmanship: If UN does Threaten Preemption and Terrorists Defy $\Rightarrow$ Preemption will occur with probability $= q$, or not with probability $= 1 - q$

The next section describes the model in detail and illustrates the extensive form of the brinkmanship game. Section 3 derives two incentive compatibility or "credibility constraints" that must be satisfied for the UN to adopt a brinkmanship strategy. The first "Effectiveness Constraint" relates to the type of terrorist organization one is up against (Hard or Soft), and the second "Acceptability Constraint" relates to risks the World might be willing to take with a policy of preemption. Section 4 offers a graphical solution and interpretation of the results. This reveals when a brinkmanship threat of preemption is credible (a set of incentive compatible Nash equilibrium solutions), and when it is not. We conclude with some policy guidance and recommendations for future research.

2 The Extensive Form of the Brinkmanship Game

Consider two players: (1) the World as represented by the UN (the "Principal" in this game), and (2) a representative Terrorist organization or rogue state (the "Agent"). The UN's objective is to identify conditions under which a brinkmanship threat

[6] http://www.hooverdigest.org/021/bdm.html 6/27/06

(much like an incentive-based contract) is incentive compatible, in the sense that it aligns the interest of the Terrorist (Agent) with the rest of the World (the Principal) – namely, to renounce the use of WMD. In reality, the UN does not know the type of terrorist organization it faces. Following Harsanyi [6], the UN views the terrorist's type as randomly determined by "Nature," drawn from a distribution of types that are common knowledge. For ease of exposition we assume two types: "Hard" and "Soft." Since the UN/World does not know which player it is up against, based upon its best information, it assigns a probability, $0 \le p \le 1$, it is playing against a "Hard" terrorist and the probability, $(1-p)$, it is playing against a "Soft" terrorist. Figure 1 illustrates the payoffs for each terrorist type (Hard and Soft), and the corresponding payoffs for the World community. This represents the extensive form of a non-cooperative brinkmanship game under incomplete information (see [4], [10]).

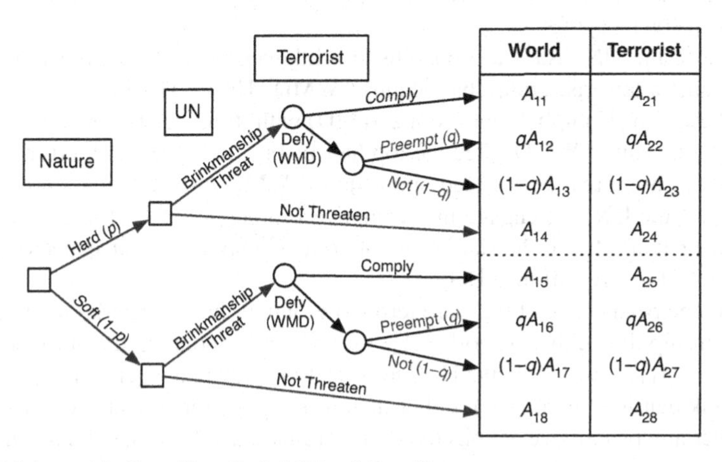

Fig. 1 Brinkmanship Game Tree: Probabilities & Payoffs

The distinction between Hard and Soft terrorists arises in terms of how each organization assesses its payoffs for various outcomes of the game. Hard terrorists are assumed to have a more optimistic view of their expected return from defying a brinkmanship threat than Soft terrorists. Consequently, a Hard terrorist group is more difficult to deter.

Given that "Nature" determines the type of terrorist it faces, the UN must decide whether or not to adopt a brinkmanship strategy (see Fig. 1). The key challenge for the UN in issuing a brinkmanship threat is that it must be credible. As Schelling [11] emphasizes: "Talk is not a substitute for moves ... Talk can be cheap when moves are not." (p. 117)

A threat aimed at deterrence will fail if it is not credible. Two "credibility constraints" define the equilibrium solution in this game: an "Effectiveness constraint" and an "Acceptability constraint." (In a different context, see [3])

To be "effective" the threatened risk of retaliation, $0 \le q \le 1$, must be high enough to at least deter the Soft terrorist from acquiring WMD, and yet still small enough (since retaliation involves a mutually harmful outcome) to be "acceptable" to the World community. Derived in Sect. 3, these two credibility (or incentive compatibility) constraints define a unique set of Nash equilibria – a restricted set of conditions under which the UN could credibly issue a brinkmanship threat of preemptive action to deter a terrorist organization from acquiring WMD.

If the UN adopts a brinkmanship strategy, then from Fig. 1 terrorists will look ahead knowing that if they acquire WMD they face a probability, q, of preemption. Expected payoffs for the terrorists depend partly on their choices to Comply or Defy, and partly on whether they are Hard or Soft (and only they know their own type).

Deriving a "Rollback Nash Equilibrium" solution of the game through backwards induction requires the UN to first focus on the Terrorist group's payoffs for each possible outcome. These are given by the far column in Fig. 1. After the unpleasantness of placing itself in the terrorist's shoes and establishing each terrorist type's best response to a brinkmanship threat ("Comply" or "Defy"), then the UN/World must evaluate its own payoffs.

Suppose the UN issues a probabilistic brinkmanship threat of preemptive action against a terrorist group that acquires WMD. Then a Hard terrorist must decide whether to "Comply" (not acquire WMD) with its associated payoff A_{21}, or to "Defy" and acquire WMD. (See the top half of Fig. 1) If the Hard terrorist under a brinkmanship threat goes ahead and acquires WMD, then there is a probability, $0 \leq q \leq 1$, the UN will engage in preemptive action, and $(1-q)$ it will not. This leads to a payoff $[A_{21}]$ if the Hard terrorist group Complies, and an expected payoff of $[qA_{22} + (1-q)A_{23}]$ if they Defy.

From the perspective of the Hard terrorist group, if the UN decides not to issue a brinkmanship threat, their payoff is A_{24} (and the corresponding threat to the world is given by A_{14}).[7] Assuming the Hard terrorist type prefers to operate in the absence of a brinkmanship threat (A_{24}), rather than to Defy the threat and "get lucky" and not suffer any preemptive actions (A_{23}), the ordinal ranking of payoffs for the Hard terrorist type is $A_{24} \geq A_{23} > A_{22} > A_{21}$.

The situation is analogous for a Soft terrorist type. Their associated payoffs appear in the bottom half of Fig. 1. Given a brinkmanship threat, Soft can choose to "Comply" (not acquire WMD) with a payoff A_{25}, or to "Defy" and acquire WMD. If the Soft terrorist type acquires WMD, there is a probability, $0 \leq q \leq 1$, the World will engage in preemptive actions and $(1-q)$ it will not. This leaves the Soft terrorist with an expected payoff $[qA_{26} + (1-q)A_{27}]$ if they Defy a brinkmanship threat. Of course, the best outcome is not to face a brinkmanship threat of preemption (A_{28}).

[7] Payoffs in the game must reflect the preferences (utility functions) of the decision makers involved to represent real world terrorism. The Nash equilibrium brinkmanship game solution derived in Sect. 4 mostly depends on ordinal not cardinal payoff values, and thus is largely independent of differences in magnitudes of payoffs across players. The magnitudes of the payoffs are not as important in determining the solution. Instead, the solution is sensitive to the order of a player's payoffs across the various possible outcomes. Some simplifying assumptions regarding relative magnitudes serve to increase tractability. This initial set of assumptions is meant to reasonably represent the order and (where possible) comparative magnitudes of the preferences of the UN and the terrorists. Section 5 explores the sensitivity of the solution to changes in the magnitudes and ordering of the payoffs, and in values of the probabilities q and p.

What distinguishes the Soft terrorist type from the Hard terrorist type is a reversal in how they rank their two worst outcomes – facing down a brinkmanship threat of preemption. A Hard terrorist would rather Defy the brinkmanship threat of preemption (pursue WMD) and suffer the possibility (with probability $0 \leq q \leq 1$) of Preemption, than to Comply and agree to renounce WMD (i. e., $A_{22} > A_{21}$). The reverse is true for the Soft terrorist. A Soft terrorist would rather Comply under a Brinkmanship Threat (i. e., not pursue WMD), than to Defy the brinkmanship threat and suffer the possibility of Preemption ($A_{25} > A_{26}$) (see Fig. 1).

Assuming the Soft terrorist prefers to operate in the absence of a UN threat (A_{28}), rather than to Defy the threat and "get lucky" and not suffer any preemptive actions (A_{27}), the ordinal ranking of payoffs for the Soft terrorist type is $A_{28} \geq A_{27} > A_{25} > A_{26}$. Thus the ordinal preference ranking for the Hard terrorist type's payoffs is

$$A_{24} > A_{23} > A_{22} > A_{21}, \tag{1}$$

whereas the Soft terrorist type's ordinal preference ranking is

$$A_{28} > A_{27} > A_{25} > A_{26}. \tag{2}$$

Conversely, from the perspective of the World community, the best outcome is for terrorists to Comply with a brinkmanship threat and renounce WMD ($A_{11}; A_{15}$) (see Fig. 1). The next worse outcome for the World community is for terrorists to retain their option to acquire WMD in the absence of a brinkmanship threat ($A_{14} < A_{11}; A_{18} < A_{15}$), followed by terrorists acquiring WMD without Preemption ($A_{13} < A_{14} < A_{11}; A_{17} < A_{18} < A_{15}$). The worst outcome is for terrorists to acquire WMD with preemption (possibly inducing terrorists to use WMD). From best to worst, the World's ordinal preference ranking of outcomes against a Hard terrorist type is

$$A_{11} > A_{14} \geq A_{13} > A_{12}, \tag{3}$$

and similarly against a Soft terrorist type is

$$A_{15} > A_{18} \geq A_{17} > A_{16}. \tag{4}$$

3 Incentive Compatibility ("Credibility") Constraints

The challenge in issuing a brinkmanship threat is that it must satisfy two "credibility constraints" – an "Effectiveness constraint" and an "Acceptability constraint." The goal is to make the probabilistic brinkmanship threat of preemption sufficiently credible and unpleasant that it deters terrorists from obtaining WMD (*Effectiveness constraint*), but not so catastrophic and repugnant to those that need to carry it out that they would refuse to exercise that option (*Acceptability constraint*).

To be "effective" the threatened risk of preemption, $0 \leq q \leq 1$, must be high enough to at least deter the Soft terrorist from acquiring WMD, and yet still small enough (since preemption involves a mutually harmful outcome) to be "acceptable" to the World community. These two incentive compatibility constraints define a set of Nash equilibrium solutions under which the UN could credibly issue a brinkmanship threat to deter terrorists from using WMD. Alternatively, it establishes the region in which a brinkmanship strategy is not a credible option, and where other actions need to be considered.

3.1 The Effectiveness Constraint

In order for a probabilistic brinkmanship threat to be "Effective," it must at least deter Soft Terrorists from acquiring WMD. In other words, the probability of preemption needs to be sufficiently high that it at least motivates Soft Terrorists to Comply. The first step is to derive the minimum threat probability, q^*, required to deter a Soft terrorist type. This is the lower bound of $q \in [0,1]$ for a threat to be effective. A brinkmanship threat of preemption with probability $1 \geq q \geq q^* > 0$ of being carried out will induce a Soft terrorist type to comply (and renounce attempts to acquire WMD) if and only if their expected value of Complying is greater than Defying, or iff:

$$E[\text{Soft Comply}] \geq E[\text{Soft Defy}] \Leftrightarrow A_{25} \geq qA_{26} + (1-q)A_{27} \tag{5}$$

From Fig. 1, the rankings in (2), and Equation (5), the set of "Effective" brinkmanship preemption threat probabilities that would deter a Soft terrorist type is given by:

$$1 \geq q \geq (A_{25} - A_{27})/(A_{26} - A_{27}) = q^* > 0. \tag{6}$$

For each possible player type (Soft or Hard), the expected return from defying the brinkmanship threat rather than complying and not acquiring WMD, is jointly determined by the threat probability and the payoffs. The inequality in (6) represents the range of threat probabilities of preemption that would be effective in deterring a Soft terrorist type. The value $q*$ represents the lower bound of preemption probabilities for a brinkmanship threat to be effective.

Uncovering the "Rollback Nash Equilibrium" conditions for the UN to issue a probabilistic brinkmanship threat involves simple backwards induction. First the UN must consider the Soft Terrorist type's decision whether to Comply or Defy faced with a brinkmanship threat given by (6). It is clear from Fig. 1, the ordinal payoff rankings given by (2), and condition (6), that the expected value of Complying for Soft is greater than Defying. Conversely, given the ordinal payoff rankings (1), Hard will Defy since

$$A_{21} < qA_{22} + (1-q)A_{23} \Leftrightarrow E[\text{Hard Comply}] < E[\text{Hard Defy}]. \tag{7}$$

(Note that, from (1) and (7), if $q = 1$, $A_{21} < A_{22}$ and if $q = 0$, $A_{21} < A_{23}$ so that $E[\text{Hard Comply}] < E[\text{Hard Defy}]$ for all $1 \geq q \geq (A_{25} - A_{27})/(A_{26} - A_{27}) = q^* > 0$)

The result from the perspective of the UN/World Community is that, given these payoffs, any probabilistic brinkmanship threat sufficient to deter a Soft terrorist type, will not deter the Hard terrorist type. So in Fig. 1, although a necessary condition for the UN to issue a brinkmanship threat of preemption is that the probability it is implemented is in the range $1 \geq q \geq q^* > 0$ (given by (6)), this is not sufficient. While Soft will Comply, Hard will Defy. So the decision that faces the UN of whether or not to issue a brinkmanship threat also rests on the probability the World is facing a Hard terrorist. Intuitively, a brinkmanship threat of preemption, with probability $1 \geq q \geq q^* > 0$ of being executed, will only be issued if there is a sufficiently low risk the World is facing a Hard terrorist type.

Given the best information available, the UN estimates it is facing a Hard terrorist type with probability, $p \in [0,1]$, and Soft with probability, $(1-p)$. The UN/World is now in a position to evaluate its expected payoffs if it issues a brinkmanship threat, E_{UN}[Threat], compared to its expected payoff if it does not, E_{UN}[No Threat]. A brinkmanship threat will only be issued as long as $E_{UN}[\text{Threat}] \geq E_{UN}[\text{No Threat}]$.

It was previously established that, for any brinkmanship threat probability in the range given by (6), Soft will Comply but Hard will Defy. Therefore, a UN brinkmanship threat is expected to result in World payoffs,

$$\begin{aligned} E_{UN}[\text{Threat}] &= pE_{UN}[\text{Hard Defy}] + (1-p)E_{UN}[\text{Soft Comply}] \\ &= p[qA_{12} + (1-q)A_{13}] + (1-p)A_{15}. \end{aligned} \tag{8}$$

Meanwhile, the No Threat option yields expected World payoffs of

$$E_{UN}[\text{No Threat}] = pA_{14} + (1-p)A_{18}. \tag{9}$$

The condition under which the UN would issue a brinkmanship threat is if $E_{UN}[\text{Threat}] \geq E_{UN}[\text{No Threat}]$. To issue a credible threat, the expected value of issuing the threat, (8), must be greater than the expected value of not issuing the threat, (9).

The question is how high the UN can set a brinkmanship threat probability of preemption (q) that is still tolerable to the World community? Since issuing a brinkmanship threat involves a very real risk of mutual harm, the higher the probability the World is facing a Hard terrorist type, p, the lower the acceptable threat probability, q. Conversely, the higher the probability of a Soft terrorist, $(1-p)$, the higher the acceptable threat probability, q. This suggests a relationship between the probability of preemption, q, and the probability of playing against a Hard terrorist type, p, or $q = f(p)$. This relationship defines the second incentive compatibility constraint under which the UN could credibly use a brinkmanship strategy to deter terrorists from using WMD – the "Acceptability Constraint."

3.2 The Acceptability Constraint

A brinkmanship threat is not credible if it is unacceptable to the party making the threat. The question is how high the brinkmanship threat probability of preemption with its possibly grim consequences can be set and still remain tolerable to the World community. Of course this depends to a large extent on the type of terrorist faced by the UN. Thus the "Acceptability constraint" involves deriving the relationship, $q = f(p)$, such that the expected value to the UN of issuing a brinkmanship threat of preemption, E_{UN}[Threat], is greater than not to threaten brinkmanship, E_{UN}[No Threat], or from (8) and (9),

$$p[qA_{12} + (1-q)A_{13}] + (1-p)A_{15} \geq pA_{14} + (1-p)A_{18}. \quad (10)$$

Assuming the cost to the World if a Hard terrorist type defies a brinkmanship threat and preemption does not occur, A_{13}, is the same as if no threat were issued, A_{14} (i. e., $A_{13} = A_{14}$), then expression (10) yields the Acceptability constraint,

$$q \leq f(p) = [(1-p)/p]z, \text{ where: } \quad 1 > z = (A_{18} - A_{15})/(A_{12} - A_{13}) > 0;$$
$$\text{and where: } q = f(p) : dq/dp < 0; d^2q/dp^2 > 0. \quad (11)$$

Combined with the Effectiveness constraint ((5) and (6)), the Acceptability constraint function given by (11) reveals conditions under which the UN can credibly issue a probabilistic brinkmanship threat of preemption – i. e., such that E_{UN}[Threat] $\geq E_{UN}$[No Threat].

4 Equilibrium Solution and Interpretation of the Results

If the UN is certain it is dealing with a Hard terrorist (i. e., $p = 1$), then from (11), no probabilistic brinkmanship threat is acceptable (i. e., $q = 0$). A brinkmanship threat in the face of a Hard terrorist is not a viable option. Figure 2 offers a graphical illustration and interpretation of the results. In graphing the solution with the probability of a Hard terrorist group (p) on the abscissa (x-axis), and the brinkmanship threat probability of preemption (q) on the ordinate (y-axis), the point $(1,0)$ defines the x-intercept of the Acceptability constraint.

To anchor the top limit of the Acceptability constraint requires deriving from (11) how low the probability of facing a Hard terrorist needs to be ($p = p^*$) in order for the UN to issue a dire threat (i. e., with certainty, $q = 1$, that it will be carried out). Letting $q = 1$ in (11) and solving for p yields,

$$p^* = 1/(1+z); \quad \text{where: } 1 > z = (A_{18} - A_{15})/(A_{12} - A_{13}) > 0 \quad (12)$$

The point $(p^*, 1)$ anchors the top limit of the Acceptability constraint which, from (11), slopes down at a decreasing rate as illustrated in Fig. 2.

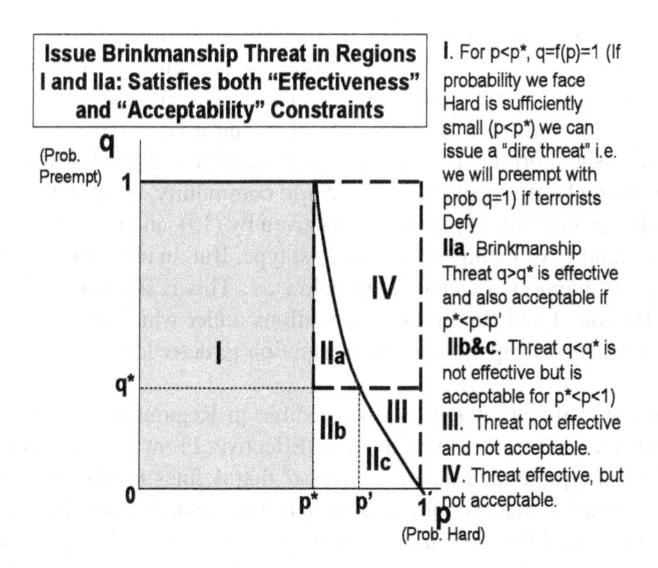
Issue Brinkmanship Threat in Regions I and IIa: Satisfies both "Effectiveness" and "Acceptability" Constraints
q
(Prob. Preempt)
1
q*
0
p*
p'
1
p
(Prob. Hard)
I
IIa
IIb
IIc
III
IV
I. For p<p*, q=f(p)=1 (If probability we face Hard is sufficiently small (p<p*) we can issue a "dire threat" i.e. we will preempt with prob q=1) if terrorists Defy
IIa. Brinkmanship Threat q>q* is effective and also acceptable if p*<p<p'
IIb&c. Threat q<q* is not effective but is acceptable for p*<p<1)
III. Threat not effective and not acceptable.
IV. Threat effective, but not acceptable.

Fig. 2 Brinkmanship Game Solution Regions

The Effectiveness constraint ((5) and (6)) requires a minimum threshold brinkmanship probability of preemption, $q^* = (A_{25} - A_{27})/(A_{26} - A_{27})$, to deter a Soft terrorist type. Substituting q^* into (11) yields the maximum associated probability of facing a Hard terrorist compatible with this threshold brinkmanship threat, or

$$p' = 1/(q^* + z) > p^*; \quad \text{where: } 1 > z = (A_{18} - A_{15})/(A_{12} - A_{13}) > 0. \tag{13}$$

The Nash Equilibrium UN strategy set of brinkmanship threat probabilities of preemption, q, are reported here and interpreted graphically in Fig. 2 in terms of the escalating probability, p, the World is facing a Hard terrorist type:

$$\text{For all } \; p \in [0, p^*], \quad q = 1 \tag{14}$$

$$\text{For all } \; p \in [p^*, p'], \quad q^* \leq q \leq f(p) \tag{15}$$

$$= [(1-p)/p][(A_{18} - A_{15})/(A_{12} - A_{13})] \leq 1$$

These results reveal conditions under which a brinkmanship threat of preemption is credible – i. e., a set of incentive compatible Nash equilibrium solutions that satisfy both the "Effectiveness" and "Acceptability" constraints. The actual region in which the UN operates is driven by its assessment of the probability the World is facing a Hard terrorist type.

The Nash Equilibrium strategy set expressed in (14) and (15) is illustrated as Regions I and IIa in Fig. 2. From (14), given a sufficiently low risk it faces a Hard terrorist type, $0 < p < p^*$, the UN can issue a "dire threat." That is, it can promise preemption with certainty ($q = 1$) if a terrorist group acquires WMD. This is illustrated as Region I in Fig. 2. However, as the risk of facing a Hard terrorist grows beyond probability, p^*, the equilibrium threat strategy changes.

Given the increased risk, $p^* < p < p'$, of facing a Hard terrorist type, the UN can credibly issue a brinkmanship threat, but not a "dire threat," since it would be too dangerous and not acceptable to the World community. Instead, the equilibrium brinkmanship probability of preemption is given by (15), and is inversely correlated with the probability of facing a Hard terrorist type. But, in order to be Effective, the probability of preemption cannot drop below $q*$. This is illustrated as Region IIa in Fig. 2. Regions I and IIa represent conditions under which the UN could credibly adopt a brinkmanship strategy of preemption to deter terrorists from acquiring WMD.

Brinkmanship threats would not be credible in Regions IIb and IIc in Fig. 2, since while Acceptable, they would not be Effective. From (15), any risk of facing a Hard terrorist type in the range $p^* < p < p'$ that defines Region IIb, corresponds to a brinkmanship probability of preemption below the minimum effective level, q^*. Similarly, any risk of facing a Hard terrorist type in the range $p' < p < 1$ that defines Region IIc, corresponds to a brinkmanship probability of preemption below q^* (see Fig. 2).

A brinkmanship threat would also not be credible if the UN found itself in Regions III and IV of Fig. 2. In Region III a brinkmanship threat is neither Acceptable nor Effective, while in Region IV, it is Effective, but not Acceptable.

In summary, Regions I and IIa represent the only two sets of conditions under which the UN could credibly issue a brinkmanship threat of preemption to deter terrorists from acquiring WMD. Outside these two regions, the UN must adopt other strategies to deter terrorist from acquiring WMD since a brinkmanship threat is either too weak to be effective, or unacceptably dangerous to the World community.

5 Conclusion

This paper examines conditions under which a world leader such as the Secretary General (or Security Council) of the United Nations (UN) might credibly issue a brinkmanship threat of preemptive action to deter sovereign states or transnational terrorist organizations from acquiring weapons of mass destruction (WMD). The model consists of two players, the United Nations (UN or World) "Principal," and a representative terrorist organization "Agent." The goal is to examine conditions under which the Principal might be able to structure a brinkmanship threat of preemption that aligns the interests of the agent with that of the principal – notably to Comply with a ban on the acquisition of WMD.

The UN (Principal) does not know which terrorist type it faces: Hard or Soft. Therefore, in choosing its brinkmanship strategy it must make a subjective assessment of the probability it is playing against a Hard terrorist type, say p, and the probability it is playing against a Soft terrorist type, $(1 - p)$. In game theory these two states of nature are captured conceptually as Nature making the first move "choosing" whether the Principal is playing against a Hard or Soft terrorist organization. In the next stage the Principal (UN) must decide whether or not to issue a Brinkmanship threat. If the Principal issues a brinkmanship threat, then the Agent (Terrorist organization) must decide whether to "Comply" and refrain from using WMD, or to "Defy" and acquire WMD. A straightforward extensive form game tree is developed to illustrate the possible outcomes.

Except in the case of a "dire threat," Brinkmanship always leaves something to chance. A Brinkmanship strategy always involves a probabilistic strategic action (preemptive retaliation in this case) that has a mutually harmful outcome. Preemption might be bad for the terrorists, but it is also bad for those that carry out the threat. The goal is to make the brinkmanship threat sufficiently effective and unpleasant that it deters terrorists from using WMD, while not being so catastrophic and repugnant to those that must carry it out that they would refuse to do so.

Two incentive compatibility or "credibility constraints" are derived that must be satisfied for the UN to adopt a brinkmanship strategy. The first "Effectiveness Constraint" relates to the type of terrorist the World is up against (Hard or Soft), and the second "Acceptability Constraint" relates to risks the World is willing to take with a brinkmanship policy of preemption. The equilibrium solution and graphical interpretation of the results reveal when a brinkmanship threat of preemption is credible (a set of incentive compatible Nash equilibrium solutions), and when it is not.

Two sets of conditions are derived under which the UN could credibly issue a brinkmanship threat of preemption to deter terrorists from acquiring WMD. Outside these two regions, the UN must adopt other strategies to deter terrorist from acquiring WMD, since a brinkmanship threat would either be too weak to be effective, or unacceptably dangerous to the World community.

Hazy information and imprecise control can generate large risks in this model. Future research might examine the uncertainty that arises in identifying the probability the World faces a Hard terrorist type, p, and the uncertainty associated with the UN being able to stick to a brinkmanship probability of preemption, q. Since, in reality, p and q are uncertain, a Monte Carlo simulation might be constructed in which the actual probability that a terrorist group is Hard, and that the actual chosen probability of preemption can be assured, are drawn from probability distributions. For example, it is possible that using Beta distributions as an illustration, and specifying cardinal payoffs, a Monte Carlo simulation could reveal how likely it is the Brinkmanship strategy would succeed under different sets of assumptions about probability distributions and payoffs.

Brinkmanship involves the strategic use of a probabilistic threat. This strategy involves exposing your rival and yourself to a gradually increasing risk of mutual harm that is not entirely within your control. The challenge is to generate a threat with a risk that is large enough to be effective and yet small enough to be acceptable. Future research could reveal the risk tolerance of each player by modeling a gradual escalation of the risk of mutual harm. In this case the UN would gradually escalate the brinkmanship risk of preemption to its equilibrium level, adjusting as it learns more about the probability it faces a Hard terrorist type from signals it receives from the terrorist organization (for example, [1])

References

1. Arce, D. and T. Sandler "Terrorist Signaling and the Value of Intelligence," British Journal of Political Science 37, (2007).
2. Blight, J. and D. Welch On the Brink: Americans and Soviets Reexamine the Cuban Missile Crisis. Hill and Wang, New York (1989)
3. Dixit, A. and S. Skeath Games of Strategy. Norton, New York (1998)
4. Friedman, J., Game Theory with Applications to Economics. Oxford University Press, New York (1986)
5. Fursenko, A. and T. Naftali, One Hell of a Gamble: The Secret History of the Cuban Missile Crisis. Norton, New York (1997)
6. Harsanyi, J., "Games with Incomplete Information Played by 'Bayesian' Players, I-III. Part I. The Basic Model", Management Science, 14/3, (1967)
7. Kagan, D. On the Origins of War and the Preservation of Peace. Doubleday, New York (1995)
8. May, E. and P. Zelikow eds., The Kennedy Tapes: Inside the White House During the Cuban Missile Crisis. Harvard University Press, Cambridge, Mass.: (1997)
9. Melese, F. and D. Angelis "Deterring Terrorists from Using WMD: A Brinkmanship Strategy for the United Nations," Defense & Security Analysis 20, (2004)
10. Rasmusen, E., Games and Information. Blackwell, Cambridge Mass. (1995)
11. Schelling, T., The Strategy of Conflict. Harvard University Press, Cambridge (2002)

Strategic Analysis of Terrorism

Daniel G. Arce and Todd Sandler

Abstract Two areas that are increasingly studied in the game-theoretic literature on terrorism and counterterrorism are collective action and asymmetric information. One contribution of this chapter is a survey and extension of continuous policy models with differentiable payoff functions. In this way, policies can be characterized as strategic substitutes (e. g., proactive measures), or strategic complements (e. g., defensive measures). Mixed substitute–complement models are also introduced. We show that the efficiency of counterterror policy depends upon (i) the strategic substitutes-complements characterization, and (ii) who initiates the action. Surprisingly, in mixed-models the dichotomy between individual and collective action may disappear. A second contribution is the consideration of a signaling model where indiscriminant spectacular terrorist attacks may erode terrorists' support among its constituency, and proactive government responses can create a backlash effect in favor of terrorists. A novel equilibrium of this model reflects the well-documented ineffectiveness of terrorism in achieving its stated goals.

1 Introduction

Terrorism is a form of asymmetric conflict where terrorists do not have the resources necessary to engage their adversaries in direct conflict. Consequently, terrorists act strategically and use violence against civilians in order to attain political, ideological or religious goals. Terrorism is a tactic of intimidation of a target audience beyond

Daniel G. Arce
School of Economic, Political and Policy Sciences, University of Texas at Dallas, 800 W. Campbell Road, Richardson, TX 75080-3021, USA, e-mail: darce@utdallas.edu

Todd Sandler
Vibhooti Shukla Professor of Economics and Political Economy, School of Economic, Political and Policy Sciences, University of Texas at Dallas, 800 W. Campbell Road, Richardson, TX 75080-3021, USA, e-mail: tsandler@utdallas.edu

N. Memon et al. (eds.), *Mathematical Methods in Counterterrorism*,
DOI 10.1007/978-3-211-09442-6_20, © Springer-Verlag/Wien 2009

that of the immediate victims. Viewed from this perspective, it is natural to study terrorism and counterterrorism through the lens of game theory, where terrorists and their targets are assumed to choose their tactics rationally, and to recognize their strategic interdependence in determining the outcomes of their actions. Note that we are not assessing whether terrorists' ultimate goals are rational or desirable; the preferences of terrorists and target governments are taken as given. What game theory presumes is that, ultimately, terrorists and target governments will select tactics and policies that minimize costs, maximize damage, effectively signal their intent, or hide their capabilities, thereby reflecting a form of procedural rationality for attaining their goals.

Two areas that are increasingly studied in the literature on terrorism and counterterrorism are collective action and asymmetric information. Collective action may refer to the efficient coordination of counterterror tactics by target governments and/or the sharing of tactical information and techniques by terrorists groups. Collective action is an issue because rational action by individual governments or terrorists may be at odds with the goals of the larger group. For example, as successful proactive counterterror policies reduce the ability of a terrorist group to attack *any* target (e. g., retaliating against a state sponsor), some governments may be willing to free ride on the policies of governments that are more likely to be attacked, leading to a case of suboptimal proactive policy at the global level. By contrast, defensive counterterror policies (e. g., air marshals) may push terrorist activity elsewhere, either to target a country's citizens on foreign soil, or to target another country altogether. Such negative externalities are often overprovided. Surveys of game theoretic models of discrete choices between proactive and defensive policies are contained in [3, 16, 17]. In Sect. 2 of this chapter, we present a unifying framework when payoffs are differentiable in these choices, thereby introducing the notion of strategic complements and substitutes to assess counterterror policy.

When nations select counterterror policies without fully knowing the policies selected by their counterparts, the corresponding game is categorized as one of *imperfect information*. Consequently, Sect. 2 contains a characterization of the dichotomy of counterterror policy under imperfect information. When this assumption is relaxed – corresponding to a situation of perfect information – one of the parties can select their policy knowing that the other party will react optimally to it. Section 2 characterizes when this leader-follower framework is desirable.

Governments may also construct strategic counterterror policy without fully understanding the intent of the terrorists whom they face. Terrorists may be primarily politically motivated, in which case concessions may be an effective counterterror policy; or they may seek maximalist objectives such as completely overturning the tenets upon which a country is governed. Politically-motivated terrorists have an incentive to mimic the attacks of more militant ones if they believe that this will quickly lead to concessions. Moreover, militant terrorists may mimic the less-damaging attacks typically associated with politically-motivated terrorists if this allows them to catch the target government unprepared in future periods. Such a scenario is known as a *signaling game*, where the government has *incomplete information* about the type of terrorists whom it is facing. In Sect 3, we introduce and analyze a signaling game that is consistent with two stylized facts normally associated with terrorism. First, terrorism is rarely successful when measured against the concessions that the terrorists seek [1, 2]. Second, even in the absence of concessions, terrorists engage in campaigns that may provide future resources, depending on how the target population reacts to the government's counterterror policy.

2 Strategic Substitutes and Strategic Complements in the Study of Terrorism

Game-theoretic models use continuous choice variables for strategic adversaries to illuminate various counterterrorism policies [17, 20]. A recurrent theme is the notion of strategic substitutes or strategic complements, which we illustrate with a number of terrorism-based examples. Some notation and definitions are required. Throughout the analysis, we assume just two players, denoted by i and j and/or 1 and 2. The mathematical analysis can be easily extended to n players, while the graphical analysis necessitates a symmetry-of-players assumption to be extended beyond two players.

Player i either maximizes a payoff function, $U_i(x_i, x_j)$, or else minimizes a cost function, $C_i(x_i, x_j)$, for i and j, $i \neq j$, where x_i and x_j are the respective agent's continuous choice strategies. Without loss of generality, we express our definitions in terms of the cost-minimization scenario. Player i's best response to agent j's choice, x_j, is equal to:

$$BR_i(x_j) = \arg\min_{x_i} C_i(x_i, x_j), \tag{1}$$

while player's j's best response to agent i's choice, x_i is equal to

$$BR_j(x_i) = \arg\min_{x_j} C_j(x_i, x_j). \tag{2}$$

The best response for player i is found by solving the implicit function,

$$\frac{\partial BR_i}{\partial x_i} = \frac{\partial C_i(x_i, x_j)}{\partial x_i} = 0 \tag{3}$$

associated with Eq. (1). A similar implicit function arises from Eq. (2) and applies to the best response for player j. The following definition is essential:

Definition 1. Strategy profile (x_i^N, x_j^N) is a *Nash equilibrium* if and only if $x_i^N \in \arg\min_{x_i} C_i(x_i, x_j^N)$ and $x_j^N \in \arg\min_{x_j} C_j(x_i^N, x_j)$.

At a Nash equilibrium, each agent's choice must be a best response to the other agent's best response, so that neither agent would unilaterally want to change its choice variable, if afforded the opportunity. When the two agents' best-response curves are displayed on a diagram, the Nash equilibrium corresponds to the intersection of these best-response curves.

Given this background, we now define the essential concepts of strategic substitutes and strategic complements [5, 6].

Definition 2. Strategies x_i and x_j are *strategic substitutes* if the slopes of the best-response functions are negative: $\partial BR_i/\partial x_j < 0$ and $\partial BR_j/\partial x_i < 0$.

Definition 3. Strategies x_i and x_j *strategic complements* if the slopes of the best-response functions are positive: $\partial BR_i/\partial x_j > 0$ and $\partial BR_j/\partial x_i > 0$.

From an economic viewpoint, strategic substitutes indicate that the other agent's action can replace the need for one's own action. Efforts to free ride by cutting down on one's contributions to an activity (e. g., air pollution abatement) as others contribute is an example. In contrast, strategic complements imply that actions by one agent encourage the other agent to act. An arms race between two adversaries reflects strategic complements, as does exploitation of an open-access resource, such as an oil pool. In the latter case, each exploiter tries to pump out more of the oil before the pool runs dry – increased efforts by one merely motivates increased efforts by the other.

2.1 Proactive Counterterrorism Measures

This first example follows from the analysis of Sandler and Siqueira [19], where two nations are attacked by the same terrorist network. This scenario is descriptive of the al-Qaida network that conducts operations throughout much of the world. As such, a country's assets – its people or property – are at risk at home and abroad. With a global terrorist threat, efforts to secure one country's borders may displace the attack abroad. Proactive measures by either country to destroy the terrorists' infrastructure, to capture their operatives, or to cut off their finances will weaken the terrorist threat for both countries and secure their assets at home and abroad.

The objective function of targeted country i consists of three cost components as it chooses its level of proactive effort, θ_i, against the common terrorist threat. First, country i incurs a proactive cost of $G(\theta_i)$ where $G'(\theta_i) > 0$ and $G''(\theta_i) > 0$. Second, country i endures an expected cost from attacks at home, which equals $\pi_i l(\theta_i)$ where π_i is the likelihood of a home attack and $l(\theta_i)$ is the loss from such an attack. Proactive measures reduce these losses by weakening the terrorists so that $l'(\theta_i) < 0$. The likelihood of an attack depends on the proactive efforts in country i and j where $\partial\pi_i/\partial\theta_i < 0$, $\partial\pi_j/\partial\theta_i < 0$, and $\partial^2\pi_i/\partial\theta_i\partial\theta_j > 0$ for i, j, and $i \neq j$. Offensive antiterrorist actions in either country reduces the risk of an attack everywhere. The cross partial indicates that there is a diminishing return to effort as both countries act. Third, proactive measures limit the expected losses to country i abroad, denoted by $\pi_j v(\theta_j)$, where $v'(\theta_j) < 0$. Thus, country i must choose its proactive level, θ_i, to minimize its costs, C_i:

$$\min_{\theta_i} C_i(\theta_i, \theta_j) = G(\theta_i) + \pi_i l(\theta_i) + \pi_j v(\theta_j). \tag{4}$$

The first-order conditions of Eq. (4) implicitly defines the best response, BR_i, of country i's choice of θ_i in terms of the level of country j's proactive response, θ_j.

To establish that this proactive choice results in strategic substitutes, we apply the implicit function rule to $\partial C_i/\partial\theta_i = 0$ to find:[1]

$$\frac{\partial BR_i}{\partial\theta_j} = \frac{-l'(\theta_i)\frac{\partial\pi_i}{\partial\theta_j} - v'(\theta_j)\frac{\partial\pi_j}{\partial\theta_i} - l(\theta_i)\frac{\partial^2\pi_i}{\partial\theta_i\partial\theta_j} - v(\theta_j)\frac{\partial^2\pi_j}{\partial\theta_i\partial\theta_j}}{\partial^2 C_i/\partial\theta_i^2} < 0. \quad (5)$$

The sign of Eq. (5) holds because the denominator is positive to satisfy the second-order conditions for a minimum, while the numerator is negative. The latter follows from the negativity of all four terms, given the model's structure. Since the slope of i's best-response curve is negative and the same holds for j's best-response curve, proactive measures are strategic substitutes.

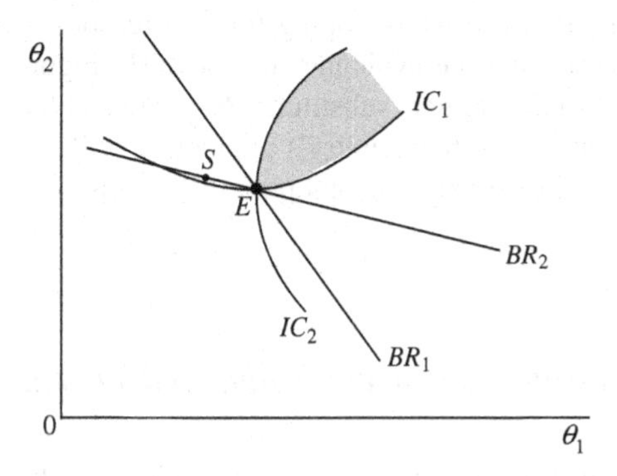

Fig. 1 Proactive measures

The downward-sloping reaction paths are illustrated in Fig. 1 for country 1 and 2. For simplicity, both paths are linearized and drawn to ensure a stable equilibrium. Reaction path BR_1 connects the minima for isocost curves of country 1 – where IC_1 is one such curve. The area above IC_1 indicates lower cost for country 1,[2] and hence greater welfare, as country 2 increases its proactive efforts for each level of θ_1. Since country 2's isocost curves are translated by 90°, 2's isocost curves are C-shaped and curves further to the east of the vertical axis denote a higher level of 2's well-being as it prospers from country 1's proactive measures. The Nash equilibrium is at E. The shaded area formed by IC_1 and IC_2 indicates welfare-improving proactive allocations by both countries, compared with the Nash equilibrium. As shown, the Nash equilibrium results in an underprovision of offensive measures, which follows because both countries fail to account for the benefits that their offensive efforts confer on the other targeted country. Next suppose that country 1 engages in leadership behavior, in which it takes the follower's (country 2's) best-response path as

[1] $\frac{\partial C_i}{\partial\theta_i} = G'(\theta_i) + l(\theta_i)\frac{\partial\pi_i}{\partial\theta_i} + \pi_i l'(\theta_i) + v(\theta_j)\frac{\partial\pi_j}{\partial\theta_i} = 0.$

[2] This follows because $\frac{\partial C_i}{\partial\theta_j} = l(\theta_i)\frac{\partial\pi_i}{\partial\theta_j} + v(\theta_j)\frac{\partial\pi_j}{\partial\theta_j} + \pi_j v'(\theta_j) < 0.$

its constraint and seeks a tangency between its isocost curve (not displayed) and BR_2 at, say, point S. At S, country 1 gains at 2's expense. Moreover, the overall level of proactive measures is even smaller than the suboptimal level at E [19]. If both countries act strategically by moving their best-response curves downwards, then both become worse off, as the underprovision of proactive measures becomes greater. Thus, game theory shows the tendency to underprovide offensive actions in a multi-target environment; hence, the need for international cooperation becomes clear.

The two-country proactive game can be extended to allow for backlash when terrorists protest proactive measures against other terrorists by launching new attacks [24]. For example, the London transport suicide bombings on 7 July 2005 supported a beleagued al-Qaida network, which had been greatly stressed since 9/11. If two countries engage in proactive responses, then the more aggressive country is anticipated to draw the backlash attack. With the introduction of backlash, the best-response curves are downward sloping for two reasons – free riding on others' proactive responses and the avoidance of backlash. Figure 1 still applies and proactive measures remain strategic substitutes. Siqueira and Sandler [24] show that strategic voting on the part of the country's population will shift the best-response curves downward and exacerbate the suboptimality resulting from free riding and backlash avoidance.

2.2 Defensive Countermeasures: Globalized Threat

Next, we turn to a situation in which each of two nations take defensive countermeasures in the hopes of shifting terrorist attacks to a different country. The desirability of this strategy depends, in part, on the country's relative interests at home and abroad. If attacks are more costly at home than abroad, then actions to transfer attacks may be especially strong. Sandler and Siqueira [19] show that the same basic cost-minimizing objective in Eq. (4) applies with some small reinterpretation. The primary difference involves the probability of attack functions, π_i and π_j, where $\partial\pi_i/\partial\theta_i < 0$, $\partial\pi_j/\partial\theta_j < 0$, $\partial\pi_j/\partial\theta_i > 0$, and $\partial\pi_i/\partial\theta_j > 0$. Greater defensive measures at home decrease the likelihood of a terrorist attack there, while these measures increase the likelihood of a terrorist attack abroad through transference, as terrorists seek out the softest target. The cross partials are now ambiguous:

$$\frac{\partial^2\pi_i}{\partial\theta_i\partial\theta_j} \underset{<}{\overset{>}{=}} 0 \quad \text{as} \quad \theta_i \underset{>}{\overset{<}{=}} \theta_j. \tag{6}$$

The expression for the slope of the best-response function is the same as Eq. (5), but its sign is typically positive, indicating strategic complements.

The orientation of the isocost curves are more difficult to pin down because there are more opposing influences in the defensive case – see [18, 19] for a full discussion. If the terrorism threat is *globalized* so that a nation experiences the same losses at home or abroad to its interests, then $l(\theta_i) = v(\theta_j)$ and $l(\theta_j) = v(\theta_i)$. In

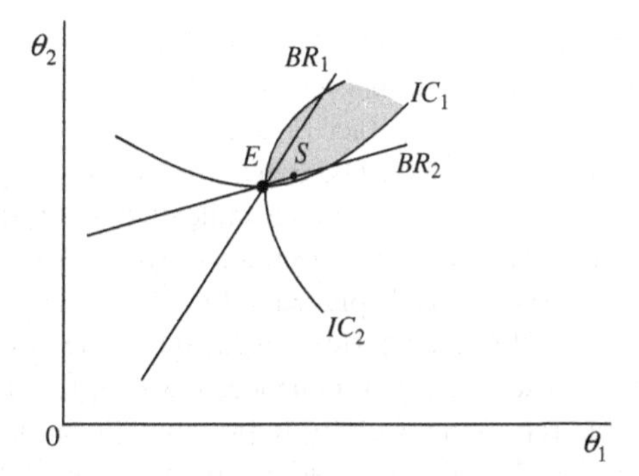

Fig. 2 Defensive countermeasures: globalized threat

this scenario, Fig. 2 applies in which the isocost curves have the same orientation as those in Fig. 1, but the *BR* curves are now upward sloping indicating strategic complements. Once again, the shaded region in Fig. 2 depicts defensive combinations that are welfare-improving compared to the Nash equilibrium. The position of these welfare-dominating defensive allocations vis-à-vis nations' independent actions indicates a tendency toward too little defensive measures when the terrorism threat is globalized. This follows because countries do not account for the benefits that homeland security affords foreign residents, visitors, and foreign property. The position of the leader-follower equilibrium at *S* (where country 1 leads) shows that strategic behavior by either or both nations can improve the allocation.

2.3 Defensive Measures: No Collateral Damage

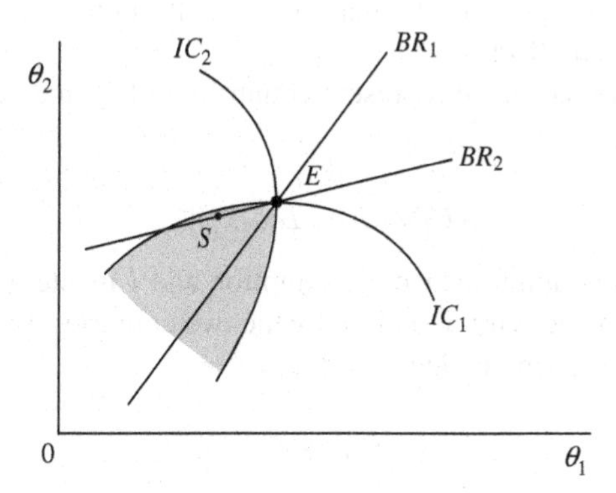

Fig. 3 Defensive race: no collateral damage

We now consider the defensive decision of two targeted countries, when attacks are host-country specific so that $v(\theta_j) = v'(\theta_j) = 0$ for $j = 1,2$. This means that a country's assets are not targeted abroad, so that there is no collateral damage on foreign interests in a venue country. In this scenario, defensive external costs imposed on other countries involve the transference of an attack; that is, an increase in θ_i augments the likelihood that the attack will be transferred to a less vigilant country. The best-response curves are still upward sloping, indicative of strategic complements; however, the isocost curves reverse their orientation, with IC_1 being hill shaped and IC_2 being a reversed *C*. This case is displayed in Fig. 3, where country 1's (2's) welfare improves as IC_1 (IC_2) shifts down (leftward), so that the other targeted country is spending less on defense. As a consequence, fewer terrorist attacks are transferred. The shaded area represents welfare-improving allocations, which highlight that nations will over spend on defense and engage in a homeland security race to become more impregnable. For this case, nations will mutually gain by coordinating defensive decisions. Leader behavior by country 1 improves the welfare of both countries and provides a second-mover advantage to the follower, whose welfare improves by relatively more.

2.4 Intelligence

Intelligence gathered on a common terrorist threat presents two targeted countries with mutual benefits and costs. If intelligence is used to weaken the terrorist group, then both countries benefit, analogous to the first case of proactive measures. This suggests a situation of strategic substitutes and downward-sloping *BR* curves, since ideally a country may be best off by free riding on the intelligence collected by the other country. Increased intelligence activity, x, by one country reduces the need for the other country to collect intelligence. Intelligence differs from proactive measures because intelligence gathering by another country may cause intelligence-collection cost, c, to the other country – i. e., $c_i(x_j)$, where $\partial c_i/\partial x_j > 0$ for $i, j, i \neq j$. This follows because more agents in the field may result in crowding and the potential for mishap and wasted effort.

To capture this scenario, we represent country $i (= 1,2)$ maximizing its welfare, U_i:

$$U_i = U_i\left[y_i, I(x_i, x_j)\right], \tag{7}$$

where y_i represents nonintelligence consumption and I is intelligence gained from intelligence-gathering activity, x_i and x_j, by the two countries. Each country or government faces the following budget constraint:

$$x_i c_i(x_j) + y_i = Y_i, \tag{8}$$

where the price of y_i is normalized to be 1 and Y_i is the country's income. The budget constraint can be substituted into the objective function to give i's maximization problem:

$$\max_{x_i} U_i\left[Y_i - x_i c_i(x_j), I(x_i, x_j)\right]. \tag{9}$$

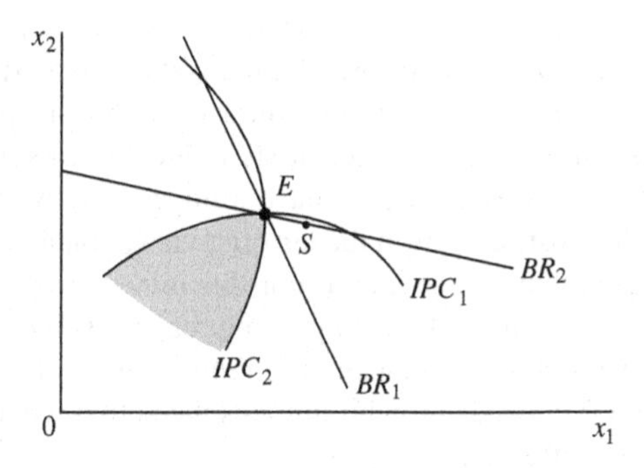

Fig. 4 Intelligence provision

Based on this maximization problem, the best-response curve can be derived and the slope shown to be negative under a wide range of scenarios. Moreover, the iso-profit curves (*IPC*) can be shaped as in Fig. 4 if there is sufficient crowding cost for intelligence collection. As before, the shaded area denotes intelligence expenditures that are more desirable than the Nash equilibrium. Strategic behavior can result in inferior outcomes at *S* or northeast of *E* if practiced by both countries.

2.5 Other Cases

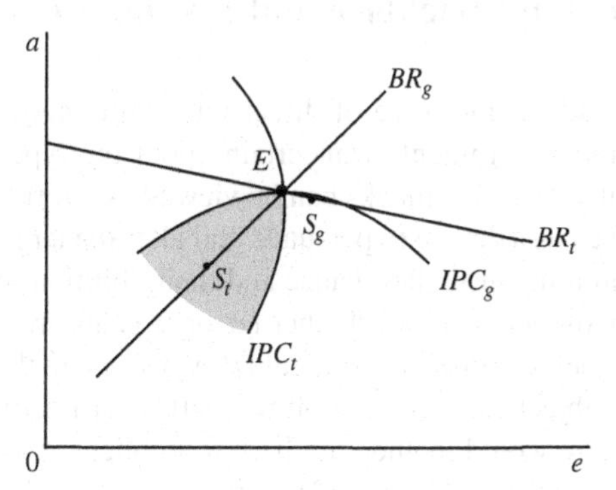

Fig. 5 Mixed case

Thus far, we examined two cases of strategic substitutes and two cases of strategic complements for various counterterrorism measures. For our final graphical example, we change the strategic agents to be the terrorists and a single targeted government. The terrorists choose a level of attacks, a, while the government exerts counterterrorism efforts, e. Siqueira and Sandler [23] present such models for which both the terrorists and the government are seeking support of members of the population. One of their models is displayed in Fig. 5, where the best-response curve of the government, BR_g, is upward sloping so that increased terrorist attacks are met with greater countermeasures. In contrast, the best-response curve of the terrorists, BR_t, is downward sloping: terrorists reduce attacks as the government augments its countermeasures. Thus, we have a mixed case where the two choice variables are strategic substitutes from the terrorist viewpoint and are strategic complements from the government viewpoint. For this mixed case, terrorist leadership results in an outcome S_t in the welfare-improving region, while government leadership does not improve *both* players' welfare. Figure 5 shows that the consequences from strategic behavior depends on who initiates the action when players view the set of choice variables differently.

A wide variety of players can be examined for continuous games applied to terrorism. In an innovative analysis, Siqueira [22] investigates strategic interaction between a political and a military wing of a terrorist group. He distinguishes four scenarios, some of which involve strategic substitutes and strategic complements. If, for example, actions by the two wings are strategic complements, then government counterterrorism actions taken against the political wing can have a double dividend by decreasing both political and military terrorist activities.

Continuous choice also applies to multi-level games, where the strategic players may change at each stage [20]. The nature of the strategic substitutes and/or complements may also change at the various stages. Clearly, the analysis can be applied to a variety of terrorism scenarios and, in so doing, enlighten policymakers.

3 Terrorist Signaling: Backlash and Erosion Effects

In this section, we address the issue of direct interaction between terrorists and a government when the government is uncertain about the type of terrorists that it confronts. Specifically, terrorist attacks can be viewed as a form of costly signaling, where violence is used as a device to persuade and alter the target audience's beliefs about terrorists' commitment to their cause and their ability to impose costs [4, 10, 13, 16]. Uncertainty often concerns whether terrorist goals are political or ideological [1, 9, 21]. We label terrorists as political (P-types) when their goals are related to concrete political objectives, such as political self-determination or eviction of an occupying force. These would include the Irish Republican Army (IRA) and Palestine Liberation Army (PLO). Abrams [2] calls this a limited objective. Here, the defining feature is that terrorism is a pure *cost* for P-types, because the resources used for terrorism represent an opportunity cost relative to their nonviolent use to achieve concessions. By contrast, militant terrorists see attacks as an intrinsic *benefit* (e. g., jihad as a religious requirement or the necessity of a Marxist struggle). Militant terrorists (M-types) have maximalist objectives corresponding to demands over ideology, either to completely transform the political system or annihilate the enemy [2]. Examples include Hamas and the Shining Path. Although some groups can be clearly characterized, many groups have idiosyncratic or a combination of limited and maximal demands (e. g., Hezbollah [2]), thereby creating uncertainty about the group's type. In such cases, there is a need for intelligence to understand the enemy.

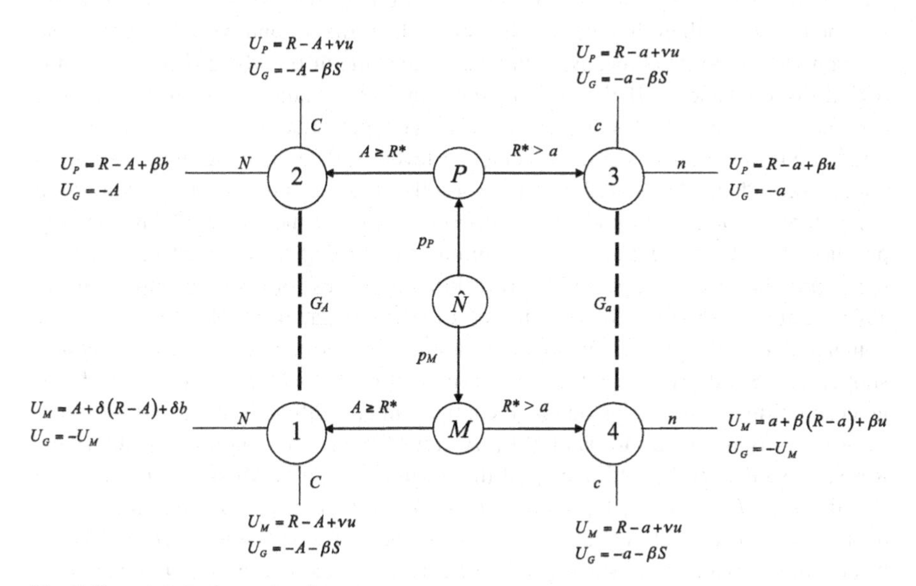

Fig. 6 Terrorist attacks as signals

Figure 6 is an example of a two-period signaling game where the government has incomplete information about its terrorist adversary. We use a two-period model because most terrorist groups are short-lived [15]. Nature moves first and selects the terrorists' type (M or P) according to a two-support probability distribution (p_M and p_P). Terrorists can either attack at or above a spectacular level, $A \geq R^*$, or below the spectacular level, $a < R^*$. *Ex ante* the government does not know the terrorist type, which is modeled as imperfect information about the move by nature ($\widehat{N}$). The dotted line labeled G_A collects the set of nodes (1 and 2) reflecting this imperfect information after a spectacular attack, A. Analogously, G_a is the information set that contains nodes 3 and 4 subsequent to non-spectacular attack, a. Following an attack (A or a), the government can either concede (C or c) or not (N or n). If the government does not concede, M-types expend all of their remaining resources in an attack, whereas P-types use their remaining resources for political goals.

Given the definition of the players and their strategies, we now turn to the payoffs. In Fig. 6, the payoffs to each terrorist type are written as U_P and U_M, and to the government as U_G. The value of any attack enters negatively into the government's payoff and, as discussed below, this value enters into the terrorists' payoff according to the terrorists' type. Let R denote the terrorists' first-period resources, common to both types and sufficient to mount a spectacular attack, $R \geq R^*$. In prior signaling models [4, 13, 14, 16], terrorists have resources exogenously available in both the first and second periods; however, part of the purpose of terrorism is to generate future resources and support [1, 8, 12]. Thus, we model terrorists' second-period resources, contingent upon their first-period actions and the response of the target government. *A priori* terrorists can expect a level of support equal to u, defined as the existing underlying support for the terrorist group. If terrorists attack below the spectacular level, then they garner this support. In the second period, P-types use this support for political means, with a baseline value of βu where β is the second-period discount factor. If the government concedes to attack a, then this support becomes vu where $v > \beta$ corresponds to a "victory" effect on discounting. If, instead, the terrorist attacks at the spectacular level, A, then the underlying support is put at risk. Indeed, Abrams [2] identifies a self-defeating component of terrorism in which large-scale attacks undermine support because of the collateral damage they produce and their potential to miscommunicate the degree to which the supporting population can be appeased through limited concessions. For example, during 1968–72 the Tupamaros [Movimiento de Liberacion Nacional (MLN)] in Uruguay conducted acts of urban terrorism so brutal that they lost any semblance of popular support [7]. Similarly, the previously maximalist *Islamic Group* has renounced violence, coming to the conclusion that the unwarranted spilling of blood and wrecking of property runs counter to Islam [26]. Abrams [2] cites the finding in a 2005 Pew Research Center global attitudes poll that in most majority-Muslim countries surveyed, support for suicide bombing (which, on average, causes twelve times more deaths than do conventional attacks) and other acts of violence in defense of Islam has declined significantly since 2002. We call this reduction in underlying support the *erosion effect*.

Of course, the purpose of terrorism is not to generate this self-defeating component, but instead to garner further support, primarily through reactions to the target government's response. Terror groups often use the politics of atrocity to produce further counter-atrocities that manifest themselves as backlash against the target government [15]. Kydd and Walter [12] classify one possible motivation for terrorism as an attempt to provoke targets into overreactions that turn public opinion into support for terrorists. Wright [25] asserts that one of Osama bin Laden's goals in implementing the attacks of 9/11 was to draw al-Qaida's opponents into responses that turn out to be repressive blunders, thereby encouraging attacks by other Muslims. Furthermore, while backlash is most closely associated with M-types, Jacobson and Kaplan [11] argue that politically motivated Palestinian terrorists have been aided by the downstream benefits of recruitment spurred by Israel's heavy-handed responses to terrorism. To capture this influence, a *backlash effect*, b, replaces u as the source of second-period resources subsequent to a spectacular if the government does not concede. The loss of u is an opportunity cost of a spectacular – i. e., erosion. Effectively, terrorists recognize that they are trading backlash support for grassroot support when they conduct a spectacular that does not lead to concessions. The net benefit of a spectacular on second-period resources is discounted by β if no concessions are received, and by δ if second-period resources are used in a second-period attack. M-types attack in the second period, whereas P-types do not. We assume that $\beta > \delta$ because a spectacular attack causes a target government to raise its defenses, thereby decreasing the effectiveness or raising the cost of a second-period attack by reducing the logistical probability of success. If terrorists must substitute into an alternative form of attack that requires the acquisition of new tactical skills, then this may also result in $\beta > \delta$, following a spectacular incident. The benefits of policies that translate into δ come at a cost of inducing a backlash effect in support of the terrorist group. If, by contrast, the government concedes, this translates into second-period cost S to the government. At nodes 1 and 4, nonconcession results in zero-sum payoffs for the government and M-type terrorists.

Definition 4. Assessment $(s_P, s_M, (s_{G_i})_{i=A,a}, (\mu_j)_{j=1}^4)$ is a perfect Bayes-Nash equilibrium (PBE) for our signaling game if and only if

1. s_P and s_M are the strategies of P-types and M-types, respectively;
2. s_{G_A} is the government's strategy at information set G_A and s_{G_a} is its strategy G_a;
3. μ_i is the government's belief that it is at node 'i' where (i) μ_i is constructed to be consistent with Bayes rule for actions that are on-the-equilibrium path, and (ii) μ_i is consistent with the expected payoff maximization for strategies that are off-the-equilibrium-path; and
4. the expected payoffs associated with this assessment are mutual best replies.

Intuitively, a PBE is a Nash equilibrium in expected payoffs. As in Arce and Sandler [4], this game contains a (separating) PBE where the government concedes to militant types who attack at the spectacular level and does not concede to political types who attack at the non-spectacular level. There are also (pooling) PBE where both types attack at level A (and the government concedes) or attack at level a (and the government does not concede).

Here, we concentrate on a new equilibrium that is a direct consequence of our introduction of backlash and erosion effects into the signaling framework. Under a spectacular attack, the government is at information set G_A and forms (conditional) beliefs that it is at node 1 (μ_1) versus node 2 (μ_2). From Fig. 6, the government does not concede subsequent to a spectacular attack if:

$$-\mu_2 A - \mu_1 [A + \delta(R - A) + \delta b] \geq -\mu_2 [A + \beta S] - \mu_1 [A + \beta S]. \tag{10}$$

Since $\mu_1 + \mu_2 = 1$, Eq. (10) implies that:

$$\beta S / [\delta(R - A) + \delta b] \geq \mu_1. \tag{11}$$

If the government's belief that it is facing an M-type, μ_1, is sufficiently low, then it does not concede subsequent to a spectacular attack. A novel result is that the greater the backlash effect, b, the more inclined is the government to concede following a spectacular attack. In contrast, the government is more apt to hold firm as the costs of concession, S, increases. Hence, understanding the potential for backlash is a key component of counterterror policy.

The government does not concede subsequent to a non-spectacular attack, a, when:

$$-\mu_3 a - \mu_4 [a + \beta(R-a) + \beta u] \geq -\mu_3 [a + \beta S] - \mu_4 [a + \beta S]. \tag{12}$$

Solving for μ_4 we have:

$$\beta S / [\beta(R-a) + \beta u] \geq \mu_4. \tag{13}$$

Under such an attack, the government's belief that it is facing an M-type, μ_4 must be sufficiently low to justify a commitment to suffer a potential second-period attack rather than face the cost of concessions, S.

Given N at G_A and n at G_a, P-types will attack at the spectacular level if

$$R - A + \beta b \geq R - a + \beta u; \text{ i.e., } b \geq u + (A-a)/\beta. \tag{14}$$

The importance of backlash is immediately apparent; if $b \leq u$ so that the backlash produced by the government's response to the spectacular does not increase support for the terrorists above the existing underlying level, then the condition in (14) cannot be met and P-types do not attack at the spectacular level (given N and n). Furthermore, M-types conduct a spectacular if

$$A + \delta(R-A) + \delta b \geq a + \beta(R-a) + \beta u. \tag{15}$$

Simplifying and solving for b, and combining this with (14), both types attack at the spectacular level if

$$b \geq \max\left\{u + \frac{(A-a)}{\beta}, \frac{\beta u + (\beta-\delta)R + (1-\beta)a - (1-\delta)A}{\delta}\right\}. \tag{16}$$

In other words, a (pooling) PBE occurs where both political and militant terrorists attack at the spectacular level and the government does not concede if Eqs. (11), (13), and (16) hold. From Bayes' rule, μ_1 equals the prior probability that terrorists are militant types (p_M); hence, Eq. (11) translates into the requirement that the prior probability of M-types is low. Because Eq. (13) corresponds to a belief that is off-the-equilibrium-path, not conceding is rational if the initial level of support, u, is low; i. e., P-types would attack at level a to conserve the resources that they have. The beliefs at G_a are therefore consistent with forward induction.

This equilibrium captures several phenomena previously absent in signaling models of terrorism. First, terrorism is rarely successful. Abrams [2] presents a scorecard of terrorist activity since 2001 and notes that terrorist objectives were obtained in only seven percent of the cases. In this equilibrium, terrorists attack at a spectacular level, but do not receive concessions. Second, an actual terrorist *campaign* occurs. M-types attack again subsequent to a spectacular, with the level of attack given as a function of the backlash to the target government's response to the spectacular. Third, this backlash effect must more than offset the current underlying support for the terrorist cause. This can be seen by examining the first term in braces on the right-hand side of Eq. (16). The size of the backlash effect must exceed the current underlying support, u, plus a multiple of the resources lost in mounting the spectacular, $(A-a)/\beta$. Terrorists must account for the degree to which spectacular attacks erode existing support. As Abrams [2] notes in his scorecard, groups that were successful in attaining their goals focused on military – rather than civilian – targets. Less erosion takes place when the target is military. Furthermore, our backlash result has a direct implication for counterterror policy, because defensive measures are less apt to promote a backlash effect whereas proactive ones are more likely to do so.

4 Concluding Remarks

This chapter has explored counterterror policy from the perspective of strategic complements – when the actions of one party cause another to act in a similar fashion – as is often the case in defensive counterterrorism policy, and strategic substitutes – when the actions of one party can replace those of another – as is the case for proactive counterterror policy. Collective action issues, relating to a dichotomy between Pareto-efficient and Nash equilibrium policies arise in these cases. Moreover, when a leader-follower structure is appropriate, our analysis identifies when policies are at cross-purposes, insofar as improving the welfare of one actor comes at the expense of the other actor. A novel insight is that in mixed models, involving both strategic substitutes and complements, this characterization may disappear, with both parties recognizing the mutual benefits or costs of leader-follower behavior. An example of the latter is government leadership when terrorists employ their tactics to garner further popular support.

The issue of how terrorist actions and government responses intersect to produce future resources and support for terrorists is one that is receiving increasing attention. The idea that a government response may generate a backlash against the government itself is embedded within a signaling game, where the government has incomplete information about the terrorists' intent. When combined with the previously unexplored idea that terrorist spectaculars that produce collateral damage may erode support for the terrorists, a trade-off is identified in which spectacular attacks occur only when government reactions produce the requisite backlash. This suggests that counterterrorism policy and intelligence should judiciously account for the net effects of backlash and erosion.

References

1. Abrams, M.: Al Qaeda's Scorecard: A Progress Report on Al Qaeda's Objectives. Stud. Confl. Terror. **29**, 509–529 (2006)
2. Abrams, M.: Why Terrorism Does Not Work. Int. Security **31**, 42–78 (2006)
3. Arce, D.G., Sandler, T.: Counterterrorism: A Game-Theoretic Analysis. J. Conflict Resolution **49**, 183–200 (2005)
4. Arce, D.G., Sandler, T.: Terrorist Signaling and the Value of Intelligence. Brit. J. Polit. Sci. **37**, 573–586 (2007)

5. Bulow, J.L., Geanakoplos, J.D., Klemperer, P.D.: Multimarket Oligopoly: Strategic Substitutes and Complements. J. Polit. Economy **93**, 488–511 (1985)
6. Eaton, B.C.: The Elementary Economics of Social Dilemmas. Can. J. Econ. **37**, 805–829 (2004)
7. Enders, W., Sandler, T.: The Political Economy of Terrorism. Cambridge University Press, Cambridge (2006)
8. Faria, J., Arce, D.G.: Terrorism Support and Recruitment. Defence Peace Econ. **16**, 263–273 (2005)
9. Hoffman, B.: Inside Terrorism. Columbia University Press, New York (1998)
10. Hoffman, B., McCormick, G.H.: Terrorism, Signaling, and Suicide Attacks. Stud. Confl. Terror. **27**, 243–81 (2004)
11. Jacobsen, D., Kaplan, E.H.: Suicide Bombings and Targeted Killings in (Counter-) Terror Games. J. Conflict Resolution **51**, 772–792 (2007)
12. Kydd, A.H., Walter, B.: The Strategies of Terrorism. Int. Security **31**, 49–80 (2006)
13. Lapan, H.E., Sandler, T.: Terrorism and Signaling. Europ. J. Polit. Economy **9**, 383–97 (1993)
14. Overgaard, P.B.: The Scale of Terrorist Attacks as a Signal of Resources. J. Conflict Resolution **38**, 452–78 (1994)
15. Rapoport, D.C.: Terrorism. In: Hawkesworth, M., Kogan, M. (eds.) Routledge Encyclopedia of Government and Politics, vol. 2, pp. 1067–1070. Routledge, London (1992)
16. Sandler, T., Arce, D.G.: Terrorism & Game Theory. Simulat. Gaming **34**, 319–37 (2003)
17. Sandler, T., Arce, D.G.: Terrorism: A Game-Theoretic Analysis. In: Sandler, T., Hartley, K. (eds.) Handbook of Defense Economics, Vol. 2 Defense in a Globalized World, pp. 775–813. North-Holland, Amsterdam (2007)
18. Sandler, T., Lapan, H.E.: The Calculus of Dissent: An Analysis of Terrorists' Choice of Targets. Synthèsis **76**, 245–261 (1988)
19. Sandler, T., Siqueira, K.: Global Terrorism: Deterrence versus Preemption. Can. J. Econ. **39**, 1370–1387 (2006)
20. Sandler, T., Siqueira, K.: Games and Terrorism: Recent Developments. Simulat. Gaming **40**, forthcoming (2009)
21. Scheuer, M.: Through Our Enemies' Eyes, revised edition. Potomac Books, Washington, DC (2006)
22. Siqueira, K.: Political and Militant Wings within Dissident Movements and Organizations. J. Conflict Resolution **49**, 218–236 (2005)
23. Siqueira, K., Sandler, T.: Terrorists versus the Government: Strategic Interaction, Support, and Sponsorship. J. Conflict Resolution **50**, 878–898 (2006)
24. Siqueira, K., Sandler, T.: Terrorist Backlash, Terrorism Mitigation, and Policy Delegation. J. Public Econ. **91**, 1800–1815 (2007)
25. Wright, L.: The Looming Tower. Knopf, New York (2006)
26. Wright, L.: The Rebellion Within. The New Yorker, June 2, 37, 16 (2008)

Underfunding in Terrorist Organizations*

Jacob N. Shapiro and David A. Siegel

Abstract A review of international terrorist activity reveals a pattern of financially strapped operatives working for organizations that seem to have plenty of money. To explain this observation, and to examine when restricting terrorists' funds will reduce their lethality, we model a hierarchical terror organization in which leaders delegate financial and logistical tasks to middlemen, but cannot perfectly monitor them for security reasons. These middlemen do not always share their leaders' interests: the temptation exists to skim funds from financial transactions. When middlemen are sufficiently greedy and organizations suffer from sufficiently strong budget constraints, leaders will not fund attacks because the costs of skimming are too great. Using general functional forms, we find important nonlinearities in terrorists' responses to government counter-terrorism. Restricting terrorists' funds may be ineffective until a critical threshold is reached, at which point cooperation within terrorist organizations begins to break down and further government actions have a disproportionately large impact.

1 Introduction

Shortly after noon on February 26, 1993, approximately 1,500 pounds of homemade explosives were detonated on the second level of the World Trade Center parking structure. Six people were killed and more than a thousand were injured. It could have been much worse; the F-350 Econoline van rented by the terrorists had

Jacob N. Shapiro
Princeton University, Department of Politics, 032 Corwin Hall, e-mail: jns@princeton.edu

David A. Siegel
Florida State University, Department of Political Science, 541 Bellamy, e-mail: dsiegel@fsu.edu

* The authors thank Jim Breckenridge, Ethan Bueno de Mesquita, James Fearon, Jeanne Giraldo, Harold Trinkunas, the participants in the CISAC Social Science Seminar, and several anonymous reviewers for tremendously helpful comments and criticisms. Any remaining errors are our own.

N. Memon et al. (eds.), *Mathematical Methods in Counterterrorism*,
DOI 10.1007/978-3-211-09442-6_21, © Springer-Verlag/Wien 2009

a cargo capacity of over 4,000 pounds. Ramzi Yousef, the leader of the attack, later testified he did not have enough money to purchase sufficient materials to build a larger bomb. He also claimed the attack was rushed because his cell ran out of money. (Levitt, 2002)

Similar underfunding problems have plagued other international terrorist organizations. Idris, the chief logistician for the 2002 Bali bombing, was arrested three weeks after the bombing. The police tracked him down through his involvement in a bank robbery intended to raise funds for subsequent operations. A detailed study of the money sent to Idris to fund the Bali bombing reveals that he received much more than he disbursed for that attack, making his subsequent involvement in risky criminal activities puzzling. (Abuza, 2003b: 54–55) Indeed, there is a recurring pattern of financially strapped operatives working for terrorist organizations that seem to have plenty of money. We seek to explain this pattern.

The puzzle for traditional perspectives on terrorist financial and logistical systems is that groups which are purportedly organized to carry out attacks often provide inadequate funds to their operatives. Standard accounts stress the efficiency with which terrorist financial networks distribute funds while operating through a variety of covert channels. They describe terrorist organizations as able to shift agilely between multiple avenues to raise and use funds with no mention of transaction costs or organizational infrastructure requirements. We are told that "al-Qaeda is notably and deliberately decentralized, compartmentalized, flexible, and diverse in its methods and targets … Al-Qaeda's financial network is characterized by layers and redundancies. It raises money from a variety of sources and moves money in a variety of manners." (Weschler and Wolosky, 2002: 6) Reports from the multinational Financial Action Task Force on Money Laundering (2004: 6), the Asia/Pacific Group on Money Laundering (2004: 32), and others provide a similar narrative. (Brisard, 2002: 6; Singapore Ministry of Home Affairs, 2003: 6)

From this perspective, it is hard to imagine why a group would underfund operational elements. As an alternative, suppose that the members of a terrorist support network, middlemen, were not uniformly driven by mission accomplishment, but that some were driven by monetary rewards. Because the network is covert, informational asymmetries abound, exacerbating the principal-agent dilemmas found in any hierarchical organization, creating numerous opportunities for middlemen to appropriate resources for personal use. In this scenario a greedy middleman need only pass on as much money to operatives as is required to achieve an acceptable number of successful attacks.[2] This rough sketch, which our model formalizes, can explain underfunding in hierarchical terrorist groups.

A game-theoretic approach is particularly attractive for dealing with this puzzle because it places the organizational dilemmas faced by terrorist groups into the starkest possible contrast. (Victoroff 2005) Developing formal models of terrorism can also help explain otherwise puzzling patterns. The best example of such work is Bueno de Mesquita (2005b), which reconciles the seemingly contradictory findings that while terrorist operatives are neither poor nor lacking in education, poverty and

[2] "Acceptable" here means the probability that the middleman is not fired or killed is sufficiently small relative to the rewards he obtains.

lack of economic opportunity are positively correlated with terrorism. In the case of our model, a formal presentation is required because the interactions are simply too complicated to understand decisively in an informal analysis.

Previous formalized rational-choice analysis of terrorism has focused on two areas: (1) the interaction between governments and terrorist groups; and (2) groups' internal dynamics. In the first area, much of the early work focused on the signaling dynamics of terrorism. Lapan and Sandler (1993) present a model in which the optimal government strategy depends on the resources of the terrorist group. The terrorist group can use the scale of attacks to send a signal about these resources and has incentives to misrepresent its resources to gain concessions from the government. Overgaard (1994) presents a more subtle signaling model in which terrorist resources are renewable between periods and the terrorists have positive alternative uses for resources, alternatives such as providing social services. Overgaard finds that if concessions are ruled out for exogenous political reasons, then only a pooling equilibrium exists and the government learns nothing from the scale of attacks. More recently, Sandler (2003) explains the under-provision of international counter-terrorism as a collective action problem. Frey and Luechinger (2004) use a simple supply and demand model to examine the relative merits of deterrence and decentralization of critical infrastructure as counter-terror policies.

In one of the first formal rational-choice analyses of terrorist groups' internal dynamics, Chai (1993) examines why people participate in covert anti-government organizations even though their chances of getting particularistic benefits out of doing so are quite slight given the low likelihood of overthrowing the government. His analysis helps explain the common requirement among European terrorist groups that new members commit a violent act as part of the recruiting process. Siqueira (2005) models the interaction between militant and political factions within a dissident movement. His work highlights how the external manifestations of competition between factions depend on the nature of the externalities each factions' actions create for the other faction. Berman and Laitin (2005) examine why radical religious groups are particularly well suited to using suicide attacks. Recent analysis of how terrorist conflicts end formalizes Crenshaw's (1991) arguments about how the internal dynamics of terrorist organizations can cause difficulties for the peace process. Bueno de Mesquita (2005c) models the problems created by differing levels of ideology within a terrorist organization trying to make peace. His analysis outlines the strategic considerations underlying the tendency of radical factions to break off from groups engaged in peace negotiations.

While the level of sophistication in organizational analysis of terrorism has increased, no one has yet explored the challenges that heterogeneity poses for terrorist support systems and how those challenges affect groups' abilities to kill people.[3] Our work begins to address this by focusing on the relationship between terrorist leaders and the middlemen they must use for security reasons.

---

[3] A more general problem which is just now being addressed in the academic literature is that most studies of terrorism fail to consider the great heterogeneity of terrorists. This problem is explored in great detail in Victoroff (2005).

Beyond contributing to a growing body of literature on the organizational dynamics of terrorist groups, our analysis has implications for important policy debates. It speaks to the treatment of possible informers, to the interpretation of the apparent success or failure of counter-terror policies, and to more fundamental questions over the role of counter-terrorist financing in the larger counter-terrorism efforts. A great deal of energy has been spent over the last five years to clamp down on terrorist financing. UN resolutions have been passed, international standards set, technical assistance programs started, and financial laws changed in over 100 countries. (Weschler and Wolosky, 2004: 9–11; Francis, 2004) Legal changes undertaken as part of this campaign have created serious threats to treasured civil liberties in Western democracies and appear to have been highly counter-productive from an intelligence standpoint. (Donohue, 2006) Asset seizures, and the threat thereof, have crippled Islamic charities that provide invaluable social services in areas of failed governance. Efforts to limit terrorists' use of alternative remittance systems have had disastrous social welfare implications in some of the world's poorest areas. (Medani, 2002)

Yet all this activity may have little positive effect. Terrorist attacks are cheap compared to the budgets of terrorist organizations, at least when fixed costs are not taken into account.[4] Of the 23 international attacks attributed to Al Qaeda since 1995, only the World Trade Center attacks appear to have cost the group even one percent of its estimated annual budget.[5] Given how hard it is to stop terrorist funds, and how cheap attacks appear to be, one must wonder whether all this effort is for naught. Our analysis informs this debate by identifying how counter-terrorist financing efforts can limit attacks without necessarily making attacks unaffordable. Specifically, seizing funds, and thereby tightening terrorists' budget constraints, can increase friction within organizations to the point where leaders choose to spend on activities other than attacks. However, before this constraint is reached, seizing funds is unlikely to reduce attacks.

The remainder of the paper proceeds as follows. Section 2 provides empirical and theoretical motivation for our model of a hierarchical terrorist group. Section 3 formally presents the model. Section 4 develops the equilibrium strategies and derives comparative statics. Readers not interested in the mathematical development can omit this section without significant loss of comprehension. Section 5 discusses the results, providing greater intuition about the model through a computational illus-

---

[4] Most estimates are that terrorist groups spend less than 10 percent of their income on actually conducting attacks. For example, estimates of the centrally managed portion of Al Qaeda's budget range from $16 million to under $50 million a year. See Second Report of the UN Monitoring Group (2002: 12, 27). The 1993 World Trade Center attack cost less than $20,000, according to its mastermind Ramzi Yousef, and the Bali bombing cost at most $35,000. (Lee, 2002: 4). See also Abuza (2003b). Given these numbers, we concur with the CIA estimate that Al Qaeda's spending on actual operations was quite small. (National Commission on the Terrorist Attacks on the United States, 2004b: 11.)

[5] On attacks, see the RAND/St. Andrews data provided by the Memorial Institute for the Prevention of Terrorism. http://www.tkb.org (2006, February 13). On costs, a brief summary is provided by Prober (2005).

tration of our predicted outcomes. Using general functional forms we find important non-linearities not revealed by the comparative statics. Section 6 concludes.

2 Motivation

Because of the covert nature of their work, terrorist networks must operate with fewer checks and balances than most financial organizations require. Indeed, the cellular structure of terrorist networks so often cited in the literature necessarily implies that leaders will be poorly informed about the actions of their subordinates. If we assume that all the members of the network are uniformly committed to the cause and that they all agree on how best to advance the group's political goals, then there is no inconsistency here. However, suppose leaders, middlemen, and operational cadres have divergent preferences over spending. Then the informational asymmetries created by the secretive nature of terrorist networks leads to myriad opportunities for spending money differently than leaders would like.

Of course, the real-world division of labor is not always so stark. The level of specialization can vary over time and between groups. Al Qaeda and affiliates used to have quite defined organizational roles with a strong distinction between support and operational roles.[6] However, since losing their refuge in Afghanistan, Al Qaeda and affiliates may have shifted to a less hierarchical system. In Madrid and Casablanca the same members appear to have engaged in logistical tasks and conducted operations.[7] Moreover, the level of specialization can be a strategic choice. Resource-poor groups must be efficient to survive, while wealthy organizations may not be concerned with inefficiencies so long as they can meet their political goals.

These subtleties aside, using an agency theory framework can help explain otherwise puzzling behaviors. Consider the following informal example where the relationship between terrorist leaders and their financial system is described in terms of a principal-agent relationship. The principals, terrorist leaders, must delegate certain tasks – raising funds and distributing them to operational elements – to their agents, the financial network. Delegation entails a risk that if the agents' preferences differ from those of the principal, the agents will not carry out their tasks exactly as the principal would like; they may "shirk" by retaining some portion of the funds intended to support an attack.

Because monitoring the agents entails a security cost, the principals can only observe the outcome of the agents' actions: whether an attack succeeds or fails. This outcome is probabilistic, and thus provides only a noisy signal as to whether the agent passed on all the funds or not. Thus there is some space for agents to skim funds, thereby reducing the principal's utility. This strategy is feasible because the principal can neither perfectly monitor nor punish the agent with certainty. Clearly

[6] For example, the planning and bomb-making for the African Embassy bombings were conducted by individuals who left the country shortly before the actual attacks.

[7] Marc Sageman, private communication, 8 Jan 2005.

an agency theory approach can help explain the observed pattern of underfunding, but is there any evidence that these kinds of problems actually occur?

Many would cite Al Qaeda as the hard case for testing the idea that terrorist organizations suffer from agency problems and other organizational pathologies. Al Qaeda reportedly had a significant vetting process for membership and is generally discussed as a remarkably cohesive group. (Testimony of FBI Agent, 2001; Jenkins, 2002: 5) However, given that Al Qaeda is a relatively small group, operating small cells, over vast distances, in areas of the world with poor communications infrastructure, it is in some sense an easy case for agency problems. Such problems should exist here if the middlemen have preferences that diverge from those of the leadership. In this light, consider the following e-mail written by Ayman Al-Zawahiri, Al Qaeda's second-in-command, to a Yemeni cell on February 11, 1999:

> ... With all due respect, this is not an accounting. It's a summary accounting. For example, you didn't write any dates, and many of the items are vague. The analysis of the summary shows the following:
>
> 1 You received a total of $22,301. Of course, you didn't mention the period over which this sum was received. Our activities only benefited from a negligible portion of the money. This means that you received and distributed the money as you pleased ...
>
> 2 Salaries amounted to $10,085, 45 percent of the money. I had told you in my fax ... that we've been receiving only half salaries for five months. What is your reaction or response to this?
>
> 3 Loans amounted to $2,190. Why did you give out loans? Didn't I give clear orders to Muhammad Saleh to ... refer any loan requests to me? We have already had long discussions on this topic ...
>
> 4 Why have guesthouse expenses amounted to $1,573 when only Yunis is there, and he can be accommodated without the need for a guesthouse? (Cullison 2004)

Al Qaeda, in fact, faces recurring difficulties due to the divergent motivations of its membership. Two examples from the East Africa Embassy bombings are instructive. Jamal Ahmed Al-Fadl, who testified in the African Embassy bombing trial, had stolen money from Al Qaeda, got caught, went on the run, and approached the U.S. government in an attempt to save himself and his family. L'Hussein Kherchtou, a member of the Nairobi team, testified for the government because he was so appalled at the un-Islamic embezzlement practiced by senior members of his team. Thus even in the hard case, we see divergent motivations creating agency problems.[8]

Agency problems over spending occur in smaller, more localized organizations as well.[9] During the Christian-Muslim violence in Poso in late 2000, a relatively senior Jemaah Islamiyah (JI) member arranged to raise funds from oil company workers to be channeled through one local militia, KOMPAK-Solo, to JI and another local militia, Mujahidin KOMPAK. The workers were so concerned about the probity of these transfers that they appointed an auditor to oversee the funds. (International Crisis Group, 2004: 9–10) This auditor's involvement exacerbated

---

[8] For a sample of captured Al Qaeda documents highlighting agency problems see Felter et al. (2006).

[9] Agency problems over strategy and tactics have bedeviled terrorist organizations since the 1890s. (Felter et al. 2006; 11–21)

tensions between Mujahidin KOMPAK and JI, and the relationship deteriorated to the point that formal authorization by senior leadership was required for members of the groups to share weapons, something they had done freely during the early months in Poso. In other words, JI and Mujahidin KOMPAK suffered from the type of agency problems our model highlights.

2.1 Game

The interaction we describe is that between a terrorist boss, who controls the flow of money and desires some impact from this expenditure, and the middlemen he hires to perform the logistical support for those who will actually carry out the operations. Upon first glance, this is simply a standard principal-agent dynamic, in which the boss principal hires the middleman agent to do a job for him, though a particularly grisly one. However, this ignores the essential covert nature of the terrorist organization, in which monitoring is all but completely limited to whether or not an attack has succeeded. Moreover, contingent payment – the means by which the principal typically obtains his desired behavior – is either infeasible because middlemen do not have the capital to finance attacks, or is not practiced.

The general pattern of funds transfer seems to be that leaders provide funding in either an initial block grant, or on a need-to-have basis. The block grant method appears to have been used for the three Thai members of JI who were carrying $50,000 to support operations when they were arrested in Phnom Penh in May 2003. (Abuza, 2003a: 194) Both the Bali bombings and the September 11 attacks followed the need-to-have model.[10] Although there is some evidence that Hambali, the key middleman for the 2003 J.W. Marriott bombing in Jakarta, received a substantial bonus following the attack,[11] we have been unable to find a single case in which a middleman or facilitator financed an attack and received payment from the group leader after the attack had succeeded, the contingent payment model.

The difficulty in monitoring agents, along with the ability of agents to go to the government, makes the application of a punishment difficult – at least in the international context – leaving little the principal can do to condition his agent. In a covert system, the agent holds an inherent threat over the organization. If he is too dissatisfied with his punishment, he can go to the authorities, as Jamal Ahmed Al-Fadl did. Because agents have exactly this option, punitive strategies should only exist where the organization can wield a credible threat of violence over the agent. It is possible for some localized groups such as the IRA to use murder to condition their agents. Yet even in such cases, punishment is problematic. One study of Palestinian groups operating in Lebanon in between 1976 and 1983 revealed that groups that punished disloyalty had to pay higher wages to their agents. (Adams, 1986: 86) It seems quite

[10] For the Bali bombing figures, see Abuza (2003b). For the September 11 attacks, see National Commission on the Terrorist Attacks Upon the United States (2004a: section 5.4). See also Roth (2004: Appendix A).

[11] Private communication, Phil Williams, April 11, 2005.

likely that financial agents operating in foreign countries, such as the Yemeni recipient of al-Zawahiri's e-mail, will be less susceptible to punishment strategies. That agent responded to being called out by quitting the network, illustrating the difficulties transnational groups face in using punishment strategies. (Cullison, 2004)

What the boss can do is refuse to use that middleman again for subsequent attacks, denying him future gains from participation. As we will show below, if middlemen are patient enough, then this incentive proves sufficient to motivate even agents solely interested in pecuniary gain to act at least partially in the boss' interests.

2.2 Actors

Our model starts from the position that a boss wants to maximize political impact. We can think of the Al Qaeda leaders whose expressed goal was to compel the United States to withold its support from apostate Arab regimes such as the Saudi royal family. (Al-Zawahiri, 2001) In Zawahiri's writings, the purpose of terrorist attacks is two-fold. First, to rally support around the movement by setting an example for others to follow. Second, to impose costs on the United States for policies that contradict the groups' goals. Both fall under the larger goal of achieving political impact.

A leader gets political impact by providing a basket of goods that includes attacks, as well as more prosaic goods such as social welfare services, ideological communications, training camps, and the like. Evidence from the occupied territories suggest that both Hamas and the PLO vary their outputs of each type of legitimacy-generating activity according to what garners the most support at a given moment in time. (Gunning, 2004: 242–243; Rees, 2001) Our model captures this substitution dynamic in a rough way. In our model, the boss simply gets utility from attacks and some disutility from spending money on attacks that could otherwise be spent on all the other goods that terrorist groups can produce. By varying the intensity of this disutility, we can account for relative preferences between attacks and other goods.

Now consider the preferences of the middlemen. We assume individuals join terrorist organizations when the utility of doing so is at least as good as that provided by their next best option. Utility is composed of two components. First, individuals get utility from the impact of their actions in furthering the group's goals, from doing what they believe is right. We simplify this to the probability that an attack succeeds. Second, individuals get utility out of monetary compensation: the money they take in that does not go to an attack. Each individual weights these two components such that the sum of the weights is one. At the extremes are individuals who are purely motivated by impact, suicide bombers perhaps, and those motivated purely by money.

There is good empirical evidence that the preferences of middlemen are not always aligned with those of leaders and operational elements. Mid-level managers of organizations such as Harakat ul-Mujahedin (HUM), a Pakistani militant group focused on Kashmir, often live luxurious lives far beyond what their followers can afford. (Stern, 2003: 213–216.) Captured PLO documents show that those who plan attacks are paid eight times as much as is given to the families of those who die carrying out the attacks. (Israeli Defense Forces, 2002) People running criminal fund-raising operations in the United States for Hezbollah drive luxury cars and live in upper-middle class neighborhoods. (Farah, 2004: 164)

This preference divergence should be expected as terrorist groups suffer from two adverse selection problems with respect to their financial and logistical operatives. The first is that those likely to survive long in terrorist networks tend to be less ideologically committed as they are less likely to volunteer for the most dangerous missions.[12] The second is that because participation as a financier or logistician is less risky than participating as a local leader or operator, middlemen in terrorist organizations will tend to be less committed. We examine each in turn.

One striking pattern that emerges from a close examination of terrorist organizations is that financial network members face dramatically lower risks than local leaders or tactical operatives. Beyond not being asked to participate in risky or inherently fatal ventures, they are less likely to be targeted by government forces. When targeted, they are less likely to be killed. And when arrested, they face more lenient treatment.

We assessed the risks of participating at different levels using Sageman's (2004) sample of 366 participants in the global Salafi jihad: Al Qaeda, affiliated organizations, and some operating outside of traditional organizations. Using open-source material we collected data on individuals' operational roles, when they left the jihad, and how they left.[13] According to these data, between 1997 and 2003, financiers were rarely killed and their chances-of-being-arrested rate were 10-20 percent lower than that of tactical operators, with 2002 being the only exception.

Even when government succeeds in capturing logisticians and other support network members, these individuals face dramatically lower consequences than operators. Only one of the 32 financiers and logisticians removed from the global Salafi jihad between January 2001 and December 2003 was killed. A particularly telling example is the Jemmaah Islamiyah (JI) cell which was broken up in Singapore in late 2001. The cell provided fund-raising services to JI and was engaged in making logistical arrangements for an Al Qaeda attack in Singapore. Of the 30 plus people arrested, the 13 engaged in direct logistical support each received two years in prison. Those engaged in fund-raising activities were released but not permitted to leave the country. (Ressa, 2003: 158–160)

With these facts in mind, consider a hierarchical organization where individuals come up through the ranks, starting out in subordinate roles and moving into man-

[12] We thank Ethan Bueno de Mesquita for pointing this out.

[13] Sageman (2004) uses data on 172 of these individuals. We conducted independent coding of operational roles for those participating between 1997 and 2003.

agement roles as local leaders, financial facilitators, or logisticians.[14] Throughout their careers, these individuals will have opportunities to volunteer for risky missions.[15] Those most likely to do so will be those who place the highest weight on impact. Thus, the longer individuals remain in the organization, and the further they move up the management structure, the more likely they are to place a heavy weight on monetary rewards.[16]

Even without these adverse selection processes, there are reasons to expect divergence. The lenient treatment observed for support network members means that the threshold level of risk acceptance and commitment required for participation in support activities is much lower than for participation in tactical roles. We make the reasonable assumptions that there is a distribution of weights in the population of potential members and that the population of true believers is limited.[17] Thus, given set wages for different activities, some individuals might participate in support activities while balking at other roles. Seeking to maximize operational capability, a rational organization would concentrate such individuals in support roles, freeing up the true believers for riskier operational duties. These personnel decisions would then lead to consistent differences in ideology between leaders and middlemen.[18]

There is good empirical evidence that terrorist middlemen often have preferences that are not aligned with their leadership. The adverse selection process and differential levels of risks faced by those filling different roles provide a mechanism for understanding why. Our model incorporates this insight explicitly into the middlemen's preferences and analyzes how this preference divergence affects groups' abilities to kill.

---

[14] This progression need not happen within one organization. For example, much of the leadership in JI have been waging jihad together, at varying levels of intensity, since the mid-1980s (International Crisis Group, 2003b: 7–9).

[15] Evidence from trial transcripts and other sources suggest that volunteerism is a primary method of selection with leaders choosing from among volunteers. The selection process for the 9/11 attacks is discussed in National Commission on the Terrorist Attacks Upon the United States (2004a).

[16] Weinstein (2005) discusses a slightly different adverse selection problem in his work on rebel groups in Sierra Leone. He posits that a wage that brings high quality recruits will also bring in low quality individuals. As leaders are unable to observe the recruits type, they face an adverse selection problem.

[17] Considering the relatively low frequency of attacks by Islamist terrorist organizations given their substantial worldwide membership, we believe this assumption is reasonable. As the Washington D.C. sniper attacks proved, sowing terror through small-scale attacks is eminently feasible and can be done with very small expenditures. Given the ready availability of fire-arms in the United States, and the thousands of prospective terrorists who have received small-arms training since Osama Bin Laden issued his first fatwa calling for world-wide attacks against the United States in 1998, the lack of frequent, small-scale attacks is puzzling. One plausible explanation is that the population of true believers is actually fairly limited.

[18] Shapiro (2008) examines these selection and recruiting dynamics in more detail.

3 Model

To bring this dynamic into sharp focus, we begin with a fairly simple model that nevertheless captures the mechanisms at work here. There are only two actors: a terrorist boss whom we refer to as "B", and a single middleman, called "M". We assume that each period of the infinitely-repeated game consists of the middleman's disbursing some fraction of the money he is given in order to fund an attack. The likelihood of a successful attack is increasing in this fraction; thus, the boss would prefer that all the money he hands to M be spent. On the other hand, if M has some innate preference for money – i. e., he cares about this in addition to the success of the attack – then he has the incentive to skim some of the funds for himself.

3.1 Game Form

Each period consists of the following sequence of events: B gives w_0 to M, M spends $w_0 - x$ on the operation, keeping x for himself, and the attack succeeds with probability $p(w_0 - x; \alpha, \beta)$ with $p(\cdot)$ increasing, eventually concave, and twice differentiable in $w_0 - x$. Here α and β define the operational environment. α provides the baseline funding level necessary to achieve a certain success rate, and so represents the difficulty of successfully completing an attack. β determines the rate at which changes in funding are translated into success probabilities, and so represents the sensitivity of the success rate to funding.[19] B observes only the outcome of the attack – whether or not it succeeds – and decides based on this whether or not to utilize M in the next period. We assume that the pool of potential middlemen is sufficiently large that B need not use a discarded M again, and that there is no cost (such as being ratted out by your former middleman) to changing M. For now, we also assume that all M are identical. This assumption is strong and will be relaxed in future work, but for now it allows us to focus on the essential aspects of the interaction before branching out into behavior across populations. The stage game just described is repeated infinitely, though in practice no particular M is retained for much of this time.[20]

---

[19] One can think of α and β as parameters of a location-scale family of cumulative density functions that defines the relationship between spending and success probability.

[20] This model is similar to that presented in Ferejohn (1986) insofar as B observes a noisy signal about M's performance, and hence his type, and has to choose retrospectively whether to hire B again in the next period. Here, however, in order to produce a signal – and achieve an appreciable chance of success – B must fund M before the signal occurs, resulting in a substantial change in both B's strategy space and M's decision calculus.

3.2 Actors

We begin with the boss, B, who is assumed to care primarily about the success probability of the attack, p,[21] but also about the amount spent on the attack. We assume a separable expected utility $E[U] = p - H(w_0)$ for each period. $H(0) \geq 0$, with $H' > 0$, $H'' \geq 0$, determines B's disutility from the loss of funds that go into the attack; w_0 is the amount given to M at the beginning of each period to use to fund the attack that period. The larger H', the more money the boss would like to allocate to other goods such as social service provision. Different groups with different goals would have different dependences of H on w_0. For example, Hamas is estimated to spend somewhere in the vicinity of 90 percent of its annual budget on providing social services, so less than 10 percent is allocated to actual attacks. (International Crisis Group, 2003a: 13) Suppose Hamas provided enough funds so that it almost always succeeded. In that case, this revealed preference would imply that w_0 spent on attacks would yield no more utility than $9 \cdot w_0$ spent elsewhere.

The middleman, M, also cares about both money and the likelihood of a successful attack, and is assumed to maximize an expected utility $E[V(x,p)] = \gamma v(x) + (1 - \gamma)p$ per period. $x \leq w_0$ is the amount M skims off the top, so that $w_0 - x$ is the actual amount used to fund each attack. For simplicity, we have again used a separable form for this utility, with $v' > 0$, $v'' \leq 0$. γ parameterizes the degree of pecuniary interest on the part of M. When $\gamma = 1$, the agent cares nothing for the cause; when $\gamma = 0$ the cause is everything. Notice that this framework makes no inherent claims about the type of middlemen a group will have. Thus, it can be used to analyze a wide variety of different organizations.

In this model we assume that B knows γ. Future will relax this assumption, but we believe it is justified here for two reasons. First, this model addresses the case of a homogeneous pool of M where we assume B has had time to develop accurate beliefs about the pool. Second, in many groups, the members work together for long periods of time. For example, most identified members and supporters of JI are also members of a long-standing informal network interested in establishing Islamic law in Indonesia. (International Crisis Group, 2002) Through their repeated interactions, we expect that the leaders would be able to develop fairly accurate beliefs about the commitment of their prospective middlemen.

Taken together, these parameters describe a particular period's payoff. We capture how utilities compare across time through the actors' discount factors. These parameters usually take one of two interpretations: either individuals value future payoffs less than present ones, or the game itself has some probability of ending each period, denying the accrual of future utility. We are agnostic between the two interpretations. In this model, the second interpretation can be understood as follows: δ is the probability an agent is not captured or killed. This probability is generated by something exogenous to this model: the skill level of the government. In a sense, values for this interpretation are already built into our model; B is guaranteed to be around in each period, while M faces a chance of being let go based on the

[21] In this section and the following one we suppress the parameters of p for notational simplicity.

outcome of his task. As such, we believe the first interpretation is the more plausible one for this model, even if it is empirically harder to identify.[22]

Defining δ_B and δ_M as the discount factors, we obtain the two utilities: $E[U] = \sum_{t=0}^{\infty} \delta_B^t (p_t - H(w_{0,t}))$ and $E[V] = \sum_{t=0}^{\infty} \mathbf{1}_{hired} \delta_M^t (\gamma v(x_t) + (1-\gamma)p_t)$, where $\mathbf{1}_{hired}$ is an indicator function that equals 1 if M has been hired that period, and 0 if not.

4 Results

As we assume that B may costlessly choose a new, identical M whenever he chooses not to retain an old one, every period looks identical to him.[23] Because B therefore faces the same payoff-related incentives in every period, we make the reasonable simplifying assumption that B utilizes a stationary strategy, one that is constant in time.[24] Since the strategy space for B consists of the amount of money paid to M, w_0, and rules for when to retain M in the face of both a successful attack and a failed one, which we will call q_S and q_F, respectively, this implies that each will be constant over time.

The retention rules, q_S and q_F, arise from the circumstances of the interaction. The lack of information for B beyond an attack's success or failure implies the form of these retention rules can only depend on the history of such information for each agent. Note that this allows for steadily increasing (or decreasing) likelihoods of retention as successes (or failures) mount. Since we are considering stationary strategies, however, we will assume the most basic of such rules: retain with probability q_F upon each failure, and retain with probability q_S upon each success. In this case, as M responds to B's actions within a completely static environment, we will assume that M's equilibrium strategy – how much to skim, x – will also be time independent. Our equilibrium concept is thus subgame perfect Nash equilibrium in stationary strategies.

We further assume that M cannot borrow money to add to the success of an attack. Of course, M also cannot skim more than is given to him. Under these assumptions we are presented with a constrained optimization problem in which B maximizes his utility, conditional on M's maximizing his utility, subject to the constraint that $x \in [0, w_0]$. The full optimization breaks the problem up into four parts: the case where all the money is skimmed, the "honest" case where no money is skimmed, the interior or "skimming" solution, which generally obtains for greedy middlemen and rich bosses focused on attacks, and the "transition" region between "honest"

---

[22] It is easier to observe survival rates than it is to estimate discount factors from observed behavior.

[23] The plausibility of the assumption of identical M varies depending on the competitive environment for a group. Where government enforcement is lax, moving funds to operational cells requires no particular skill, and so most any agent can fulfill the task. When a group is under significant government pressure, moving funds can require substantial machinations, and thus the population of adequately skilled middlemen who care sufficiently about the cause may be quite small.

[24] We view this as reasonable because the only restriction it imposes is that B cannot condition on non-payoff-related histories.

and "skimming" regimes, in which no skimming occurs, but the boss nevertheless must consider his agent's divergent interests.

In the next subsection we derive solutions for the equilibrium actions of both Boss and Middleman and briefly discuss how changing the assumptions of the model would alter this equilibrium. In the one following we examine how the equilibrium outcome changes with the model's parameters, introducing general functional forms for p, H, and v to aid our intuition. Finally, Sect. 5 adds a budget constraint for the boss in order to explore the model's meaning in real-world terms and highlight some important nonlinearities not captured fully in the comparative statics provided in this section.[25]

4.1 Equilibrium Strategies

Ignoring constraints for the moment, B solves the maximization problem:[26]

$$\max_{w_0, q_S, q_F} p - H(w_0), \tag{1}$$

subject to:

$$\max_{x} \gamma v(x) + (1-\gamma)p + \delta_M (p q_S + (1-p) q_F) C(x^*, w_0, q_S, q_F; \gamma, \delta_M), \tag{2}$$

where $C(x^*, w_0, q_S, q_F; \gamma, \delta_M)$ is M's continuation value.[27] Note that C depends on the future actions of M in equilibrium. We solve the game using backward induction, beginning with the middleman's problem, which is to choose the x^* that maximizes:

$$C = \frac{\gamma v(x) + (1-\gamma)p}{1 - \delta_M (p q_S + (1-p) q_F)}. \tag{3}$$

The form of C is suggestive: increasing q_S and decreasing q_F would both seem to increase C's dependence on p. Since dependence on p correctly aligns the incentives of B and M, it is in B's best interests to do condition M's utility on p as much as is possible. Thus we would expect that, in equilibrium, B would always rehire upon a

---

[25] The budget constraint is not included in this section as here we only consider the interior solution of the constrained maximization problem.

[26] B's optimization problem is set up as a static one because: (1) he chooses an identical M whenever one is fired; and (2) we are considering stationary strategies. These imply that every period is identical to B, and so there is no incentive for B to alter his strategy after successes or failures. B's discount factor thus does not impact the analysis.

[27] We can separate M's utility given at the end of the previous section into terms corresponding to M's present utility and M's expected future utility, given that M will be playing his best response (the equilibrium value x^*), since we have assumed stationarity. Comparing this separated utility with Equation (2) implies that $C(x^*, w_0, q_S, q_F; \gamma, \delta_M)$ equals $(\gamma v(x^*) + (1-\gamma) p(w_0 - x^*)) \sum_{i=0}^{\infty} \delta_M^i (p(w_0 - x^*) q_S + (1 - p(w_0 - x^*)) q_F)^i$. Since the summand is less than one, this yields (3) when summed.

success, and always fire upon a failure. We show this formally with the following two lemmas:[28]

Lemma 1. *x^* is decreasing in q_S, and the optimal $q_S^* = 1$.*

Lemma 2. *x^* is increasing in q_F, and the optimal $q_F^* = 0$.*

We can therefore eliminate the retention rules from further analysis.[29] This simplifies (3), as well as the first-order condition that arises from it:[30]

$$\frac{v'(x^*)}{p'} - \frac{\delta_M p v'(x^*)}{p'} - \delta_M v(x^*) = \frac{1-\gamma}{\gamma}. \tag{4}$$

(4) implicitly defines x^*. Recall that p is also a function of x^* and w_0; we have left off its dependence for readability.

Noting that $\frac{dp}{dw_0} = \frac{\partial p}{\partial w_0} + \frac{\partial p}{\partial x^*}\frac{dx^*}{dw_0}$, we can use the equilibrium value in (4) with (1) to find the second first-order condition:

$$p'(1 - \frac{dx^*}{dw_0}) = H'(w_0). \tag{5}$$

With (4) and (5) in hand, we can now find the full equilibrium, which consists of the pair (x^*, w_0^*), along with the p resulting from these. There are four regimes in which an equilibrium might lie: (i) the "breakdown" region where M prefers to appropriate all the funds he is given; (ii) the interior "skimming" region where skimming is sustainable in equilibrium; (iii) the "transition" region between (ii) and (iv) in which M takes nothing in equilibrium, but would were he to be given more funds; and finally (iv) the "honest" region where M takes nothing regardless of how much he is given. Cutoffs between the regimes are in general complex functions of the parameters. As we are interested primarily in γ, however, we will only determine cutoff values in this parameter, but one should note that this is only a one-dimensional picture of the cutoffs in the larger parameter space.

First consider the case where the constraint $x = w_0$ binds, so M prefers to take the money and run. In this regime B would obtain the same minimal success probability – $p(0;\alpha,\beta)$ – as he would by not funding M at all, and so in equilibrium no money

[28] Proofs for these and all other results in this section can be found in the Mathematical Appendix.

[29] It is important to keep in mind that this simplification arises due to the homogeneity of the pool of middlemen and the lack of switching costs. Future work will relax these assumptions, resulting in more complex retention rules.

[30] In an infinitely repeated game, the continuation value is identical to the value of the game at its outset. Given our assumption of stationarity we may thus maximize (3) directly. A slightly more intuitive approach, but one that entails more notation, would be to calculate a first-order condition from M's full utility, separated into present and future components, as discussed in a previous footnote. This yields $\gamma v'(x^*) - (1-\gamma)p'(w_0 - x^*) - \delta_M(p'(w_0 - x^*)q_S - p'(w_0 - x^*))q_F) \times (\gamma v(x^*) + (1-\gamma)p(w_0 - x^*))[(1 - \delta_M(p(w_0 - x^*)q_S + (1 - p(w_0 - x^*)q_F))]^{-1} = 0$. The total expected future payoff, contained in M's continuation value, is thus explicitly compared to the marginal present benefit. Utilizing Lemmas 1 and 2, multipling through by $\frac{(1-\delta_M p)}{\gamma p'}$, and simplifying yields the same (4) as we obtained by maximizing (3), as it must.

changes hands. While the equilibrium in this case is therefore trivial, *when* this breakdown of cooperation occurs is of vital importance. We find that this occurs in situations where the money granted to each M is limited, usually due to a budget constraint, and so we put off further discussion of this case until the next section.

Next, consider the regime in which neither constraint on M binds, so there is some potential for skimming. In order to analyze (5) in this case, we must first implicitly differentiate the equilibrium value x^*. The required derivative is:

$$\frac{dx^*}{dw_0} = \frac{v'(x^*)[\delta_M (p')^2 + (1-\delta_M p)p'']}{(1-\delta_M p)[v''(x^*)p' + v'(x^*)p'']}. \tag{6}$$

Combining (5) and (6) yields the condition:

$$H'(w_0^*) = \frac{(p')^2[v''(x^*)(1-\delta_M p) - \delta_M v'(x^*)p']}{(1-\delta_M p)[v''(x^*)p' + v'(x^*)p'']}. \tag{7}$$

(4) and (7) implicitly define (x^*, w_0^*) in the "skimming" region.

Now consider the "transition" region. Here x^* is zero, but B cannot simply provide his optimal level of funding, and so achieve his optimal chance of success, as he can in the "honest" regime described below. Here M will skim if B pays some w greater than the equilibrium value of w_0, the point at which M's utility from starting to skim balances exactly against the lowered probability of success this skimming would cause. More funding would allow M to skim while also increasing p; mathematically this means that $\frac{dx^*}{dw_0} > 0$ in this regime. Thus γ_1, the cutoff between the "skimming" and the "transition" regimes, is the highest value of γ such that $x^* = 0$, and is obtained by solving (4) with $x = 0$ for γ. This is:

$$\gamma_1 = \left[1 + v'(0)\frac{[1-\delta_M p_1]}{p_1'} - v(0)\delta_M\right]^{-1}, \tag{8}$$

where p_1 is the equilibrium value of the interior solution obtained by decreasing the value of γ until M just stops skimming. The probability of success, $p_1(w_0)$, in the "transition" region is given implicitly by (4) with $x = 0$. It is worth noting again that the existence of the conflict of interest depresses funding levels even though no actual skimming takes place. We will see this graphically in Sect. 5.

Finally, consider the case where the constraint $x = 0$ binds M regardless of how much money he is given: M is sufficiently motivated by the cause so as to give up graft entirely. In this case x^* is unchanging regardless of how much funding M receives – it is simply 0 always – and so (5) becomes

$$p_0' = H'(w_0). \tag{9}$$

This specifies p_0, the equilibrium value of the success probability in the honest regime.

Since B maximizes his own utility without regard to M's desires here, p_0 is the ideal success rate for B, taking into account the other potential uses of his money. Not surprisingly, we find that this best possible situation for B occurs in general when middlemen are less desirous of money, at least in relation to the mission's success. Define γ_0 as the cutoff between "transition" and "honest" regimes. This is the point at which both $x^* = 0$ and $\frac{dx^*}{dw_0} = 0$, and can be found by substituting p_0 for p_1 in (8).

Together, (4)–(9) provide equations that specify both x^* and w_0^* analytically in all four regimes, and give the cutoffs in γ between the last three as well. In the next subsection we will build an intuition for the model's behavior by signing comparative statics of interest under the assumption of specific, empirically reasonable functional forms. Before doing so, we briefly consider the effect certain variations of the model would have on this equilibrium.

There are two slight alterations to the present model that can be productively examined without going too far afield: a change to the way in which B pays M, and a change to the signal that B receives. We examine the logic of each in turn.[31]

Presently the boss is constrained to fund middlemen through what may be thought of as a block grant: both "salaries" and operational resources are taken from the same sum. While we justified this assumption in Sect. 2, it is not unreasonable to ask what would change if a payment, possibly contingent on the success of the attack, were offered in addition to the normal level of funding.

For the noncontingent payment, the answer is very little. When a change does occur, it is uniformly negative for the boss, implying that such a payment would never be used in equilibrium. This occurs because a noncontingent payment is indistinguishable from graft for M: increasing such a payment decreases graft equally much. Thus, this method cannot improve the outcome for B, but can hurt him because an M who would not want to engage in graft would also prefer the noncontingent payment be spent on operations. Since B would rather have all the funding be usable for an attack – unless a different funding mechanism could spur less graft – there is no reason for B ever to use this mechanism. Essentially, the model mirrors what we see in reality: graft is often an implicit salary.[32] Notice, though, that it is a salary only imperfectly defined; the middleman's attributes dictate what he gets, as opposed to the boss's choice.

In contrast, the existence of a contingent payment alters the outcome substantially. By providing a payment that depends on the success of the attack, the boss provides strong incentives to the middleman to minimize graft, since now success yields money in addition to its other benefits. Thus, while such a payment is not used in the regime where no graft takes place, it is used substantially when graft would otherwise be seen, with just enough of a payment so as to make M skim basically nothing off of the top of the block grant. Because receiving such a payment depends on p, which is decreasing in x, the net effect of contingent payments is to further align the motivations of the middlemen with those of B. Allowing such payments lets the boss achieve roughly the same outcome at less cost, lessening the budget constraint at which cooperation breaks down. The fact that we do not observe such

[31] Because they add little to the overall, analysis we do not include the analytical and computational results that underlie this logic. They are available from the authors upon request.

[32] Graft as an implicit salary could explain high levels of corruption within Palestinian militant organizations prior to the Oslo Accords.

payment is testament either to the fact that contingent payments entail extra communications, and thus a security cost, or, less positively, to our inability to discern terrorist organizations' behavior.

In a different vein, we ask: What if the boss can roughly determine the level of graft in M, perhaps by noting recent changes in his middlemen's lifestyles? Though the mathematics quickly get complex, we can discern the broad strokes of this change's effect from the form of M's utility, equation (3). Adding a signal of the level of M's skimming impacts only B's decision as to whether or not to fire a particular M in equilibrium[33], effectively making δ a function of x. Since this $\delta(x)$ will be decreasing in x – implying that agents who skim more have more of a chance of being discovered and fired – such a signal of skimming will provide an additional incentive for agents not to skim, with roughly the same impact that an equivalent decrease in δ would have. By not including such a signal, we are thus treating the worst case scenario for the boss.

4.2 Comparative Statics

As we saw in the previous subsection, the way in which M responds to changes in funding levels is of central importance to the decision-making of both Boss and Middleman. The equation specifying this in the interior regime (when x is constrained it clearly does not vary) is given in (6), which tells us whether middlemen increase or decrease their skimming when the level of funding is increased. As the next result makes clear, the answer depends upon several factors.

Proposition 1. *Assume that p is increasing and concave over the region in which an interior solution obtains and that $v(x)$ is increasing. Then there exists an $\varepsilon > 0$ such that for all $\delta_M \leq \varepsilon$, x^* is increasing in w_0.*

Proposition 1 tells us that, when agents do not care much about the payoffs from future periods (a setting we can mimic either by setting $\delta_M = 0$ or $C = 0$), increasing the level of funding increases the amount skimmed. This is intuitive: a greedy and short-sighted agent has little motivation to spend money on an attack that provides a low level of utility, even if it should succeed. Unfortunately, we cannot say more than this; (6) can be both positive and negative in general, depending on γ, δ_M, and the functional form of p. This indeterminacy propagates through due to the centrality of M's skimming decision, and we find that, without further assumptions, we cannot sign relevant comparative statics.[34]

In order to circumvent this indeterminacy, we must make additional assumptions on our functions v, H, and p. In what follows, we assume what we believe are

[33] Lowering the level of w_0 is less efficient, as it results in a lower level of p for B than would otherwise be obtainable by increasing the likelihood of a firing.

[34] In fact, even when $\delta_M = 0$, the sign of many comparative statics is in general indeterminate.

reasonable functional forms for each of the three, and derive a full range of comparative statics using these. Before doing so, we briefly introduce our functional forms, accompanied by empirical justifications of our choices whenever possible.[35]

For p, we assume that an attack succeeds with probability

$$p(w_0 - x; \alpha, \beta) = \frac{e^{\beta(w_0 - x - \alpha)}}{1 + e^{\beta(w_0 - x - \alpha)}}, \tag{10}$$

where α and β define the difficulty of successfully completing the attack and the sensitivity of the success rate to funding, respectively. This form for the probability function has three distinct advantages. First, it matches our intuition: for low levels of spending, success rates increase slowly; at moderate levels of spending, each additional unit of spending greatly enhances the probability of success; once spending levels reach a certain level, there are diminishing marginal returns in terms of success. Second, it asymptotes nicely at 0 and 1, so we do not need to truncate it artificially. Third, using this form the parameters of our model can be estimated using a standard logistic regression, although doing so requires data on how much is being spent on attacks and on the rate of both successful and failed attacks.[36]

For v, we assume M's overall level of wealth is large enough that his decisions regarding any one attack are made within the linear section of his utility curve.[37] Thus we define $v(x) = bx$.

Finally, for H we again make the empirically reasonable assumption that the range of money B pays out to M in equilibrium is a relatively minor fraction of B's overall wealth.[38] This allows us to treat B's utility function as linear in this range,

---

[35] It should be noted that, though we cannot sign derivatives for the fully general case, our results do not rest entirely on the exact functional forms of p, H, or v. Holding the other two functions constant, as long as p is always increasing and concave over the region for which a "skimming" equilibrium exists, we can reproduce nearly all the results below. Most of the comparative statics continue to hold as well as long as H is a convex function. The only real difference is that the boss is less likely to increase funding to greedy agents, so that p is no longer constant with γ in equilibrium. Letting v be non-linear, but concave, results in greater complexity, and here we can only say that the middleman still reacts to changes in the parameters in the same way, holding funding constant, only he now values the future more due to his risk aversion. Because we believe the simplifications we use are reasonable and do not feel that the significantly increased algebraic complexity entailed in relaxing them yields additional insight, we do not report these more general results here. They are available upon request.

[36] Several datasets provide information on terrorist successes and failures in the Palestinian-Israeli conflict. Unfortunately no time-series data exist on the costs of terrorist attacks. For data on failed and foiled attacks from October 2000 to August 2004, see Merari (2004). For data that includes successful attacks, Israeli interdictions, and targeted killings by Israel, see Kaplan, Mintz, Mishal, and Samban (2005).

[37] Linearity is a common assumption, is extremely tractable, and no other functional form, such as $v(x) = \sqrt{x}$, is obviously better. Moreover, we rarely see individuals getting rich by supporting terrorism, or using money from supporting terrorism as their sole source of income. Where they do, it is on the fundraising end through skimming money from fundraising ventures. For examples of this, see Adams (1986: 85, 103).

[38] Recall we showed in Sect. 1 that attacks appear to consume a small portion of groups' overall resources.

so that $H(w_0) = cw_0$. Using these three functions with (4)–(9) yields the following lemma.

Lemma 3. *In the interior, "skimming" region, and thus in the "transition" and the "honest" regions as well, $p > \frac{1}{2}$. Further, $\frac{dx^*}{dw_0} > 0$ whenever $\delta_M < \frac{2p-1}{p^2}$, and less than or equal to zero otherwise.*

Lemma 3 states that, when there is no budget constraint, B will not fund M unless the probability of success is at least one-half. Further, there are numerous circumstances in which M's skimming becomes *less* of an issue the *more* money he receives from B. The reason for this perhaps counterintuitive result can be seen in M's utility function, given in (3). M discounts the future according to both δ_M and the probability that he will be rehired, which here is just p.[39] The form of the utility is therefore very sensitive to changes in p, producing increasing returns for M's decreasing the amount he skims in certain regions. Due to the functional forms of p and M's utility, these increasing returns occur when p is close to $\frac{1}{2}$, and when there are high discount factors. In this region, therefore, a small increase in funding for M can lead to a substantial increase in the likelihood that an attack would succeed, a result we discuss further in the next section. There we also discuss the impact of a budget constraint in more depth. Such a constraint actually makes B's decision-making easier in the region in which it binds, in that B's decision devolves to whether or not to fund with the entire budget or with nothing at all.

Proposition 2. *Increasing the disutility B obtains for funding attacks – c – decreases the level of funding B offers M and decreases the probability of success of the attack in both "skimming" and "honest" regions, while having no impact on either in the "transition" region. Increasing c does alter the range in γ over which the transition region obtains, though, increasing both γ_0 and γ_1. (Recall that γ_0 is the cutoff between "transition" and "honest" regimes, and γ_1 is the cutoff between "skimming" and "transition" regimes).*

The statement that increasing B's disutility for funding decreases funding in equilibrium is hardly surprising. Lemma 3 makes the rest of the statement less obvious, however, since the amount of skimming can be either increasing or decreasing in funding depending on δ_M. Proposition 2 cuts through the indeterminancy and states assuredly that the more B values actions other than attacks, the less likely attacks are to succeed.

The results involving the transition region are more complex, but may be understood by recalling that this region occurs when B prefers not to increase his funding, because then M would start skimming. Thus the equilibrium values of p and w_0 are independent of c in this region. Despite this, c does have an effect; increasing c means that B's optimal funding level is lowered, increasing γ_0 as he is

---

[39] Note that p is only equivalent to the probability of being rehired by virtue of Lemmas 1 and 2. Unlike the rehiring probabilities q_S and q_F, p depends explicitly on both x and w_0 in equilibrium. M therefore will not react to a high probability of being rehired by skimming more, as this would cause his chance of being rehired to decrease, resulting in a less favorable outcome for M.

able to achieve this lower level with more greedy agents. Further, as he wants less to subsidize skimming, γ_1 increases as well.

Proposition 3. *Increasing M's value for money or M's greed – b or γ – in the "skimming" region increases both the funding B offers M and the amount M skims, while leaving the probability of the attack's success constant. In the "transition" region, increasing both decreases both funding and success probability, while in the "honest" region, altering either parameter has no effect on funding or the probability of success. In addition to these within-region variations, raising b lowers both cutoffs γ_0 and γ_1 so that less greedy agents are required to ensure better scenarios for B.*

Again the statement of the proposition contains more and less surprising results. While it is intuitive that increasing M's greed or value for money should yield additional skimming, it is by no means obvious that B should simply replace the extra money lost to skimming, so as to keep the likelihood of success the same.[40] In the absence of a budget constraint, utilizing greedier agents does not alter the outcome of the attempted attack in the "skimming" region. What it *does* alter, as we shall see in the next section, is what level of budget constraint binds, and thus the point at which cooperation within the organization breaks down.

Proposition 4. *Increasing the degree to which M values the future, his discount rate δ_M, has an indeterminate effect on both skimming and funding levels in the "skimming" region, but increases the probability of success of an attack. It raises both funding and the probability of success in the "transition" region, and has no effect on either in the "honest" region, though it does raise γ_0, allowing more greedy agents to be used in the "honest" region.*[41]

The feedback mechanisms discussed after Lemma 3 make signing comparative statics involving δ_M difficult: increasing M's patience leads to lower levels of skimming, which, depending on other parameters, can lead to decreased funding, which can then lead to increased skimming, and so on. Despite this, the effect on the mission's chance of success is clear: a more patient M implies a greater likelihood of success.

Proposition 5. *In all regions in which funding is provided, creating a more difficult environment for terrorist activity through increasing α increases levels of funding by the same amount, leaving skimming and the probability of success unchanged. Increasing the sensitivity of the success probability to the net level of funding through increasing β decreases skimming and has an indeterminate, though usually negative, effect on funding in the "skimming" region. Increasing β increases the success probability in equilibrium in all regions in which funding is provided and also raises both γ_0, allowing more greedy agents to be used in the "honest" region, and γ_1, so that more agent types fall into the "transition" region.*

---

[40] This is due to B's assumed linear preference for money, as discussed briefly in footnote 35.

[41] Increasing δ_M also increases γ_1 for all $\delta_M > 1/2$.

The difference between the equilibrium dependence on each type of environmental parameter clarifies our choice of functional form for p. The first, α, has no effect on the likelihood of success until the budget constraint binds,[42] but increasing α does directly decrease the slack in B's budget. Thus we would expect a non-linear impact of increasing α: no effect until the budget constraint is reached, and a big effect afterward as cooperation breaks down. This is what we see in the next section. In contrast, β has a more gradual impact, in that increasing it makes the cost of underfunding greater for M and provides better information to B about the likelihood of M's failing to fund the attack sufficiently.

5 Discussion

Thus far we have shown that an equilibrium of the game exists and illustrated analytically how it changes when the parameters shift. Here we flesh these results out by exploring computationally the effect of a budget constraint on the Boss-Middleman interaction[43] utilizing the specific functional forms discussed in Sect. 4.2. This illustration reveals important nonlinearities in terrorists' responses to government action under very reasonable assumptions about functional forms and parameterizations.

We begin by choosing values for the operational environment, α and β. Since estimating these parameters for a particular conflict is impossible without data on the costs of attacks, we arbitrarily set $\alpha = 600$ and $\beta = .005$. With $\beta = .005$ groups have a 10 percent success rate when funding per attack is at \$300.

Given extant data on Al Qaeda and affiliated groups such as Jemaah Islamiyah (JI), we were able to use realistic values for δ_M. Here we consider the interpretation that δ represents the probability that the game will not be ended by an exogenous shock such as government action. We found that the average survival rate for middlemen between 1997 and 2003 was approximately 83 percent.

We set c, the boss' marginal disutility for spending on attacks, so that B's utility is approximately zero for the case of p near 1 with $\gamma = 0$, so that zero utility corresponds to receiving a particular payoff from the least greedy agent. This yields an estimate of $c = 1/(4\alpha)$, which is low enough that the boss would want to fund an attack up to his ideal level if skimming by the middlemen were not an issue. We find this reasonable, as bosses with higher values of c than this would be uniformly less willing to fund attacks in all cases, and thus be less of the type we are studying here. Nevertheless, we do discuss what happens if c increases below.

We use a similar strategy to calculate b. Since we have a different parameter, γ, to describe M's relative utility for increasing success probabilities and skimming money, all b must do is put dollar amounts on the same footing as probabilities, which are bounded by zero and one. Accordingly, we calculate b by equating M's

[42] This again is due to the linearity of B's disutility for money.

[43] The budget constraint places a limit on how much the boss can allocate to funding the middleman. The computational procedure for finding the equilibrium does not consider funding levels above the budget constraint.

utility over a range of parameters, setting a perfectly greedy middleman's equilibrium utility at $p = \frac{1}{2}$, equal to a perfectly committed M's utility, also at $p = \frac{1}{2}$. This yields an estimate of $b = \frac{\beta\delta}{8(1-\delta/2)}$. While not perfect, we know from the previous section that positive changes to this parameter have consistent positive effects on the equilibrium levels of funding and skimming for $p > \frac{1}{2}$, and so our exact choice is not important.

To summarize, we use the following parameters to explore the model: [44]

- $\alpha = 600$
- $\beta = .005$
- $\delta_M = .83$
- $c = 0.00042$
- $b = 0.00089$

Using these estimates, we solve numerically for the constrained maximization problem described in the previous section. This amounts to maximizing M's utility over the range $x \in [0, w_c]$, where w_c is the budget constraint in effect, and then maximizing B's utility over the range $w_0 \in [0, w_c]$, for a range of γ.[45] Allowing w_c to vary from zero to a point high enough such that it would not bind anyone's actions – in this case, to $1,500 – provides us with the ability to explore the entire range of potential interactions, from the point at which M would rather just take the money and run, to the point where B has achieved a sufficient success rate and chooses to spend the rest elsewhere. The results of this are summarized in Figures 1–4.

Figure 1 shows the amount B spends as a function of his budget constraint and how venal his middleman is. Figure 2 shows the amount M skims, and Figure 3 shows the probability that the attack will succeed, both as a function of w_c and of γ.

There are two major things to note here. First, the equilibrium level of funding in Fig. 1 – that which occurs at a budget constraint of $1,500, and sometimes considerably less – does not linearly increase with venality. While that is what we would intuitively expect, as greedier middlemen might require more money to overcome their increased skimming and ensure a reasonable level of success, Fig. 2 shows why it is not the case. Below some level of venality, it no longer pays for M to skim at all, and the boss can eke out an additional 10 percent in the success rate by upping the amount of money he pays to M relative to what he would be willing to do if skimming were an option for M. The existence of this cutoff in γ – derived analytically in the previous section – leads to a clear policy implication: state action that serves to remove the most venal of the available middlemen can lead to a sizable *increase*

---

[44] We did not choose set values for γ, as that is our key causal variable, and in all the illustrations we either explore its full range or estimate results for $\gamma = \{.25, .5, .75, .9, .1\}$. We do not consider the perfectly committed middleman, as the case is trivial in this model. In future models where the boss wants to achieve a specific level of impact, rather than maximizing his success probability, we will consider the case where a boss gets more impact than he wants by hiring a perfectly committed agent, an agent for whom $\gamma = 0$.

[45] Essentially, w_c serves as an endpoint in the players' strategy space in the computational search for an equilibrium.

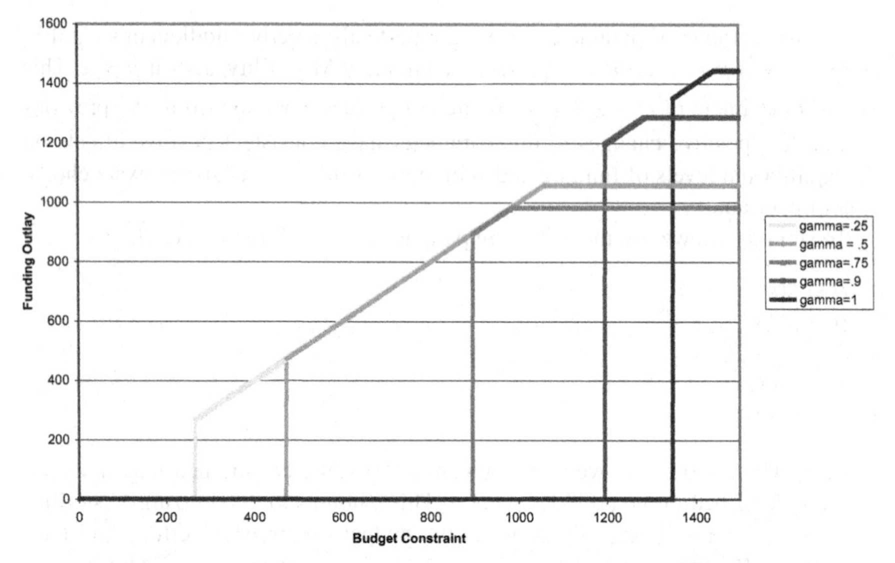

Fig. 1 Funding Outlay as a Function of Budget Constrain ($\beta = 0.005$, $\alpha = 600$)

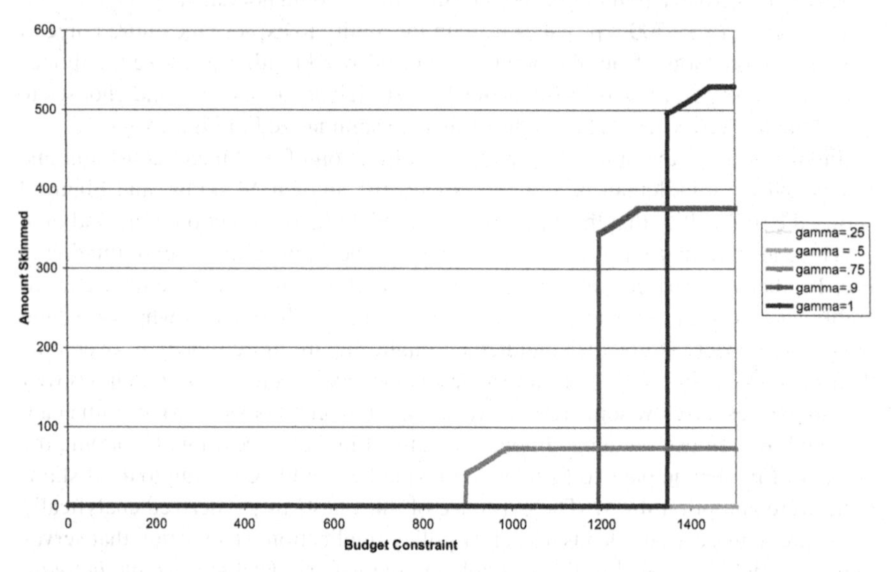

Fig. 2 Amount Skimmed as a Function of Budget Constraint ($\beta = 0.005$, $\alpha = 600$)

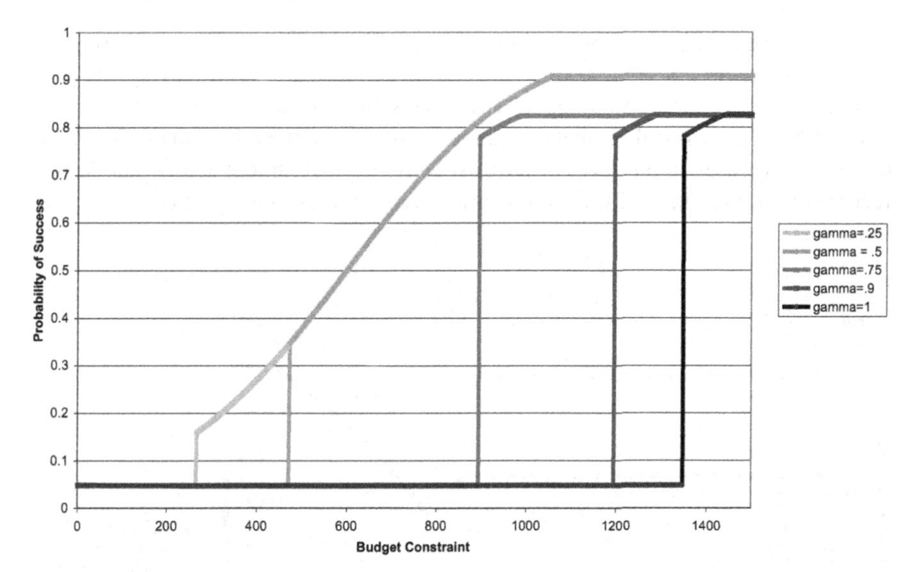

Fig. 3 Probability of Success as a Function of Budget Constraint ($\beta = 0.005$, $\alpha = 600$)

in the likelihood of terrorist success if those middlemen are replaced by more committed types. Increasing the risks to middlemen is likely to have a similar impact by pushing less committed individuals to give up the cause. This suggests that efforts to limit terrorists' finances should be focused on interdicting funds, rather than on capturing financiers.

Second, the behavior of each quantity of interest is decidedly nonlinear in the budget constraint. Figure 3 clearly illustrates this dynamic. For all levels of middleman greed, there is a constraint level below which the chance of an attack's succeeding becomes small due to the breakdown of cooperation between boss and middleman.[46] Essentially, there is a point at which B prefers to spend all the money on non-attack goods rather than on achieving the low success probability he can get given his budget and M's greed. However, once the constraint exceeds that level – which will be different for different values of the other parameters – the chance of success undergoes a discontinuous increase, up to 80 percent in the case of the more venal agents. Past this point, increasing the level of funds available to B causes a more gradual, nearly linear increase in all the variables, until an equilibrium is reached, whereupon B no longer desires to increase the likelihood of success. This leads to a second clear policy implication: interdicting terrorists' funds can provide dramatic decreases in the likelihood of an attack, even in cases where previous reductions in available funds seemed to have a more gradual impact.

Note that government actions which make the operational environment more challenging, increasing α, should have effects similar to tightening the budget

[46] It is only nonzero due to the form of the success probability and the particular parameter estimates in use.

constraint. Thus we expect important nonlinearities in the returns to government counter-terrorism. For some budget constraints and levels of α, it may appear that counter-terror policies are having no effect. However, once government makes the environment hard enough so that the budget constraint starts to bite, there should be dramatic decreases in the level of terrorist activity. Government actors that do not recognize the existence of this nonlinearity may cease eventually effective actions too soon.

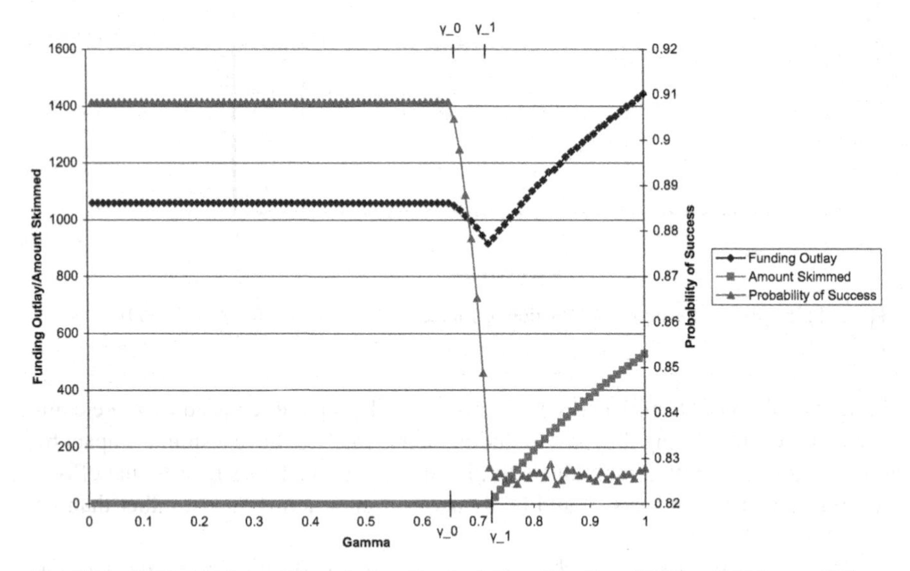

Fig. 4 Equilibrium Values with no Budget Constraint as a Function of γ ($\beta = 0.005$, $\alpha = 600$)

Figure 4 displays results of a different sort, mirroring the comparative statics described in the previous section. Here we examine how the equilibrium values of p, x, and w_0 vary with γ when no budget constraint binds. We note several things. Past the cutoff point given earlier in Equation (8), which for these parameters yields $\gamma_1 = 0.72$, the equilibrium values of both skimming and funding come from the interior "skimming" solution described in the previous section. Accordingly, they increase at the same rate with venality, and the probability of success therefore remains constant. Below this point, the situation is much different. M does not skim, and so the boss ups his financial outlay to achieve a higher success rate. This is the "transition" region discussed in the previous section, and here the success rate continues to climb with further decreases in γ until a second cutoff is reached, at $\gamma_0 = 0.65$. At this point B has achieved his optimal level of funding given his other uses for money – the level he would choose if he did not require M – and further decreases in γ have no effect on the equilibrium.[47]

[47] Figure 4 also reinforces the point that removing the most venal middlemen may lead to sudden increases in terrorists' success rates.

Because we are primarily interested in γ and the budget constraint, due in large part to their clear relation to policy, we do not produce similar figures for all the other parameters in the model. Thanks to the work of the previous section, however, we know what varying each of the remaining parameters would entail.

Increasing c decreases the probability of a successful attack, and shifts the cutoffs between regions to the right. It also decreases the area over which a non-zero equilibrium obtains, and a large enough c can cause cooperation to vanish entirely. Thus increasing the relative utility of non-violent uses of money can have a disproportionate effect on reducing the rate of successful attacks. This finding thus provides an explicit organizational mechanism by which encouraging groups using terrorist tactics to enter the political arena – or to engage in social service provision – can reduce the level of violence.

Increasing b has almost the same effect as increasing venality, and so is of little additional interest. From the previous section, increasing δ_M increases the probability of success of an attack as long as M's skimming or potential skimming has an effect on B's funding levels. Further, increased patience mitigates increased greed, to some extent, in that increasing δ_M generally causes more greedy agents not to skim in equilibrium. We infer from the form of the utility for M, (3), that increasing δ_M results in a tighter budget constraint being necessary for cooperation to collapse.

Increasing α has no effect on the probability of success when there is no budget constraint, but it does cause an existing constraint to bind sooner, leading to the outcomes described above. Finally, increasing β increases the success probability in equilibrium, and reduces the types of M who choose to skim in equilibrium, shifting the skimming region to the right in Fig. 4. Computational work indicates that increasing β also produces a wider range of budgets in which cooperation can occur. This is because an increase in β makes the transition from a high probability of success to a low probability more abrupt, effectively increasing the informativeness of an attack's success or failure which, much like the additional signal of skimming discussed in the previous section, is generally good for B.

6 Conclusion

We have presented a model of a hierarchical terrorist organization in which leaders must delegate financial and logistical tasks to middlemen for security reasons. However, these middlemen do not always share their leaders' interests. In particular, the temptation always exists to skim funds from any financial transaction. To counteract this problem, leaders can threaten to punish the middlemen. Because logisticians in terrorist organizations are often geographically separated from leaders, and because they can defect to the government if threatened, violence is rarely an effective threat. Therefore, leaders must rely on more prosaic strategies to solve this agency problem; we focused on leaders' ability to remove middlemen from the network, denying them the rewards of future participation.

Because our model is inherently hierarchical, its explanatory scope is limited to those groups which operate through a defined chain of command. As such, the model's applicability to Al Qaeda and related groups may be diminishing. Prior to the loss of their Afghan sanctuary, these groups operated through defined hierarchies with relatively clear distinctions between operational and supporting roles.[48] More recent attacks have followed a different, more bottom-up pattern. The London, Madrid, and Casablanca bombings were locally initiated and in all three cases some individuals undertook both logistical and operational roles. While this pattern is apparent in some groups, the majority of terrorist organizations worldwide remain hierarchically organized, and thus amenable to the principal-agent analysis we have conducted.

Our analysis yielded several important policy implications. The first is that the removal of the most venal middlemen, those who are the easiest to identify and will be most vulnerable to government incentives, actually makes the terrorist boss' problem easier and can lead to a jump in the number of successful attacks. This is in line with others' findings that as moderates leave an organization, it often becomes more violent (Bueno de Mesquita, 2005a). The second is that reducing a group's available funds below a certain threshold can have a dramatic impact, even if previous reductions yielded only gradual effects. The third is that a similar threshold effect exists for counter-terror efforts that make it harder to conduct attacks. Such efforts may yield no result until they make the environment hard enough that it is not worth it for a group to attempt any attacks. The fourth is that efforts to restrict funding to terrorist organizations are unlikely to reduce attacks unless they can reach a threshold which depends on the operational environment and the nature of the individuals in the group. This finding should give pause to policy-makers who must balance spending on counter-terrorist financing efforts against other, less socially damaging, counter-terror efforts. Finally, our model strongly suggests that increasing the value of non-violent uses of money, perhaps by encouraging entry into the political process, can have a disproportionate effect by making it impossible for leaders and middlemen to cooperate.

Overall, our main finding is that the level of greed among middlemen is central to the internal dynamics of terrorist groups. All of the boss' problems become harder when they have to rely on less ideologically committed individuals. While this result is by no means counter-intuitive, its implications deserve consideration. Rather than differentiate terrorist organizations solely on the basis of their strategic goals, or ideological style, our theories should take into account their internal make-up. Specifically, we should seek to understand how the motivations of members who do not volunteer for the riskiest activities impact groups' behavior. Doing so will lead to a more thorough, more effective understanding of terrorism.

[48] Felter et al. (2006) examine captured Al Qaeda documents detailing the specific goals and responsibilities for those filling different roles within Al Qaeda.

Mathematical Appendix

This appendix contains proofs of all propositions in the text. The first three results utilize the techniques of monotone comparative statics.[49]

Proof of Lemma 1: In the general case, the first-order condition that implicitly specifies x^* is $\frac{\partial C}{\partial x} = 0$, which is

$$\frac{(\gamma v' - (1-\gamma)p')(1-\delta_M(pq_S + (1-p)q_F)) - \delta_M(p'q_S - p'q_F)(\gamma v + (1-\gamma)p)}{(1-\delta_M(pq_S+(1-p)q_F))^2} = 0. \tag{11}$$

The cross-partial of M's objective with respect to x and q_S is

$$\frac{-\delta_M \gamma(pv' + vp')}{(1-\delta_M(pq_S+(1-p)q_F))^2} + \frac{2\delta_M p \frac{\partial C}{\partial x}}{(1-\delta_M(pq_S+(1-p)q_F))^2}. \tag{12}$$

The first term is strictly negative by assumption, while the second is zero by (11). Thus, x^* is decreasing in q_S. Since there is no cost to B of increasing q_S and a positive benefit to decreasing x, B maximizes his utility with $q_S^* = 1$. □

Proof of Lemma 2: We follow the same logic as in Lemma 1. The cross-partial of M's objective with respect to x and q_F is

$$\frac{\delta_M(\gamma(v'(1-p)+p'v)+(1-\gamma)p')}{(1-\delta_M(pq_S+(1-p)q_F))^2} + \frac{2\delta_M(1-p)\frac{\partial C}{\partial x}}{(1-\delta_M(pq_S+(1-p)q_F))^2}. \tag{13}$$

The first term is strictly positive by assumption, while the second term is zero by (11). Thus, x^* is increasing in q_F. Since there is no cost to B of increasing q_F and a positive benefit to decreasing x, B maximizes his utility with $q_F^* = 0$. □

Proof of Proposition 1: Assume that p is increasing and concave over the region in which an interior solution obtains, and that $v(x)$ is increasing. Taking into account Lemmas 1 and 2, $\frac{\partial C}{\partial x} = 0$ becomes

$$\frac{(\gamma v' - (1-\gamma)p')(1-\delta_M p) - \delta_M p'(\gamma v + (1-\gamma)p)}{(1-\delta_M p)^2} = 0. \tag{14}$$

The cross partial of this with respect to w is

$$\frac{-\gamma\delta_M v'p' - (1-\gamma+\gamma\delta_M v)p''}{(1-\delta_M p)^2} + \frac{2\delta_M p' \frac{\partial C}{\partial x}}{1-\delta_M p}. \tag{15}$$

The second term is zero by (14). The first is positive – implying that x^* is increasing in w_0 – whenever

$$\gamma(v'p' + vp'')\delta_M < -(1-\gamma)p''.$$

---

[49] Specifically, Theorem 3 in Ashworth and Bueno de Mesquita (2006).

The term on the right hand side of this inequality is always positive by assumption. If the term on the left hand side is negative, then the inequality is true for all δ_M. If it is positive, then by continuity there must exist a maximal $\varepsilon > 0$ such that the inequality holds whenever $\delta_M \leq \varepsilon$. □

The remaining results for the "skimming" region utilize total differentiation of (4) and (7), with the appropriate functional forms inserted. Simplified, (4) and (7) become

$$\frac{1-\delta_M p}{\beta p(1-p)} - x^* \delta_M - \frac{1-\gamma}{\gamma b} = 0, \tag{16}$$

$$\frac{\delta_M \beta p^2 (1-p)^2}{(1-\delta_M p)(2p-1)} - c = 0. \tag{17}$$

The total derivative for w is computed directly from (17); that for x is computed using this result, along with the partial derivative of (16) and the fact that $\frac{dx^*}{dv} = \frac{\partial x^*}{\partial v} + \frac{\partial x^*}{\partial w_0^*}\frac{dw_0^*}{dv}$, where v is any parameter. Subtracting the latter from the former dictates how p changes, since it is increasing in the difference $w_0^* - x^*$. For the "transition" region we directly differentiate (16) with $x^* = 0$. For the "honest" region we directly differentiate (9).

Determining the effects of parameter variation on the cutoffs γ_0 and γ_1 is somewhat more complex, due to the fact that they each depend upon the equilibrium value of p, and the derivatives of p with respect to the parameters are discontinuous at the cutoffs in many cases. Thus, to sign these comparative statics we look instead at relative changes in p between regions, relying upon the definition of the "transition" region as that in which $p < p_0$ and $x^* = 0$. If p is decreasing faster in some parameter in the honest region than in the transition region, then the level of greed, γ_0, required for M to be willing to skim, and hence for B to have to take this potential into account, will increase. Otherwise it decreases. Likewise, if p is decreasing faster in the skimming region than in the transition region, the level of greed for which choosing w_0 such that $x^* = 0$ is optimal for B will increase. Otherwise it decreases. Only the actual derivatives are given below; full derivations can be obtained from the authors.

Proof of Lemma 3: The form of (17) directly implies the first part of the lemma: since c is positive, $(2p-1) > 0$ if the "skimming" region is to have a solution. The same is true for the "transition" and the "honest" regions by definition. The second part is derived similarly, using the simplified version of (6):

$$\frac{dx^*}{dw_0} = \frac{1-2p+\delta_M p^2}{(1-\delta_M p)(1-2p)}. \tag{18}$$

With $p > \frac{1}{2}$, the denominator of (17) is negative, so $\frac{dx^*}{dw_0} > 0$ whenever $1-2p+\delta_M p^2 < 0$, and less than or equal to zero otherwise. □

Proof of Proposition 2: For c, the total derivatives are:

$$\frac{dw_0^*}{dc} = \left[\frac{(1-\delta_M p)^3(2p-1)^3}{\delta_M^2\beta^2p^3(1-p)^3}\right]\frac{1}{Q}, \tag{19}$$

$$\frac{dx^*}{dc} = \left[\frac{(1-\delta_M p)^2(2p-1)^2}{\delta_M^2\beta^2p^3(1-p)^3}\right]\frac{-1+2p-\delta_M p^2}{Q}, \tag{20}$$

where $Q = -2+(6+\delta_M)p-3(2+\delta_M)p^2+4\delta_M p^3$, which is always negative for $p > \frac{1}{2}$. We will use Q throughout. (20) is of indeterminant sign. (19) is negative for $p > \frac{1}{2}$, as is the difference between (19) and (20), implying that p is decreasing in c as well in the "skimming" region. In the "transition" region, the equilibrium value of p is independent of c from (16). In the "honest" region, differentiating (9) yields $\frac{dp_0}{dc} = -\frac{1}{2\beta}\left(\frac{1}{4}-\frac{c}{\beta}\right)^{-\frac{1}{2}} < 0$. As p is unchanging in c in the transition region, and decreasing in c in both other regions, both cutoffs are increasing in c. □

Proof of Proposition 3: The total derivatives for b are

$$\frac{dw_0^*}{db} = \frac{(1-\gamma)}{\gamma b^2\delta_M}, \tag{21}$$

$$\frac{dx^*}{db} = \frac{(1-\gamma)}{\gamma b^2\delta_M}, \tag{22}$$

and those for γ are

$$\frac{dw_0^*}{d\gamma} = \frac{1}{\gamma^2 b\delta_M}, \tag{23}$$

$$\frac{dx^*}{d\gamma} = \frac{1}{\gamma^2 b\delta_M}. \tag{24}$$

All are positive for $p > \frac{1}{2}$. The differences of each pair are zero, so p does not change with b or γ. In the "transition" region, the derivative with respect to b is: $\frac{dp(w_0)}{db} = \frac{-(1-\gamma)\beta p^2(1-p)^2}{\gamma b^2(-1+2p-\delta_M p^2)} < 0$, since (18) is positive in this region, and with respect to γ is: $\frac{dp(w_0)}{d\gamma} = \frac{-\beta p^2(1-p)^2}{\gamma^2 b(-1+2p-\delta_M p^2)} < 0$ for the same reason. In the "honest" region, the probability of success depends on neither b nor γ. As p is decreasing in the transition region and constant in the skimming and honest regions, both cutoffs are decreasing in b. □

Proof of Proposition 4 The partial derivative of x^* with respect to δ_M is

$$\frac{\partial x^*}{\partial \delta_M} = \frac{-p}{\beta(1-\delta_M p)(2p-1)}\left[1+\beta x^*(1-p)\right].$$

This is negative for $p > \frac{1}{2}$, but neither of the following two total derivatives can be signed:

$$\frac{dw_0^*}{d\delta_M} = \left(\frac{1-\gamma}{\delta_M^2 b\gamma}\right)+\left(\frac{-1}{\delta_M^2\beta p(1-p)}\right)\left[\frac{-1+2p-(2-\delta_M)p^2}{Q}\right]. \tag{25}$$

$$\frac{dx^*}{d\delta_M} = \left(\frac{1-\gamma}{\delta_M^2 b\gamma}\right) + \left(\frac{-1}{\delta_M^2 \beta p(1-p)}\right)\left[\frac{-1+(2+\delta_M)p-2(1+\delta_M)p^2+2\delta_M p^3}{Q}\right]. \tag{26}$$

In each case the second term is always negative, while the first is always positive. The difference of the two derivatives is positive, though, implying that p is increasing in δ_M in the "skimming" region. In the "transition" region, $\frac{dp(w_0)}{d\delta_M} = \frac{p^2(1-p)}{-1+2p-\delta_M p^2} > 0$ since (18) is positive in this region. In the "honest" region, the probability of success is independent of δ_M. p is increasing in δ_M in all regions but the honest one. Thus γ_0 is increasing in δ_M. To discern if the same is true for γ_1, one needs to compare the relative rates of increase in δ_M. For all $\delta_M > 1/2$, $p(w_0)$ increases more rapidly than p, implying that in this range γ_1 is also increasing. □

Proof of Proposition 5: The total derivatives for α are:

$$\frac{dw_0^*}{d\alpha} = 1, \tag{27}$$

$$\frac{dx^*}{d\alpha} = 0. \tag{28}$$

Increasing w^* and α by the same amount leaves p unchanged in the "skimming" region. α has no effect as well in the "honest" and "transition" regions, and both cutoffs are unchanging in it. The total derivatives for β are:

$$\frac{dw_0^*}{d\beta} = \frac{1}{\beta^2}\ln\left(\frac{p}{1-p}\right) - \frac{(1-\delta_M p)(-2-\delta_M+4\delta_M p)}{\beta^2\delta_M Q}, \tag{29}$$

$$\frac{dx^*}{d\beta} = \frac{(1-\delta_M p)(2-\delta_M)}{\beta^2\delta_M Q}. \tag{30}$$

Although the first is usually negative, it goes positive for very high ($\sim .96$) p. The second is negative for all $p > \frac{1}{2}$. Subtracting the second from the first yields a positive result, so increasing β increases the probability that an attack will be successful in the "skimming" region. This is also true in the both the "transition" and "honest" cases, since: $\frac{dp(w_0)}{d\beta} = \frac{p(1-p)(1-\delta_M p)}{\beta(-1+2p-\delta_M p^2)} > 0$ and $\frac{dp_0}{d\beta} = \frac{c}{2\beta^2}\left(\frac{1}{4}-\frac{c}{\beta}\right)^{-\frac{1}{2}} > 0$. As p is increasing in β in all three regions, we must compare relative rates to discern the effect on both cutoffs. $p(w_0)$ increases more rapidly than both p and p_0, implying that both γ_0 ad γ_1 are increasing. □

References

1. Abuza, Zachary. (2003a) Funding Terrorism in Southeast Asia: The Financial Network of Al Qaeda and Jemaah Islamyiah. *Contemporary Southeast Asia* **25**:169–199.
2. Abuza, Zachary. (2003b) Funding Terrorism in Southeast Asia: The Financial Network of Al Qaeda and Jemaah Islamiyah. *NBR Analysis* **14**:5.

3. Adams, Jim. (1986) *The Financing of Terror*. New York: Simon and Schuster.
4. Al-Zawahiri, Ayman. December 2, 2001. Knights Under the Prophet's Banner, part 11. Al-Sharq al-Awsat (London). Foreign Broadcast Information Service (FBIS) translation at http://www.fas.org/irp/world/para/ayman_bk.html (2005, March 1).
5. Ashworth, Scott and Ethan Bueno de Mesquita. 2006. Monotone Comparative Statics for Models of Politics. *American Journal of Political Science* 50:214–231.
6. Asia/Pacific Group on Money Laundering. (2004) *APG Annual Typologies Report*. Sydney: Asia/Pacific Group on Money Laundering.
7. Berman, Eli and David Laitin. (2005) Hard Targets: Theory and Evidence on Suicide Attacks. NBER Working Paper 11740.
8. Bounty for Murder Offered by Iran's Revolutionary Guards, Syria, Hizballah. February 26, 2004. DEBKA-Net-Weekly. http://www.debka-net-weekly.com (2004, March 15).
9. Brisard, Jean-Charles. (2002) Terrorism Financing: Roots and Trends of Saudi Terrorism Financing. Report Prepared for the President of the Security Council United Nations. New York: JCB Consulting.
10. Bueno de Mesquita, Ethan. (2005a) Conciliation, Commitment, and Counter-Terrorism. *International Organization* **59**:145–176.
11. Bueno de Mesquita, Ethan. (2005b) The Quality of Terror. *American Journal of Political Science* **49**:515–530.
12. Bueno de Mesquita, Ethan. (2005c) The Terrorist Endgame: A Model with Moral Hazard and Learning. *Journal of Conflict Resolution.* **49**:237–258.
13. Chai, Sun-Ki. (1993) An Organizational Economics Theory of Antigovernment Violence. *Comparative Politics* **26**:99–110.
14. Crenshaw, Martha. (1991) How Terrorism Declines. *Terrorism and Political Violence* **3**:69–87.
15. Cullison, Alan. (2004) Inside Al Qaeda's Hard Drive. *The Atlantic Monthly* **294**:55–65.
16. Donohue, Laura K. (2006) Anti-Terrorist Finance in the United Kingdom and United States Part II: How Effective? *Michigan Journal of International Law* **27**:303–435.
17. Farah, Douglas. (2004) *Blood From Stones: The Secret Financial Network of Terror*. New York: Broadway Books.
18. Felter, Joseph, . (2006) *Harmony and Disharmony: Exploiting al-Qa'ida's Organizational Vulnerabilities*. Westpoint, N.Y.: Combating Terrorism Center.
19. Ferejohn, John. (1986) Incumbent performance and electoral control. *Public Choice* **50**:5–25.
20. Financial Action Task Force. (2004) *Report on Money Laundering and Terrorist Financing Typologies: 2003–2004.* Paris: Financial Action Task Force.
21. Francis, David R. April 8, 2004. The war on terror money. Christian Science Monitor.
22. Frey, Bruno S., and Simon Luechinger. (2004) Decentralization as a disincentive for terror. *European Journal of Political Economy* **20**:509–515.
23. Gunaratna, Rohan. (2002) *Inside Al Qaeda*. New York: Columbia University Press.
24. Gunning, Jeroen. (2004) "Peace with Hamas?" *International Affairs* **80**:233–255.
25. International Crisis Group. (2002) "Al-Qaeda in Southeast Asia: The case of the 'Ngruki Network' in Indonesia." *ICG Indonesia Briefing* 1.
26. International Crisis Group. (2003a) "Islamic Social Welfare Activism In The Occupied Palestinian Territories: A Legitimate Target?" *International Crisis Group Middle East Report* 13.
27. International Crisis Group. (2003b) "Jemaah Islamiyah in South East Asia: Damaged but Still Dangerous." *ICG Asia Report* 63.
28. International Crisis Group. (2004) "Indonesian Backgrounder: Jihad in Central Sulawesi." *ICG Asia Report* 74.
29. Israeli Defense Forces, Ministry of Parliamentary Affairs. (2002) *The Involvement of Arafat, PA Senior Officials and Apparatuses in Terrorism against Israel, Corruption and Crime*. http://mfa.gov.il/mfa/go.asp?MFAH0lom0. (2005, March 14).
30. Jenkins, Brian Michael. (2002) *Countering Al Qaeda*. Santa Monica, C.A.: RAND.
31. Kaplan, Edward H., Alex Mintz, Shaul Mishal, and Claudio Samban. (2005) What Happened to Suicide Bombings in Israel? Insights From a Terror Stock Model. *Studies in Conflict and Terrorism* **28**:225–235.

32. Lapan, Harvey E., and Todd Sandler. (1993) Terrorism and Signaling. *European Journal of Political Economy* **9**:383–397.
33. Lee, Rensselaer. (2002) *Terrorist Financing: The U.S. and International Response.* Congressional Research Service Report for Congress, RL31658.
34. Levitt, Matthew A. (2002) "The Political Economy of Middle East Terrorism." *The Middle East Review of International Affairs (MERIA) Journal* **6**:49–65.
35. Medani, Khalid. (2002) Financing Terrorism or Survival? Informal Finance, State Collapse and the US War on Terrorism. *Middle East Report* **32**:2–9.
36. Merari, Ariel. (2004) "Suicide Terrorism in the Context of the Israeli-Palestinian Conflict", Paper presented at National Institute of Justice Suicide Terrorism Conference (Washington, D.C.), October 25–26.
37. National Commission on Terrorist Attacks upon the United States. (2004a) *The 9/11 Commission Report.* New York: W.W. Norton and Company, Inc.
38. National Commission on Terrorist Attacks upon the United States. (2004b) *Staff Statement no. 15: Overview of the Enemy.*
39. Overgaard, Per Baltzer. (1994) The Scale of Terrorist Attacks as a Signal of Resources. *Journal of Conflict Resolution* **38**:452–478.
40. Prober, Joshua. (2005) Accounting for Terror: Debunking the Paradigm of Inexpensive Terrorism. Washington Institute for Near East Policy, Policy Watch #1041.
41. Rees, Matt. 23 April, 2001. How Hamas-Hezbollah Rivalry is Terrorizing Israel. Time.
42. Ressa, Maria A. (2003) *Seeds of Terror.* New York: Free Press.
43. Roth, John. (2004) *Monograph on Terrorist Financing: Staff report to the Commission.* Washington, D.C.: National Commission on the Terrorist Attacks Upon the United States.
44. Sageman, Marc. (2004) *Understanding Terror Networks* Philadelphia: University of Pennsylvania Press.
45. Sandler, Todd. (2003) Collective Action and Transnational Terrorism. *World Economy* **26**:779–802.
46. Second Report of the [UN] Monitoring Group, Pursuant to Security Council Resolution 1390. Sept. 19, 2002. New York: United Nations.
47. Shapiro, Jacob N. (2008) "The Greedy Terrorist: A Rational Choice Perspective on Terrorist Organizations' Inefficiencies and Vulnerabilities." In *Terrorist Financing in Comparative Perspective*, edited by Harold Trinkunas and Jeanne K. Giraldo. Stanford: Stanford University Press.
48. Singapore Ministry of Home Affairs. (2003) *The Jemaah Islamiyah Arrests and the Threat of Terrorism.*
49. Siqueira, Kevin. (2005) Political and Militant Wings within Dissident Movements and Organizations. *Journal of Conflict Resolution* **49**:218–236.
50. Stern, Jessica. (2003) *Terror in the Name of God: Why Religious Militants Kill.* New York: HarperCollins.
51. Testimony of FBI Agent John Anticev on Odeh. February 2001. United States of America v. Usama bin Laden, et al. 5 (7) 98 Cr. 1023, 27, 1630–1638.
52. Victoroff, Jeff. (2005) The Mind of the Terrorist: A Review and Critique of Psychological Approaches. *Journal of Conflict Resolution* **49**:3–42.
53. Weinstein, Jeremy. (2005) Resources and the Information Problem in Rebel Recruitment. *Journal of Conflict Resolution* **49**:598–624.
54. Weschler, William F., and Lee S. Wolosky. (2002) *Terrorist Financing: Report of an Independent Task Force Sponsored by the Council on Foreign Relations.* New York: Council on Foreign Relations.
55. Weschler, William F., and Lee S. Wolosky. (2004) *Update on the Global Campaign Against Terrorist Financing.* New York: Council on Foreign Relations.

Part VI
History of the Conference on Mathematical Methods in Counterterrorism

Personal Reflections on Beauty and Terror

Jonathan David Farley

Abstract The history of the Conference on Mathematical Methods in Counterterrorism is briefly described.

1 Shadows Strike

It was a Tuesday morning, and I awoke into a nightmare. The phone rang; a friend spoke. "Turn on the television," she said.

You couldn't see it at first. This was before anyone knew it had been caught on camera. All you could see was a plane disappearing behind a building, and a burst of hellfire.

Terrorism is the watchword of the day, and the fear – regardless of whether the threat is real or imagined – requires an antidote.

2 The "Thinking Man's Game"

The opening line of the Oscar-winning movie *A Beautiful Mind* is: "Mathematicians won the war." Winston Churchill recognized Alan Turing, the mathematician who had mastered the German codes, as the man who had perhaps made the single greatest individual contribution to Hitler's defeat. Bletchley Park is now a place out of legend. During the Cold War, research in game theory heated up, even as the first frosts descended on the Soviet East.

Now there is a new war. What is the new mathematics?

Jonathan David Farley
Institut für Algebra, Johannes Kepler Universität Linz, 4040 Linz, Austria, e-mail: lattice.theory@gmail.com

N. Memon et al. (eds.), *Mathematical Methods in Counterterrorism*,
DOI 10.1007/978-3-211-09442-6_22, © Springer-Verlag/Wien 2009

At Los Alamos National Laboratory, the lab that built The Bomb, Cliff Joslyn's team used formal concept analysis, a branch of applied lattice theory, to mine data drawn from hundreds of reports of terrorist-related activity, and to discover patterns and relationships that were previously in shadow.

Formal concept analysis is a way of determining non-obvious implications between data. It could potentially help discover links between people connected to terrorist cells: individuals who share many of the same characteristics are grouped together as one node, and links between nodes in this picture, called a *concept lattice*, indicate that all the members of a certain subgroup with certain attributes also have other attributes in common. For instance, you might group together people based on what cafés, bookstores and churches they attend, and then find out that all the people who go to a certain café also attend the same church (but maybe not vice versa).

Lattice theoretical ideas developed at the Massachusetts Institute of Technology (MIT) tell us the probability that we have disabled a terrorist cell based on how many members we have captured and what rank they hold in the organization: Assume that terrorist plans are formulated by the leaders of a cell, and are transmitted down through the chain of command to the foot soldiers, who carry out those plans. Further assume that the message only needs to reach one foot soldier for damage to result. We endeavor to block all routes from the leaders to the foot soldiers by capturing some subset of the agents. (The agents we remove need not be leaders or foot soldiers.) Such a subset is called a *cutset* in a *partially ordered set* or *poset*.

Boston-area student Lauren McGough experimentally tested the accuracy of this model. She simulated the way that commands pass from the leaders of cells to foot soldiers by organizing fifteen of her classmates as a "binary tree": each person, except the eight foot soldiers at the bottom, had two immediate subordinates. The "leader" was assigned a message or "command" to pass on to her subordinates (for instance, "Look for a red flower with blue thorns," or, my favorite, "Twinkle-toes says hello"). These individuals then passed the message on to their direct subordinates, and so on, until the message reached the foot soldiers, like a game of "telephone". To simulate the capture of terrorists in a cell, three people were randomly removed from the binary tree for each of fifty trials. Each of the three was told not to pass on the message that was sent out for that specific trial (unless it had already reached him or her). McGough's findings showed that, if anything, the predictions of the model concerning the likelihood that a cell would be disrupted were too conservative. The point of all this is that you could calculate how much it would take to turn a comprehensive plan into a botched operation. Of course, the model assumes we know the structure of the cell to begin with, but elucidating that structure is the job of law enforcement, not mathematicians (although the theory, with its perhaps flawed assumptions, can account for gaps in our knowledge of the structure of a terrorist cell by making assumptions about how the "perfect" terrorist cell must be organized).

There is the ever-present threat of a dirty bomb being carried across the borders of the US or Europe. Which border do you guard? Which border do you want the terrorist to think is weak? Phoenix Mathematics, Inc., co-founded by lattice theorist Stefan Schmidt and later joined by Tony Harkin, is using reflexive theory – a branch of mathematical psychology developed by Vladimir Lefebvre and the Soviet military – to devise a quantitative way to help border patrols allocate personnel, and spread disinformation to the adversary. The US Army Research Lab, the US State Department, and a major defense contractor have been interested in applications of reflexive theory to military matters as well.

3 The Elephant: Politics

I am perhaps one of only two mathematicians in the world working on counterterrorism who have actually been targeted by terrorists (the other being fellow algebraist and former Iraqi Oil Minister Ahmed Chalabi). I became interested in developing mathematical methods in counterterrorism around the time I was forced to flee the US state of Tennessee, leaving many of my possessions behind, after receiving a few dozen death threats. (Gordon Gee, now president of Ohio State University, has written about the single death threat he received in the same episode; Vanderbilt University student Nia Toomer received another.) I arrived at MIT and saw a flier for a talk entitled, "Modeling the Al Qaeda Threat." The speaker was a Gordon Woo of Risk Management Solutions.

It was – like all his talks, I would learn – stimulating, and it led me to draft my first paper on this topic. If you will remember the climate in 2003, my mother and brother were afraid that even writing about counterterrorism would lead to reprisals. Regardless, I published the article in the journal *Studies in Conflict and Terrorism* and, a year later, Mel Janowitz, associate director of the Center for Discrete Mathematics and Theoretical Computer Science (DIMACS), Tony Harkin, Stefan Schmidt and I organized the first in a series of Conferences on Mathematical Methods in Counterterrorism.[1] Later Bernard Brooks joined us as an organizer. I am particularly glad we chose Nasrullah Memon, the driving force behind this volume, as one of our speakers. (I also appreciate the efforts of editors David Hicks and Torben Rosenørn, as well as Claus Atzenbeck of Aalborg University for his help with layout and typesetting, and Stephen Soehnlen for the support of Springer Verlag.) As with this volume, my intention for the conferences (not always realized) was to have speakers talk about mathematical methods that could actually be applied, and not merely discuss theory, although interesting theoretical questions could conceivably arise.

For this reason, I have been pleased to have as speakers or participants at our conferences US Air Force majors, US Army colonels, a division head of the US Army Research Lab, and representatives of the Royal Dutch Defense Academy,

---

[1] Success breeds competition. In 2006, after initially promising to fund our third conference, the so-called "Center for Advanced Defense Studies" instead organized a competing conference at almost exactly the same time as ours in the same city: It was called "the Conference on Mathematical Models in Counterterrorism," to distinguish it from our "Conference on Mathematical Methods in Counterterrorism."

the Canada Border Services Agency, and the Jamaican Constabulary Force. The representatives from the last two organizations were actually using mathematics in their jobs: Kevin Blake, an Assistant Superintendent of Police who heads of a unit of Operation Kingfish (which has been credited with reducing the amount of drug trafficking going through Jamaica), not only carries a gun but draws posets of criminal networks ferrying drugs and weapons between Jamaica and Haiti.

Although it may seem odd for a mathematician to be concerned with this, the media reaction to our work has been quite good, with television coverage by Fox News and CNBC Europe, radio coverage by Air America Radio, US National Public Radio, and Public Radio International, newspaper coverage by the *New York Times*, the *San Francisco Chronicle*, the *Associated Press*, and *The Times Higher Education Supplement*, magazine coverage by *The Economist* and *Seed Magazine*, and on-line coverage by *Science News* and MIT's *Technology Review*. The hit television crime drama *Numb3rs* even began drafting a script dealing with reflexive theory, and the American television series *Medium* aired an episode about a mathematician who was fighting terrorists.

The reaction from funding agencies, on the other hand, has been less than expected: other than the welcome support of the Proteus Group and the US Army War College, we have seen little financial support and no sustained interest by US (or other) government agencies in our enterprise, despite face-to-face meetings with the Jamaican Minister of National Security, a former senior official of the US National Security Agency and a former senior official of the CIA, the director of Homeland Security for one of America's largest ports, a former US ambassador to the European Union, a former governor of the US state of New Mexico, a US Air Force general, two US Navy admirals, at least three tycoons, and the presidents of the Massachusetts Institute of Technology, the California Institute of Technology (Caltech), and the Rochester Institute of Technology.

One reason, I believe, is the inability of some people to imagine that mathematics other than cryptography could really be useful in counterterrorism. One MIT professor openly lampooned the idea of using mathematics for counterterrorism, asking me if I was going to use a fixed point theorem to catch Osama bin Laden. (He thus revealed also a lack of originality, as there is an old joke about using math to catch a lion.) Another MIT professor actually laughed at the mere mention of the word "poset."

But I suspect that a bigger reason is politics. When I tried to garner support for a European Institute for Mathematical Methods in Counterterrorism in Vienna, an Austrian algebraist told me that this could not (or should not) be created there, as Austria was a neutral country.

At a dinner held at the home of the Austrian ambassador to the United States, a senior officer of the US National Academies asked me about using mathematical psychology for counterterrorism. But when he uncovered a 2001 article I had written for the British newspaper *The Guardian*, despite the fact that I myself always assiduously separated my interactions with mathematical colleagues from personal politics, communication ceased.[2]

But this is not about politics. Mathematicians as world citizens can and should reject anyone's demand to be "with them or against them," and soberly get on with the business of pursuing cheaper but more rational methods to save civilian lives, whether it be in London, Bali, Nairobi, or New York.

4 Toward a Mathematical Theory of Counterterrorism

Contributors Bert Hartnell and Todd Sandler were writing about mathematics, counterinsurgency and counterterrorism decades before it became "fashionable." But as Fred Roberts, director of DIMACS, has indicated, since 2001, tremendous amounts of information have been gathered regarding terrorist cells and individuals potentially planning future attacks. There is now a pressing need to develop new mathematical and computational techniques to assist in the analysis of this information, both to quantify future threats and to quantify the effectiveness of counterterrorism operations and strategies.

Progress on these problems requires the efforts of researchers from various disciplines such as mathematics, computer science, political science, and psychology. By having researchers from diverse disciplines come to one place to conduct their research, greater progress will be made in developing scientific and analytical tools to deal with the problem of terrorism. Hence, to facilitate the invention of new tools, the exchange of new ideas, and the dissemination of new results to users in the intelligence and law enforcement communities, we have endeavored to publish this volume.

Can mathematics contribute the way it did in World War II? The "war on terror" is a much more haphazard, unpredictable operation. In some ways the problems are more complicated: we generally knew, then, where the German panzer divisions were. But there are clear benefits to trying: mathematics can help take some of the guesswork out of the decision-making process (if only by replacing laypeople's guesses with mathematicians').

I do not know if the conference co-organizers, the other editors, or the volume contributors will agree with all, or even any, of the thoughts I have expressed above. But I believe they will all agree with me when I say that it's high time we chose brains over brawn: Against terror, beauty may succeed where brute force has failed.

[2] On November 18, 2001, as a result of the *Guardian* article, I was invited to be the second speaker at what was then the largest peace rally in the United Kingdom, with a crowd estimated at 100,000; amongst the other speakers were various Members of Parliament. On September 11, 2002, I was invited by BBC World News Television to discuss the issues in the article in special anniversary coverage live from Ground Zero in New York City.